Science, Technology and Innovation in BRICS Countries

The term BRICS (Brazil, Russia, India, China and South Africa) is gaining global attention both in scholarly and popular discourse. BRICS countries are crucial in terms of their vast areas and huge population and have massive economic potential. These countries are also categorized as developing countries and are aspiring to be considered as developed countries. There is commonality among these countries in that they have similar issues and problems, which may require common solutions. *Science, Technology and Innovation in BRICS Countries* examines whether more emphasis on Science Technology and Innovation (STI) capability building could be the solution to these countries' economic upgradation and poverty reduction.

This book is a collection of various STI issues of BRICS economics, and will be of interest to general readers and scholars working in this field, as well as policy-makers all over the globe.

The contributions come from various scholars across the world who have published their BRICS economics research in a special issue of the *African Journal of Science, Technology, Innovation and Development*.

Swapan Kumar Patra is Career Advancement Research Fellow at Tshwane University of Technology, Pretoria, South Africa. He holds a PhD in Science Policy from the Centre for Studies in Science Policy, School of Social Science, Jawaharlal Nehru University, New Delhi, India.

Mammo Muchie is a DST/NRF SARChI Chair Research Professor in Innovation Studies at Tshwane University of Technology, Pretoria, South Africa. He is the founder of the *African Journal on Science, Technology, Innovation and Development*. He received the first SARChI Research Professorship Chair on Innovation Studies in South Africa in 2008.

Science, Technology and Innovation in BRICS Countries

Challenges, Issues, Opportunities

Edited by
Swapan Kumar Patra and Mammo Muchie

LONDON AND NEW YORK

First published in paperback 2024

First published 2020
by Routledge
4 Park Square, Milton Park, Abingdon, Oxon OX14 4RN

and by Routledge
605 Third Avenue, New York, NY 10158

Routledge is an imprint of the Taylor & Francis Group, an informa business

Publisher's Note
The publisher has gone to great lengths to ensure the quality of this reprint but points out that some imperfections in the original copies may be apparent.

Disclaimer
Every effort has been made to contact copyright holders for their permission to reprint material in this book. The publishers would be grateful to hear from any copyright holder who is not here acknowledged and will undertake to rectify any errors or omissions in future editions of this book.

British Library Cataloguing in Publication Data
A catalogue record for this book is available from the British Library

Typeset in Times
by Newgen Publishing UK

ISBN: 978-0-367-44280-4 (hbk)
ISBN: 978-1-03-283907-3 (pbk)
ISBN: 978-1-00-300877-4 (ebk)

DOI: 10.4324/9781003008774

Contents

Citation Information

The chapters in this book were originally published in the *African Journal of Science, Technology, Innovation and Development*, volume 9, issue 5 (December 2017). When citing this material, please use the original page numbering for each article, as follows:

For any permission-related enquiries please visit:
www.tandfonline.com/page/help/permissions

Notes on Contributors

Dinesh Abrol is Coordinator in the South Asia Sustainability Hub and Knowledge Network (SASH&KN) at the Transdisciplinary Research Cluster on Sustainability Studies, Jawaharlal Nehru University, New Delhi, India, and is formerly Chief Scientist at the Council of Scientific and Industrial Research National Institute of Science, Technology and Development Studies (CSIR-NISTADS).

Amitkumar Singh Akoijam holds his PhD from the Centre for Studies in Science Policy, Jawaharlal Nehru University, New Delhi, India. He currently teaches at the University of Delhi.

Angathevar Baskaran is Associate Professor in the Department of Development Studies, University of Malaya, Malaysia. He is also Senior Research Associate at SARChI (Innovation Studies), Tshwane University of Technology, Pretoria, South Africa, and Editor-in-Chief of the *African Journal of Science, Technology, Innovation and Development*.

Chux U. Daniels is Research Fellow in Science, Technology and Innovation Policy at the Science Policy Research Unit (SPRU), University of Sussex, UK. He is the Director of the Transformative Innovation Policy (TIP) Africa Hub research project at SPRU and Visiting Fellow at the University of Pretoria, South Africa.

S. R. Khalimova is Senior Researcher in the Institute of Economics and Industrial Engineering at the Siberian Branch of the Russian Academy of Sciences, Novosibirsk, Russia.

V. V. Krishna is Professorial Fellow in the Faculty of Arts and Social Sciences at the University of New South Wales, Sydney, Australia. He is a former Professor of Science Policy in Jawaharlal Nehru University, India. He has recently been elected as Fellow of the Royal Society of New South Wales, Australia, for his contribution to the field of science, technology and society studies.

V. A. Krykov is Doctor of Sciences (Economics) and Director of the Institute of Economics and Industrial Engineering, Siberian Branch of the Russian Academy of Sciences, Novosibirsk, Russia.

Kashmiri Lal works at CSIR-NISTADS, New Delhi, India. He earned his graduate degree from Birla Institute of Technology and Science, Pilani, India, and holds an MBA from the Faculty of Management Studies, University of Delhi, India.

Sergio Leão holds a PhD in Economics from Pontificia Universidade Catolica, Rio de Janeiro, Brazil.

Ju Liu is Senior Lecturer in the Department of Urban Study, Malmö University, Sweden. She holds a PhD in Innovation Study, a master's degree in Business Administration and a bachelor's degree in Electric Engineering.

MME 'Tshidi' Mohapeloa is Associate Professor in the Business School at Rhodes University, Grahamstown, South Africa, where she coordinates a postgraduate programme in Enterprise Management (PDEM).

Mammo Muchie is a DST/NRF SARChI Chair Research Professor in Innovation Studies at Tshwane University of Technology, Pretoria, South Africa. He is the founder of the *African Journal on Science, Technology, Innovation and Development*. He received the first SARChI Research Professorship Chair on Innovation Studies in South Africa in 2008.

Oluwayemisi Adebola Oyekunle completed her PhD at Tshwane University of Technology Business School, Pretoria, South Africa. For the past twenty-five years, she has worked as a human resources lecturer and researcher in reputable international institutions/organisations.

Swapan Kumar Patra is Career Advancement Research Fellow at Tshwane University of Technology, Pretoria, South Africa. He holds a PhD in Science Policy from the Centre for Studies in Science Policy, School of Social Science, Jawaharlal Nehru University, New Delhi, India.

N. P. Pokhilenko is Doctor of Sciences (Geology and Mineralogy), Academician of the Russian Academy of Sciences and Professor and Academician of the International Academy of Ecology, Man and Nature Protection Sciences.

Julio Raffo is Researcher in the Economics and Statistics Division of the World Intellectual Property Organization (WIPO).

N. Yu. Samsonov is Candidate of Sciences (Economics) and corresponding member of the International Academy of Ecology, Man and Nature Protection Sciences. From 2009 to 2019, he was Senior Researcher at the Institute of Economics and Industrial Engineering, Siberian Branch of the Russian Academy of Sciences, Novosibirsk, Russia.

Swarup Santra is Assistant Professor in the Department of Economics, Satyawati College, University of Delhi, India. He completed his studies at the University of Calcutta and Jawaharlal Nehru University and has been teaching macroeconomics and econometrics for nine years.

Nidhi Singh is Scientist 'C' at the Department of Health Research, Ministry of Health and Family Welfare, Government of India, New Delhi. She holds a PhD from Jawaharlal Nehru University, New Delhi, India.

A. V. Tolstov is Doctor of Sciences (Geology and Mineralogy) and Academician of the Russian Academy of Natural Sciences and the International Academy of Ecology, Man and Nature Protection Sciences. He discovered the Tomtor niobium-rare-earth deposit and the Mayskoye diamond deposit.

Olga Ustyuzhantseva is Director of the Centre for Policy Analysis and Studies of Technologies at the Tomsk State University, Russia. For the last five years, she has studied the grassroots innovation phenomenon based on the cases of India and Russia.

Hui Yan is Lecturer at the School of Management, Shanghai University, China, and Vice-Director of Shanghai Competition Ecology Research Centre. She holds a PhD from Aalborg University, Denmark.

Wei Yao is Associate Professor and Assistant to the Dean at the Institute of China's Science, Technology and Education Policy Strategy, Zhejiang University, China. His research interests are engineering creativity development, social entrepreneurship, innovation management and inclusive innovation.

Graziela Ferrero Zucoloto works at the Institute of Applied Economic Research, Rio de Janeiro, Brazil.

Science, technology and innovation in BRICS countries: Introduction

Swapan Kumar Patra ⓘ and Mammo Muchie ⓘ

The BRICS (Brazil, Russia, India, China and South Africa) economies have been the subject of many scholarly studies in recent years, as well as featuring in the popular media. This special issue of the African Journal of Science, Technology, Innovation and Development (AJSTID) called for original research contributions in the various aspects of Science Technology and Innovation (STI) studies in the BRICS group of emerging economics. The issue is composed of 11 research papers on different STI topics relevant to the BRICS countries.

Introduction

The term 'BRIC' (Brazil, Russia, India and China) was coined in 2001 by Jim O'Neill, a former Goldman Sachs economist anticipating that these emerging economics would be global economic powerhouses in the mid-twenty-first century (O'Neill 2001). It was estimated that by the 2040s the BRICs economies together could be larger than the G6 (group of six European Union member states: France, Germany, Italy, Poland, Spain and the United Kingdom) in US dollar terms (Wilson and Purushothaman 2003). A decade later, South Africa was added to this group of countries and the term henceforth became 'BRICS'. Today, the term has become very common with BRICS economies increasingly playing an important role in the global economic landscape. In fact, with BRICS countries accounting for approximately 40% of the total world population, more than 25% of the earth's surface area and almost 25% of the world GDP, it is predicted that over the next 50 years BRICS economies will be a determining factor in the global economy.

These countries share common socioeconomic and demographic conditions, for example, a young population, rich natural resources, crucial geographical locations and so on. A large number of skilled workers and a young population are two advantages these emerging economies have. These five countries together formed an association to foster mutual growth and development. Although, BRICS member countries are undergoing rapid growth and show potential, they face many mutual developmental issues in the area of science, technology and innovation (STI). Despite their commonalities, the flow of technology and knowledge among the BRICS economies still falls far behind the inflows from the global North. Moreover, the origin of these groups of countries is rooted in macroeconomic analyses of national conditions. It remains to be seen, therefore, whether these countries hold similarly strong prospects at the microeconomic level, especially in relation to STI.

The BRICS economies have been the subject of many studies. This special issue of the African Journal of Science, Technology, Innovation and Development (AJSTID) called for original research contributions in the various aspects of STI studies in BRICS group of emerging economics. The special issue is composed of research papers on various topics related to BRICS countries in general, as well as individual and cross-country studies covering the following: national, regional, and technological innovation systems, technological capability building, science technology and sustainable development, the role of these economies in the globalization of R&D, university-industry relations, science, technology and innovation policy, science and technology collaboration in and among these countries, traditional knowledge systems, creative industries and so on.

With the call for papers for this special issue, 11 papers were selected after the double-blind peer reviews. The papers may be broadly segmented into three groups. Three papers deal with various BRICS issues; two papers deal with two country cases and six papers focus on various issues specific to individual BRICS countries, namely: Brazil (1 paper), India (2 papers), South Africa (2 papers) and Russia (1 paper).

BRICS countries are presently undergoing substantial structural changes towards becoming more industrialized economies. Along with economic growth, energy utilization, carbon emission and its effects from these countries are also of global concern. Swarup Santra's paper explores the impact of economic transformation on the environmental quality of BRICS countries. Using econometrics models on OECD panel data set during 2005–2012, on the relevant variables, the paper reveals that innovation in environment-related technology has a sound impact on the sustainable performance of BRICS countries. The study notes that green technological innovation reduces these countries' energy absorption and CO_2 emissions. Further, it indicates that green technological innovation may improve production-based energy productivity and production-based CO_2 emission productivity.

Chux C. Daniels, Olga Ustyuzhantseva and Wei Yao deal with the role of innovation in inclusive development (IID) and link innovation to public policy in BRICS countries. Based on the national systems of innovation (NSI) framework, these authors examine the roles played by various triple helix (TH) actors (university, industry and government) in IID activities across BRICS countries.

Their study observes that there is a significant gap in scholarly literature in advancing knowledge of innovation as a mechanism for inclusive development. Perhaps the gap is due to the inappropriate conceptualization of innovation as a mechanism to include the wider society in socioeconomic and development activities.

Patents are an important indicator to map the innovation activities for an entity. Kashmiri Lal's study on design patents examines the profile of design patents among BRICS economies, using the United States Patents and Trademark Office (USPTO) data for the period 2002–2011. The study observes significant Chinese activity in design patents, followed by Brazil, India, South Africa and Russia. Most of the design patents in the BRICS region are the result of single inventor's activity. Joint patents as an indicator of collaborative research show very little inter-BRICS country collaboration.

Outward foreign direct investment (OFDI) from developing countries has undergone important quantitative and qualitative changes over the last decade (Gammeltoft 2008). Using the case study method, Angathevar Baskaran, Ju Liu, Hui Yan and Mammo Muchie explore the factors driving OFDI in emerging MNEs (EMNEs) and the patterns of knowledge transfer of selected MNEs from three of the BRICS economies (India, China and South Africa). The study observes that there are various complex aspects of OFDI by EMNEs. This phenomenon perhaps cannot be explained with existing FDI theories. Hence, this study proposes a theoretical model that integrates both 'latecomer strategies for catch up' and the 'traditional FDI' models to explain the FDI phenomenon.

Since the 1990s, MNEs have been off-shoring their R&D activities to developing Asian countries, particularly in India and China (Krishna, Patra, and Bhattacharya 2012). With this recent trend, foreign R&D by MNEs is becoming important and has attracted attention from all over the globe. Swapan Kumar Patra deals with foreign R&D units in India and China, examining their locations and motives for off shoring crucial R&D which is usually located in their home base. The study observes that both 'market driven' and 'technology driven' factors are the major motives for MNEs to invest in R&D in these two emerging economies. Firms prefers R&D locations in India and China, where there are knowledge centers with an abundant supply of qualified and high skilled labor available at comparatively low cost.

Six papers deal with various burning issues in specific BRICS member countries. Jawaharlal Nehru National Solar Mission (JNNSM) is one of the most major initiatives undertaken by the Government of India in the recent years 'to promote ecologically sustainable growth while addressing India's energy security challenge'. The article by Amitkumar Singh Akoijam and V. V. Krishna presents the current solar energy situation in India and explores the ways in which various actors, agencies and policies shape this initiative from the innovation system perspectives. The study observes that productive R&D institutions and universities, with supportive policy initiatives, have significantly increased knowledge generation in this area (measured in terms of number of research papers published in relation to solar energy) in recent years

The paper by Graziela Ferrero Zucoloto, Julio Raffo and Sergio Leão entitled 'Technological appropriability and export performance of Brazilian firms' evaluates the strategies of Brazilian manufacturing firms in their use of intellectual property (IP) and its impact on their export performance. The paper discusses the relationship between innovation and exports in Brazil, showing that innovative Brazilian firms tend to export more than non-innovative firms, confirming the main literature findings.

Cultural and creative industries can play a strategic role in South African development, and the government has realized the role of these industries in the economic growth of cities and rural areas. The paper by Oluwayemisi Adebola Oyekunle, 'The contribution of creative industries to sustainable urban development in South Africa', addresses the evolution and development of these industries in the South African context. Taking the case of two South African cities, the paper observes that the role of creative industries in South Africa's urban development plan may be an essential component of broader development plans in spite of its many shortcomings.

Nidhi Singh and Dinesh Abrol's article brings out policy measures and the related institutional infrastructure of in-vitro diagnostics (IVDs) in India. The study observes that R&D investment in IVDs is very small and a stable market mechanism is yet to form. The major portion of diagnostics needs is still met through imported technologies. Further, the paper notes that the collaboration between public R&D institutions and large domestic firms is the major factor in defining the features of the national innovation system in the case of IVDs.

Nikolay Samsonov, Alexander Tolstov, Nikolay Pokhilenko, Valery Krykov and Sophia Khalimova deal with the possibilities of Russian hi-tech rare earth products meeting the industrial needs of BRICS Countries. Their paper speculates on the future prospects of the new Russian scientific and technological production sector. From rare earth materials, an effective innovative technological chain 'ore processing – getting highly liquid REM-products' could integrate Russian products into the global market. These products may then compete with two dominant BRICS suppliers, i.e. China and Brazil.

The final paper in this category is by Tshidi Mohapeloa and deals with entrepreneurial mindset development in the non-profit social sector in the South African context. Analyzing literature data from the Scopus database, the paper finds that entrepreneurial mindset development not only includes the way of thinking, skills and knowledge but also reflections in attitudes and behavioral patterns. The study recommends that a targeted educational curriculum, starting at school level and continuing up to university level, is required to foster entrepreneurial development with a social focus and 'Ubuntu' principles.

Disclosure statement

No potential conflict of interest was reported by the authors.

Funding

This work was supported by South African Research Chairs Initiative (SARChI) Innovation Studies, Tshwane University of Technology, Republic of South Africa.

ORCID

Swapan Kumar Patra ⓘ http://orcid.org/0000-0002-0825-7973

Mammo Muchie ⓘ http://orcid.org/0000-0003-4831-3113

References

Gammeltoft, P. 2008. "Emerging Multinationals: Outward FDI From the BRICS Countries." *International Journal of Technology and Globalisation* 4 (1): 5–22.

Krishna, V. V., S. K. Patra, and S. Bhattacharya. 2012. "Internationalisation of R&D and Global Nature of Innovation: Emerging Trends in India." *Science, Technology and Society* 17 (2): 165–199.

O'Neill, J., 2001. "Building Better Global Economic BRICs." Global Economics Paper No. 66, Goldman Sachs. S01–11. Available at http://www.goldmansachs.com/our-thinking/archive/archive-pdfs/build-better-brics.pdf

Wilson, D. and Purushothaman, R. 2003. "Dreaming with BRICs: The Path to 2050." (Vol. 99). New York, NY: Goldman, Sachs & Company. 1–23.

The effect of technological innovation on production-based energy and CO_2 emission productivity: Evidence from BRICS countries

Swarup Santra ⓘ

Collectively, Brazil, the Russian Federation, India, China and South Africa (known as BRICS) are emerging as an economic superpower. In 2014, BRICS countries accounted for approximately 40% of the total world population, 30% of the total earth surface and almost 20% of the world's economic output. Along with increasing economic growth in BRICS countries, energy usage and related carbon emissions are drawing increasing attention as a result of increasing international concern about climate change. Environmental policy has two types of role to play in the economic process. On one hand, firms can adopt or purchase existing 'cleaner' technology; on the other hand, they can invest in R&D for inventing new 'cleaner' technology. A panel data econometrics analysis of the relevant variables taken from the OECD data set for the period 2005–2012 reveals that innovative environment-related technology has had a sound impact on the sustainable performance of BRICS countries, with the green technological innovations having helped firms and countries, as a whole, to reduce their energy absorption and CO_2 emissions. These technologies have positively helped to improve production-based energy productivity and production-based CO_2 emission productivity.

Introduction

In pursuance of economic development, regional integration has emerged as the new economic order. On almost every continent, or even across the continents, countries form regional blocs. Five emerging economies, Brazil, the Russian Federation, India, China and South Africa, from four continents formed a different form of cooperative bloc, BRICS. BRICS is not a bloc of regional integration; it is a bloc of emerging economies formed to enhance their 'strategic position' in an almost unipolar world scenario. Since the 1990s, the thrust of economic power has been gradually shifting towards BRICS countries. These countries are emerging as boom markets and popular global investment destinations. Rapid and high demand growth in recent years has been one of the main reasons behind the rise of BRICS countries. In 2014, BRICS countries accounted for approximately 40% of the total world population, 30% of the total earth surface and almost 20% of the world's economic output. Merely those figures are sufficient to say that 'Globally and politically, the influence of the BRICS is rapidly increasing' (Tian 2015).

BRICS countries are also important on the global climate change negotiation platform. With the stringent environment policies in developed countries and the growing markets in developing countries, many production units in developed countries are shifting to developing countries, mostly in the BRICS bloc. This has resulted in increased concerned about environmental degradation, especially in the context of climate change, in BRICS countries. Hence, along with increasing economic growth in BRICS countries, energy usage and related carbon emissions are drawing increasing attention as a result of increasing international concern about climate change.

According to the World Resources Institute (WRI), in 2013 China, India, Russia, Brazil and South Africa ranked first, fourth, fifth, sixth and twelfth, respectively, in global greenhouse gas (GHG) emissions (Tian 2015). However, energy features are different in each BRICS country, as is the way each one deals with its CO_2 gas emissions. For example, while Brazil is ranked eighth in total energy consumption globally and tenth in energy producution, the country's emissions represent only 5% of total global emissions, mainly due to the fact that a large portion of the country's electricity is generated from renewable resources (Freitas, Dantas, and Iizuka 2011; Shaw 2016). China, on the other hand, is a major global player in coal production and consumption. In addition, more than 80% of electricity generated in China comes from coal-based power plants. Consequently, in 2011, China was responsible for 29% of total global carbon consumption (Pao and Tsai 2011; Shaw 2016). India and the Russian Federation account for about 6% and 5% of total global carbon emissions, respectively. However, the difference between Russia and India's energy consumption scenarios is that almost 54% of energy consumption in Russia is from natural gas, whereas almost 52% of energy consumption in India comes from coal (Pao and Tsai 2011; Tian 2015). Almost 78% of South Africa total emissions come from its power sector.

To curb their respective emissions levels, BRICS countries have separately developed and implemented relevant policies. Brazil, for example, under the Kyoto Protocol, promised to reduce its carbon emission by 36–39% of its 1990 level by 2010 (Freitas, Dantas, and Iizuka 2011; Shaw 2016) and in 2013 set up a national Climate Change Fund. China has taken a stand to invest heavily in renewable energy projects through its Clean Development Mechanism (CDM) fund projects (Freitas, Dantas, and Iizuka 2011). China also launched its Pilot Carbon Trading Scheme in 2013, with help from the European Union (Shaw 2016). In 2007 China launched a national Climate Change Programme for mitigating GHG emissions, and fixed a target of a 17% reduction in carbon intensity by 2015 (Freitas, Dantas, and Iizuka 2011). India has taken a national 'Action Plan on Climate Change' to reduce emission intensity, namely an almost 20–25% emissions intensity cut of GDP by 2010 based on its 2005 level (Freitas, Dantas, and Iizuka 2011; Shaw 2016).

The sixth BRICS summit in Brazil in 2014, with its slogan 'Inclusive growth: sustainable solution', focused on social inclusion and sustainable development (Fabbri and Ninni 2014). A New Development Bank (NDB) was set up to address the issues of financial assistance for infrastructure and sustainable development. The UNFCCC (United Nations Framework Climate Change Convention) established a fund is called the Green Climate Fund (GCF) in 2010 at its meeting held in Cancun (Fabbri and Ninni 2014; Pao and Tsai 2011). Many developed and developing countries have shown a willingness to contribute to the GCF. BRICS countries' NDB instrument is also a part of the GCF as promised by these countries in 2014 at the UNFCCC in Bonn (Fabbri and Ninni 2014; Pao and Tsai 2011). The GCF has been used to promote low-emission and climate-resilient technology and its own developmental goals, and has provided developing countries with financial support to fight against climate change (Lantz and Feng 2006; Tian 2015). There is however not much empirical evidence available that indicates a direct correlation between technological innovation, energy production and consumption, and CO_2 efficiency. This paper therefore aims to answer the question whether or not environment-related technological innovation is improving energy efficiency and CO_2 emission efficiency.

The objective of this study is to investigate whether or not environment-related technological innovation can positively influence production-based energy efficiency and CO_2 emission efficiency. To achieve this objective, the paper focused on BRICS countries. Energy efficiency in the production sector was measured by production-based energy productivity, output per unit of energy usage, as referenced in OECD reports. CO_2 emission efficacy was measured by energy-related CO_2 emission productivity, output per unit of CO_2 emission.

Having looked at the background and emergence of BRICS countries in the introduction of this paper, as well as their importance in global climate change negotiation platforms and their counter-acting policies, the next section reviews the literature related to the topic of this paper. By studying the works of different scholars in this section, this study builds analytical arguments in line with the objectives of the paper. Methodological issues are dealt with in the third section. A panel data analysis is used to determine the findings. The description of data sources is also a part of this section. The results and their interpretation are covered in the fourth section, while some concluding remarks are incorporated in the last section.

Review of literature
Pollution, market failure and environmental policy
When an economy produces its required goods and services, it also generates pollution as a by-product, which is a negative externality to society. This negative externality reduces the total social welfare of the economy by incurring a cost to society (Byrne et al. 2011; Jaffe, Newell, and Stavins 2005; Löschel 2002). A free-market economy fails to correct this market failure (Jaffe, Newell, and Stavins 2005; Requate and Unold 2003). In such cases, government intervention is desirable to correct this market failure. Governments can use the command and control (CAC) policy to make rules and regulations to mitigate such market failure. Alternatively, governments can develop and implement suitable economic policy instruments (Pao and Tsai 2011; Requate and Unold 2003). Environmental regulations enforce environmental laws and set the standard for emission levels that are permissible in terms of social cost (Aghion, Hemous, and Veugelers 2009; Jaffe, Newell, and Stavins 2005; Markewitz et al. 2012). On the other hand, environmental policies – through taxation and subsidization – can try to push firms to internalize the cost of pollution into their production decisions (Byrne et al. 2011; Carraro and Siniscalco 1994; Löschel 2002). Taxing highly polluting firms and subsidizing less pollution firms are instruments for environmental policy (Goulder and Mathai 2000; Jaffe, Newell, and Stavins 2005; Socolow et al. 2004) that can minimize the social cost created by pollution. The internalization of green tax payments, made for generating pollution, may force firms to reduce their pollution levels. Increasing levels of green tax payments may also induce firms to incorporate less-polluting technologies in long-run.

Induced technological development
Technological enhancement towards green technology can reduce pollution generation without decreasing the level of output (Carraro and Siniscalco 1994; Requate and Unold 2003). New green technology can help firms to cut their energy consumption and CO_2 emissions. In other word, it can help to increase the productivity of the energy used (i.e. production-based energy productivity) and the productivity of CO_2 emissions (i.e. production-based CO_2 emission productivity) (Goulder and Mathai 2000; Lantz and Feng 2006; Pao and Tsai 2011; Socolow et al. 2004). In this context, firms would be willing to incur costs on purchasing new, environment-friendly technologies (Goulder and Mathai 2000). Carraro and Siniscalco (1994) asserted that environmental policy plays two types of role in the economic process. On the one hand, firms can adopt or purchase existing 'cleaner' technology; on the other hand, they can invest in research and development (R&D) in order to invent new 'cleaner' technology (Byrne et al. 2011; Requate and Unold 2003; Socolow et al. 2004). Their motivation for the latter is often increased because, due to the current patent systems and hence the high cost structure of existing technology, existing clean technology is not diffused to many firms. Firms thus try to invest in R&D in order to obtain new technological innovations which would consume less energy and release less emissions (Aghion, Hemous, and Veugelers 2009; Requate and Unold 2003). Therefore, in the long run, firms understand that R&D investment is a prerequisite to obtaining environment-related technologies (Carraro and Siniscalco 1994; Jaffe, Newell, and Stavins

2005). Thus, an effective environmental policy designed to combat pollution levels through taxing the polluting firms can encourage them to innovate and adopt environment-friendly technology (Aghion, Hemous, and Veugelers 2009; Goulder and Mathai 2000; Markewitz et al. 2012), without reducing their level of output.

BRICS and environmental policies

According to the Global Environment Outlook (UNEP 2013) and the OECD Forecast (OECD 2014), the major challenges for BRICS countries are the increasing level of energy use and the high level of energy intensity in the production process. However, BRICS countries have not been taking many serious steps to decouple energy inputs from their production systems, 'partly because of the displacement effects[1] and the delocalization of firms and sectors to emerging countries' (Fabbri and Ninni 2014, 12; Pao and Tsai 2011). Therefore, carbon and energy productivity have become the most important factors responsible for GHG emissions in emerging economies, like those of BRICS (Fabbri and Ninni 2014; Markewitz et al. 2012; Socolow et al. 2004). Production-based emission productivity, as measured by GDP per unit of GHG emitted, and demand-based emission productivity, as measured by real income per unit of GHG emitted, are the two best ways to monitor the stabilization progress of the atmospheric concentration of GHG emissions (Fabbri and Ninni 2014; Socolow et al. 2004). Therefore, there is an urgent need for the evaluation of the impacts for environmental policies, such as environmental regulation, energy taxation/subsidies, technological innovation, increasing investment in R&D for inventing new environment-related technologies, etc. (Fabbri and Ninni 2014; Markewitz et al. 2012; Pao and Tsai 2011).

Wang and Feng (2014) showed in their study in China that environmental regulation has a great influence on total-factor energy efficiency. Weak regulation results in having low energy efficiency in production sectors, whereas strong regulation is the determining factor in achieving high energy efficiency (Goulder and Mathai 2000; Löschel 2002). The emission scenarios push the economy to take countermeasures, such as developing environmental policies that enable designs to potentially increase the green technology base in industrial sectors. (Aghion, Hemous, and Veugelers 2009; Fabbri and Ninni 2014). A sharp decline in energy intensity or a rising trend of energy productivity in many production sectors in BRICS countries has been evident from recent data (Goulder and Mathai 2000). A feasible explanation for the increasing energy productivity or emission productivity (or reducing energy intensity/emission intensity) is the diffusion of renewable energy sources across BRICS countries (Fabbri and Ninni 2014; Pao and Tsai 2011).

However, no studies have been conducted to investigate the impact of innovative, environment-related technology on energy intensity and emission intensity in production sectors in BRICS. The aim of this paper is to investigate that impact, despite the limitation of not being able to relevant long-run historical data.

Methodology and sources of data

It has been assumed that an economy has an aggregate production function. The total output (or real GDP) of any economy depends on the level of its labour forces, capital accumulation, raw materials and natural resources, and, of course, on the level of technology. Usage of energy is also a very important factor of production. While a country is producing its output, it is also producing 'pollution' as a by-product (Pao and Tsai 2011). In climate change analysis, it has become convention that the pollutant is proxied by CO_2 emission in absence of proper information on the actual GHGs. Conventionally, the use of emission intensity is generally familiar in climate change analysis. However, in this paper, the variable of production-based CO_2 emission productivity is considered due to the conformity to the refence of the sources of data for this paper. On the other hand, for sustainable energy use, energy productivity and energy intensity are also important aspects of an economy. However, in this paper production-based energy productivity has been considered for the investigation.

Data sources and variables

Production-based CO_2 emission productivity (in short, CO_2 productivity or CO2PRD), and production-based energy productivity (in short, energy productivity or ENGPRD) are considered for the performance parameters in response to combating climate change. On the other side, per capita innovation of environmental-related technology is considered as the policy variable for firms or for a country, in a macro sense. The data was collected from the Green Growth Indicators of Statistics from the OECD for the period 2005 to 2012 for BRICS countries only. The details of the variables are described in Table 1.

Production-based energy productivity (ENGPRD) is calculated as gross domestic product (GDP) per unit of total primary energy sources (TPES), i.e. (US$ in 2005 PPP/TPES). It reflects the efforts to improve the energy efficiency. The higher the value of energy productivity, the lower are the uses of energy per unit of GDP and the higher is the energy efficiency. The case for production-based CO_2 emission productivity (CO2PRD) is similar. The CO2PRD variable is measured as the GDP per unit of energy-related CO_2 emission (US$/Kg). The total CO_2 emission, here, includes the CO_2 emission from combustion of oil, coal, natural gas and other fuels. The country-specific estimates of CO_2 emission were taken from the International Energy Agency's (IEA) database, where the IPCC default emission factors were used through the *Revision of the Revised 1996 IPCC Guidelines for National Greenhouse Gas Inventories* (IPCC 2003).

The explanatory variable for innovation was taken from the information on development of environment-related technology in percentage form. The per capita innovation (INVPC) was considered as a policy variable generated by dividing the number of environment-related technological innovations by the number of total population (see Table 1). The other two dependent variables are energy productivity and emission productivity and the main explanatory variable in per capita innovation. Country-specific scatter diagrams were drawn between energy productivity

Table 1: Description of variables.

Variables	Full Name	Description
Dependent variables		
ENGPRD	Production-based energy productivity	GDP (US$ 2005) per unit of total primary energy sources (TPES)
CO2PRD	Production-based CO_2 emission productivity	GDP (US$ 2005) per unit (kilogram) of energy-related CO_2 emissions
Explanatory variables		
INVPC	Innovation per capita	Per capita innovation of environment-related technologies
ENGINT	Energy intensity	Per capita total primary energy use (TPES/Population)
RENGSS	Renewable energy resources	Renewable energy supply as percentage of TPES
RGDP	Real gross domestic product	Real GDP per capita (US$ 2005)

and per capita innovation (Figure 1) and between emission productivity and per capita innovation (Figure 2) to get a rough conception of the associations of these variables.

In Figure 1, the scatter diagrams between energy productivity and per capita innovation are for all five BRICS countries. It shows that the differential patterns of association between these two variables are not same. For China and India, the association is clearly positive. For the Russian Federation, the association is almost positive. However, for Brazil the association is not clear, whereas for South Africa it is clearly negative. Figure 2 depicts the association between CO_2 emission productivity and per capita innovation. For Brazil and China, the association is vividly positive, while for India it is more or less positive. For the Russian Federation, the picture is not clear, whereas for South Africa it is negatively associated. These two sets of scatter diagrams provide country-specific scenarios of the association between dependent and independent variables through which it is easier to conceptualize the specification of the regression model.

The other related variables would be used to control the environment as control variables. In this paper, energy intensity, supply of renewable energy and real GDP were considered for other explanatory variables. Energy intensity (ENGINT) is measured by the ratio of TPES to total population, i.e., per capita energy usage of any country. The lower the value of the overall energy intensity of an economy, the greater the energy efficiency in the production and consumption sectors of that economy. So, overall energy intensity may have some impact on energy productivity and CO_2 emission intensity. The percentage share of renewable energy out of total primary energy sources (RENGSS) may have some positive impact on energy productivity and CO_2 emission productivity. The larger share of renewable energy sources is an indicator of higher level of energy and CO_2 emission efficiency. The per capita real GDP (RGDP) plays a crucial role; it reflects the level of economic activity in the economy. In other word, the explanatory variable, per capita real GDP, shows the strength of a country which indirectly shows the ability of the economy to pursue for environment-related technological innovation. The descriptive statistics for all such variables are provided in Table 2 to facilitate a better understanding of them, country-wise.

From Table 2, it can be noted that the average energy productivity (ENGPRD) of Brazil is highest of the BRICS countries. India and China stand second and third. Similarly, Brazil's average emission productivity for the highest, followed by India and China. Russia, on the other

hand, has the highest per capita innovation, with China a close second. India leads with the most efficient per capita energy use in its economy, with Brazil and China second and third, respectively, in their per capita energy efficiency. Brazil leads in terms of the supply of renewable energy, followed by India. Finally, in terms of real per capita GDP, Russia is first, followed by Brazil and South Africa in second and third places, respectively (see, Table 2).

Model specification

$$ENGPRD = F(INVPC, \ ENGINT, \ RENGSS, \ RGDP) \tag{1}$$

$$CO2PRD = F(INVPC, \ ENGINT, \ RENGSS, \ RGDP) \tag{2}$$

where ENGPRD is the production-based energy productivity measured by GDP (US dollars, 2005) per unit of total primary energy sources (TPES); CO2PRD is the production-based CO_2 productivity measured by GDP (US dollars, 2005) per unit (kilogram) of energy-related CO_2 emissions; INVPC is the per capita innovation of environment-related technologies; ENGINT is the per capital total primary energy use (TPES/population); RENGSS is the renewable energy supply as a percentage of TPES; RGDP is the real GDP per capita (US dollars, 2005).

In the above two equations, (1) and (2), ENGPRD and CO2PRD were considered as the dependent variables, for the respective equations. The explanatory variable, INVPC, is assumed to be the main explanatory variable to be investigated. The other variables such as ENGINT, RENGSS and RGDP are included in both the equations as control variables. The econometric specification for the study is discussed in the next sub-section.

Pooled regression modelling

$$\begin{aligned}(ENGPRD)_{it} = \ & \alpha + \beta_1(INVPC)_{it-3} \\ & + \beta_2(ENGINT)_{it} + \beta_3(RENGSS)_{it} \\ & + \beta_4(RGDP)_{it} + \varepsilon_{it}\end{aligned} \tag{3}$$

$$\begin{aligned}(CO2 \ PRD)_{it} = \ & \alpha + \beta_1(INVPC)_{it-3} \\ & + \beta_2(ENGINT)_{it} + \beta_3(RENGSS)_{it} \\ & + \beta_4(RGDP)_{it} + \varepsilon_{it}\end{aligned} \tag{4}$$

where α is the intercept coefficient; and $\beta_1, \beta_2, \beta_3, \beta_4$ are

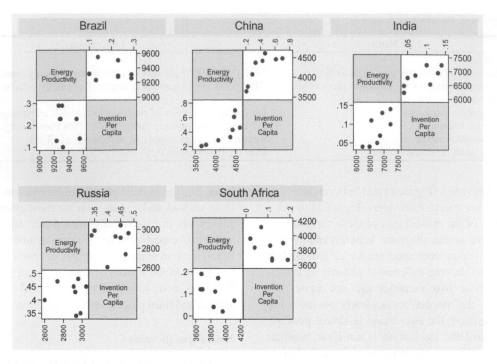

Figure 1: Scatter diagram between energy productivity and per capita innovation.

the slope coefficients for INVPC, ENGINT, RENGSS and RGDP; and, ε_{it} is the random error term.

Least-squares dummy variable (LSDV) regression model

The data set is the balanced panel data with five countries and nine time periods. To capture the country-specific heterogeneity, a fixed effects (FEs) panel data model is used. Here the slope coefficients are considered as constant, but the intercept varies across the countries. And, the variation in intercept term can be capture by introducing the country-specific dummy variables. That is why this method is also known as least-squares dummy variable (LSDV) regression model.

$$
\begin{aligned}
(\text{ENGPRD})_{it} = \; & \alpha + \beta_1 (\text{INVPC})_{it-3} \\
& + \beta_2 (\text{ENGINT})_{it} + \beta_3 (\text{RENGSS})_{it} \\
& + \gamma_1 (\text{Dummy\_Brazil})_{it} \\
& + \gamma_2 (\text{Dummy\_China})_{it} \\
& + \gamma_3 (\text{Dummy\_India})_{it} \\
& + \gamma_4 (\text{Dummy\_Russia})_{it} + \varepsilon_{it} \quad (5)
\end{aligned}
$$

$$
\begin{aligned}
(\text{CO2PRD})_{it} = \; & \alpha + \beta_1 (\text{INVPC})_{it-3} \\
& + \beta_2 (\text{ENGINT})_{it} + \beta_3 (\text{RENGSS})_{it} \\
& + \beta_4 (\text{RGDP})_{it} \\
& + \gamma_1 (\text{Dummy\_Brazil})_{it} \\
& + \gamma_2 (\text{Dummy\_China})_{it} \\
& + \gamma_3 (\text{Dummy\_India})_{it} \\
& + \gamma_4 (\text{Dummy\_Russia})_{it} + \varepsilon_{it} \quad (6)
\end{aligned}
$$

There are five countries in BRICS, so to avoid the dummy

variable trap only four dummy variables were included here. South Africa is assumed as the reference category, arbitrarily, as the choice of reference category cannot change the ultimate interpretation of the results. Here, γ_1, γ_2, γ_3, and γ_4 are the differential intercept dummies.

Analysis and interpretation
Production-based energy productivity

The energy productivity variable (ENGPRD) has been regressed on INVPC with three-years lag periods; the other explanatory variables are ENGINT, RENGSS, and RGDP. The data has been pooled for all the BRICS countries. A robust regression has been run to avoid the heteroscedasticity problem, through the econometric software, STATA. The result is documented in column (1) in Table 3.

From the pooled regression analysis (Table 3, column 1), it can be said that per capita innovation of environmental-related technology (INVPC (−3)) is a significant factor for energy productivity (ENGPRD). For BRICS countries, one unit increase in environment-related technological innovation per population three years ago could lead to almost 1208 US\$ (in 2005 PPP) of GDP per unit level of TPES. In addition, it is also evident that the renewable energy supply as a percentage of TPES (RENGSS) has a significance impact on energy productivity.

Since, BRICS is not an homogeneous entity, pooling some heterogeneous units into one may lead our results to heterogeneity bias. Apart from these included variables in the pooled regression analysis, there may be other country-specific attributes which are not addressed properly. A scatter diagram of ENGPRD and INVPC, as shown in Figure 3, gives a rough idea of the country-specific association of these two variables.

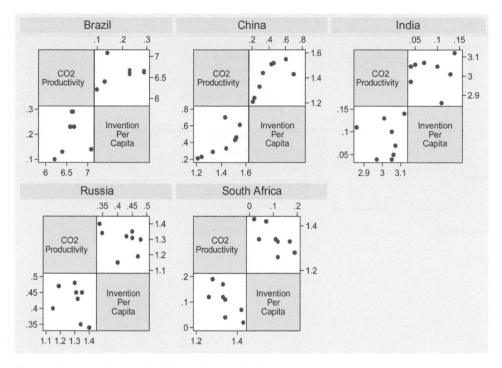

Figure 2: Scatter diagram between CO_2 productivity and per capita innovation.

Clearly, as can be seen in Figure 3, these BRICS countries do not show the same patterns of association. Not only does each country-specific scatter diagram show the difference in levels, as shown in Figure 3, but the differences in slope of the country-specific trend lines are also very vividly depicted. South Africa, China and the Russian Federation are on same level in the case of energy productivity, whereas Brazil and India's level of energy productivity are higher than those of the rest of the BRICS countries, irrespective of the level of per capita innovation. In fact, Brazil's energy productivity is the highest among BRICS countries. On the other hand, for per capita innovation (INVPC), the percentage values for Brazil, India and South Africa are much lower than those of China and the Russian Federation. In addition to that, the variation of per capita innovation for China is varied widely over the targeted time horizon.

To overcome this country-specific heterogeneity, the country-specific dummy variables were incorporated in least-squares dummy variable (LSDV) methods. The result is presented in Table 3, column (2). Here, the

Table 2: Descriptive statistics of variables.

		ENGPRD	CO2PRD	INVPC	ENGINT	RENGSS	RGDP
Total	Mean	5441.663	2.725	0.245	2.231	19.474	9620.59
	S.D.	2373.763	2.035	0.172	1.505	14.851	3840.04
	Min.	2603.520	1.150	0.020	0.460	2.410	2863.24
	Max.	9549.940	7.080	0.700	5.170	45.800	15374.15
Brazil	Mean	9315.338	6.523	0.205	1.319	43.087	12287.51
	S.D.	137.848	0.328	0.073	0.107	1.951	967.14
	Min.	9094.740	5.900	0.100	1.170	39.490	10840.83
	Max.	9549.940	7.080	0.290	1.480	45.800	13486.81
China	Mean	4268.863	1.420	0.408	1.729	12.011	7457.23
	S.D.	357.627	0.131	0.177	0.286	1.007	1759.6
	Min.	3644.630	1.210	0.210	1.350	10.760	4908.43
	Max.	4611.700	1.550	0.700	2.180	13.680	10011.12
India	Mean	6844.104	3.026	0.085	0.543	29.081	3733.39
	S.D.	379.202	0.074	0.040	0.060	2.546	591.84
	Min.	6234.420	2.860	0.040	0.460	26.310	2863.24
	Max.	7325.180	3.120	0.140	0.620	32.960	4536.45
Russia	Mean	2903.213	1.310	0.421	4.827	2.668	14028.37
	S.D.	142.873	0.090	0.053	0.236	0.183	1155.45
	Min.	2603.520	1.150	0.340	4.530	2.410	11822.33
	Max.	3044.770	1.430	0.480	5.170	2.880	15374.15
South Africa	Mean	3876.798	1.348	0.105	2.737	10.526	10596.45
	S.D.	186.598	0.059	0.059	0.096	0.418	445.93
	Min.	3665.800	1.260	0.020	2.600	9.750	9746.84
	Max.	4180.420	1.430	0.190	2.920	11.030	11190.09

Table 3: Result of regression analysis – effects of technological innovation on production-based energy productivity.

	Linear regression (robust) [Pooled] **ENGPRD** (Energy productivity) [GDP (US $)/TPES] (1)	**LSDV** regression (robust) [Fixed effects model] **ENGPRD** (Energy productivity) [GDP (US $)/TPES] (2)
Constant	**2556.102**	**4244.35**
	(14.42***)	(5.87***)
INVPC (−3)	**1207.953**	**871.942**
(Per capita innovation)	(4.59***)	(2.92***)
ENGINT	−102.1226	−180.2193
(Energy intensity)	(−0.89)	(−1.00)
RENGSS	**157.0861**	5.8136
(% Renewable energy/TPES)	(18.52***)	(0.17)
RGDP	−0.00473	
(Real per capita GDP)	(−0.11)	
Dummy_Brazil		**4900.287**
		(4.75***)
Dummy_China		210.4615
		(0.96)
Dummy_India		**2656.122**
		(4.76***)
Dummy_Russia		**−776.8416**
		(−2.29**)
No. of observations	30	30
R-square	0.9884	0.9953
F-Test	$F(4,25) = 588.29$***	$F(7,22) = 1161.85$***
Prob > F	0.0000	0.0000
Root MSE	274.32	186.82

The values in parentheses are t-values.
*** stands for significance at the 0.01 level, ** stands for significance at the 0.05 level and * stands for significance at the 0.1 level.

category for South Africa is considered as the reference category or benchmark for the dummy variable analysis.

The per capita innovation is a statistically significant explanatory variable for explaining the variation of energy productivity. The positive sign of the coefficient of INVPC shows that an increase in per capita innovation of environmental-related technology positively increases energy productivity, GDP (US dollar, 2005) per unit of total primary energy sources (TPES). However, the other two explanatory variables, ENGINT and RENGSS, came out as statistically not significant in influencing energy

productivity. In addition, the coefficient of dummy variable for China is not statistically different from zero. This implies that the intercept for China is the same as that of South Africa, i.e., 4244.35 US dollar (2005) per unit of TPES, which can be seen from Figure 3. Further, the positive coefficients for the dummy variables for India and Brazil indicate that energy productivity is the highest in Brazil, followed by India, both of which are more than that of South Africa and China. The negative coefficient of the dummy variable for Russia reveals that its energy productivity is the worst among all BRICS countries.

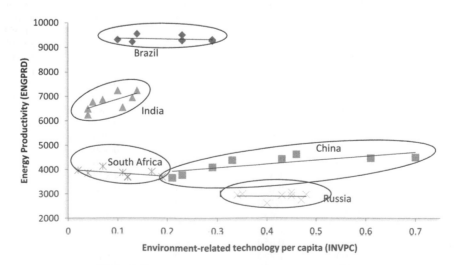

Figure 3: Scatter diagram of ENGPRD and INVPC.

Table 4: Result of regression analysis – effects of technological innovation on production-based CO_2 emission productivity.

	Linear regression (robust) [Pooled] **CO2PRD** (CO_2 Emission productivity) [GDP (US $)/ Kg of CO_2] (1)	**LSDV** regression (robust) [Fixed effects model] **CO2PRD** (CO_2 Emission productivity) [GDP (US $)/ Kg of CO_2] (2)
Constant	−1.8647	−0.87297
	(−17.09***)	(−1.30)
INVPC (−3)	1.0733	0.4571
(Per capita innovation)	(4.08***)	(2.22**)
ENGINT	0.2458	−0.02889
(Energy intensity)	(3.19***)	(−0.21)
RENGSS	0.1589	0.14016
(% Renewable energy/TPES)	(29.45***)	(3.69***)
RGDP	0.0000763	0.000074
(Real per capita GDP)	(3.49***)	(1.87*)
Dummy_Brazil		0.3730
		(0.30)
Dummy_China		0.02995
		(0.31)
Dummy_India		−0.28445
		(−0.57)
Dummy_Russia		0.7378
		(3.05***)
No. of observations	30	30
R-square	0.9961	0.9977
F-Test	$F(4,25) = 2241.56***$	$F(8,21) = 1429.38***$
Prob > F	0.0000	0.0000
Root MSE	0.136343	0.11283

The values in parentheses are t-values.
*** stands for significance at the 0.01 level, ** stands for significance at the 0.05 level and * stands for significance at the 0.1 level.

Production-based CO_2 emission productivity

To find the impact of environment-related technological innovation on CO_2 emissions, a pooled regression was done. The result is given in Table 4 in column (1).

All the t-values of all the coefficients are statistically significant. This implies that all the coefficients, including the intercept, are statistically different from zero – they are statistically significant in explaining the variation of CO_2 productivity, CO2PRD. The value of R-square is 0.9961; it shows that the model is highly fitted, which is justified by the high value of the F-statistics. If the per capita INVPC, before three years, increases by one unit, then

CO_2 emission per GDP would decrease or GDP per CO_2 emission would increase; in other word, CO_2 productivity would increase by 1.075 US dollar (2005).

To capture country-specific heterogeneity, as was done in the case of energy productivity, the four dummy variables for five BRICS countries were included in the regression analysis as control variables. This was done because the scatter diagram in Figure 4 reveals a picture of heterogeneity across the countries, in association between the CO2PRD variable and the INVPC variable. Figure 4 shows almost the same type of pattern of scatter as Figure 3.

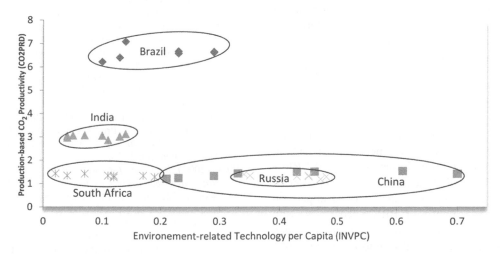

Figure 4: Scatter diagram of CO2PRD and INVPC.

The least-squares dummy variable (LSDV) regression analysis was done with the help of STATA command. The result is documented in Table 4, column (2). The high value of R-square (0.9977) confirms that the regression line is well fitted. A significant F-statistics (1429.38) validates the above statement. Almost 99.77% of the variation of CO2PRD variables can be explained by the included explanatory variables. To avoid the heteroscedasticity problem, the robust regression was run. Here, South Africa was considered as the reference category, as it was also considered in the previous analysis for energy productivity. However, all the coefficients of dummy variables came out as statistically insignificant, except that of Russia. Per capital real GDP has a positive effect on CO2PRD. The higher the percentage of renewable energy supply (RENGSS), the greater the productivity of CO_2 emissions (CO2PRD); a positive sign of the coefficient of RENGSS variable reflects its conformity. The coefficient of ENGINT is negative; this indicates that the higher the energy intensity, the larger the emission per GDP would be, which means lower CO_2 emission productivity. However, the result shows that the coefficient of ENGINT is statistically significant. Lastly, for per capita innovation (INVPC) with three year lags periods, it seems that it is a statistically significant factor for influencing CO_2 emission productivity with its desirable sign. One unit increase in INVPC can increase CO_2 productivity by 0.4571 US dollar (2005) of GDP per kg of CO_2 emissions.

Concluding remarks

This study was undertaken to investigate the effects of per capita innovation of environment-related technologies on production-based energy productivity and production-based CO_2 emission productivity, with the help of data sets collected for BRICS countries. The results of the analysis indicate that the experiences of BRICS countries from 2005 to 2012 reveal that technological innovation (on a per capita basis) had some effects on energy consumption and CO_2 emission in their production sectors. Technological innovation had positive effects on production-based energy productivity in BRICS countries and, consequently, a negative effect on energy consumption in their production sectors. In other words, technological innovation resulted in using lower amounts of energy to produce the same level of output. Therefore, production sectors of firms can improve their energy efficiency if they pursue technological innovation for green technology. The results also indicate that technological innovation had positive effects on CO_2 emission productivity. In other words, technological innovation produced relatively less CO_2 while producing the same level of output. Technological innovation directly and indirectly reduced CO_2 emission per unit of output, thereby improving the CO_2 emission situation of BRICS countries. Regarding other factors, the development of environment-related technology had a sound impact on the sustainable performance of BRICS countries, with green technological innovation helping firms and each country to reduce their energy absorption and CO_2 emissions. This positively helped to improve production-based energy productivity and production-based CO_2 emission productivity.

Policy implications also can be drawn from the experiences of BRICS countries. The primary implication is that to improve energy productivity (or to reduce energy intensity) and CO_2 emission productivity (or to reduce CO_2 emission intensity), BRICS economies should develop and adopt policies that support technological innovation for sustainable technologies.

This study has some limitations, mainly related to data comparability and model specifications. Since BRICS countries are different in various dimensions, obtaining a data set which conformed to inter-country comparability was a challenging issue. In addition, the time taken for this analysis was relatively short in relation to the time gap between technological innovation and technological absorption in production units. Further, there is much scope for investing similar research with more data sets such as expenditure in research and development, technological diffusion and spillover effects, national and international environmental tax policies and foreign direct investment. Finally, country-specific analyses may yield richer, more detailed explanations of relevant issues.

Note

1. Because of the stringent environmental policies in developed countries, many energy intense and polluting firms have been displaced to emerging countries like Brazil, India and China where the environmental laws are not so strong.

ORCID

Swarup Santra (iD) http://orcid.org/0000-0002-9127-3834

References

Aghion, P., D. Hemous, and R. Veugelers. 2009. "No Green Growth Without Innovation." *Bruegel Policy Brief,* 7: 1–8.

Byrne, R., A. Smith, J. Watson, and D. Ockwell. 2011. "Energy Pathways in Low-Carbon Development: From Technology Transfer to Socio-Technical Transformation, Energy Pathways." STEPS Centre, The Sussex Energy Group, University of Sussex.

Carraro, C., and D. Siniscalco. 1994. "Environmental Policy Reconsidered: The Role of technological Innovation." *European Economic Review* 38: 545–554.

Fabbri, Paolo, and Augusto Ninni. 2014. "Environmental Problems and Development Polices for Renewable Energy in BRIC Countries, Federalismi.it, Rivista di Diritto Pubblico Italiano, Comparato, Europeo." October 29, 2014.

Freitas, I. M. B., E. Dantas, and M. Iizuka. 2011. "The Kyoto Mechanisms and the Diffusion of Renewable Energy Technology in the BRICS." *Energy Policy.* doi:10.1016/j.enpol.2011.11.055.

Goulder, L. H., and Koshy Mathai. 2000. "Optimal CO2 Abatement in the Presence of Induced Technological Change." *Journal of Environmental Economics and Management* 39: 1–38.

IPCC (2003). *Revision of the Revised 1996 IPCC Guidelines for National Greenhouse Gas Inventories*, IPCC Expert Group Scoping Meeting Report, Geneva, Switzerland. http://www.ipcc-nggip.iges.or.jp/meeting/pdfiles/2006GLs_scoping_meeting_report_final.pdf

Jaffe, Adam B., Richard G. Newell, and Robert N. Stavins. 2005. "A Tale of Two Market Failures: Technology and Environmental Policy." *Ecological Economics* 54: 164–174.

Lantz, V., and Q. Feng. 2006. "Assessing Income, Population, and Technology Impacts on CO_2 Emissions in Canada: Where's the EKC?" *Ecological Economics* 57: 229–238.

Löschel, Andreas. 2002. "Technological Change in Economic Models of Environmental Policy: A survey." *Nota di Lavoro, Fondazione Eni Enrico Mattei* 4: 105–126.

Markewitz, P., W. Kuchshinrichs, W. Leitner, J. Linssen, P. Zapp, R. Bongartz, A. Schreiber, and T. E. Muller. 2012. "Worldwide Innovations in the Development of Carbon Capture Technologies and the Utilization of CO_2." *Energy & Environmental Science* 5: 7281–7305.

OECD. 2014. *OECD Environmental Outlook to 2050: The Consequence of Inaction*. Paris: OECD Green Growth Indicators.

Pao, Hsiao-Tien, and Chung-Ming Tsai. 2011. "Multivariable Granger Causality between CO_2 Emissions, Energy Consumption, FDI (Foreign Direct Investment) and GDP (Gross Domestic Product): Evidence from a Panel of BRIC (Brazil, Russian Federation, India, and China) Countries." *Energy* 36: 685–693.

Requate, Till, and Wolfram Unold. 2003. "Environmental Policy Incentives to Adopt Advanced Abatement Technology: Will the True Ranking Please Stand Up?" *European Economic Review* 47: 125–146.

Shaw, Nathan. 2016. "BRIC by BRIC: How Our Emerging Markets are Tacking Carbon Emissions?" *Energy and Renewables, Sustainable Business Toolkit*. Accessed March 23, 2016. http://www.sustainablebusinesstoolkit.com/bric-countries-carboon-emissions/.

Socolow, R., R. Hotinski, J. B. Greenblatt, and S. Pacala. 2004. "Solving the Climate Problem – Technologies Available to Curb CO_2 Emissions." *Environment: Science and Policy for Sustainable Development* 46 (10): 8–19.

Tian, Huifang. 2015. "Gathering Momentum for BRICS Cooperation on Climate Change." *Chinese Academy of Social Sciences*. May, 2015. Retrieved from: http://www.nkibrics.ru/system/asset_docs/data/5568/7b17/6272/693b/d153/0000/original/Huifang_Tian_Session7.pdf?1432910615.

UNEP. 2013. "Annual Report." http://www.unep.org/annualreport/2013/docs/hr_ar2013.pdf.

Wang, Z., and C. Feng. 2014. "The Impact and Economic Cost of Environmental Regulation on Energy Utilization in China." *Applied Economics* 46: 3362–3376.

Innovation for inclusive development, public policy support and triple helix: perspectives from BRICS

Chux U. Daniels ⓘ, Olga Ustyuzhantseva ⓘ and Wei Yao ⓘ

This paper investigates the role of innovation in (inclusive) development – subsequently referred to as innovation for inclusive development (IID) and the links to public policy in BRICS (Brazil, Russia India, China and South Africa) countries. To achieve this aim, the authors examine the roles played by Triple Helix actors (THA), namely university, industry and government in IID activities across BRICS countries, drawing on the national systems of innovation (NSI) framework. The findings indicate that: (1) significant gaps exist in literature useful in advancing our knowledge of innovation as a mechanism for inclusive development; (2) BRICS countries focus, mostly, on innovation in the broad sense, with less attention paid to IID, the essence of this paper. One reason for this gap may lie in the inability to conceptualize and theorize innovation as a mechanism for including the wider society in socio-economic and development activities, or the lack of appreciation of the potential roles that innovation can play in development; (3) there is absence of specific public policies and policy support for IID in BRICS; and, (4) paucity of empirical evidence needed to critically analyse and explain the roles that THA in BRICS play in innovation ecosystems.

Introduction

Developing countries increasingly face the problem of growing economic inequality and lack of access to innovation for a large segment of their population which is outside the formal economy and formal innovation system (Cozzens 2010; Prahalad 2012). In BRICS (Brazil, Russia, India, China and South Africa) the size of the 'informal economy'[1] is estimated to account for 32 to 83% of non-agricultural employment (ILO Department of Statistics 2012). Innovation (as a mechanism) for inclusive development (IID) potentially offers one avenue to address the growing challenges of, for example, poverty, inequality and exclusion of large segments of the population from national socio-economic and development activities (Kraemer-Mbula and Wamae 2010; Marcelle 2014; UNCTAD 2014).

IID operates on two principles: enabling participation of marginalized people in the mainstream economy and their involvement in innovative and development activities. In some countries, India for instance, the informal sector reveals extensive innovative activities resulting in effective solutions to development challenges facing rural people and communities. Some of these challenges are unmet by innovations in the formal sector as defined by the national systems of innovation (NSI) framework (Freeman 1987; Lundvall 1992; Nelson 1993). Empirical evidence suggest that the informal sector has the potential to generate various types of innovations (Anderson and Markides 2007; NIF 2013; Brem and Wolfram 2014), such as, radical, frugal, social, and grassroots; some of which have been regarded by scholars in science, technology and innovation (STI) domain as innovations geared towards inclusive development (IID). Such innovations help to integrate excluded people into national mainstream economic activities and formal innovation systems. Relatedly, IID generates specific innovation environment,

challenges, and prospects for conventional NSI actors: university, industry and government – as captured in the triple helix (TH) framework. In this paper, we focus on BRICS.

Establishing conceptual clarity

In establishing conceptual clarity, we use inclusive innovation, innovation, inclusion and IID in the following manner. 'Inclusive innovation' comprises 'initiatives that serve the welfare of lower-income groups, including poor and excluded groups. While growth dynamics have lifted many people out of poverty, they have not eliminated poverty and exclusion, which continue to affect millions of people' (OECD 2015b, 9). In this paper, we focus on 'innovation' in a broad sense, i.e. the application of a new or improved product, process, service, or organizational or marketing strategy that addresses a specific societal challenge or challenges. By 'inclusive' we refer to the concept of 'opening-up' or 'broadening out' innovation, innovation activities, project, services, products and processes, in the broadest possible sense without any form of exclusion. Against this backdrop, therefore, IID as used in this paper refers to innovation in its broad sense that facilitates the participation of the widest possible spectrum of the society in socio-economic development activities. This is different from designing and applying (innovation) products (e.g. M-PESA, Tata Nano) or services (e.g. microcredits) that are specifically targeted at the poor, marginalized or low-income segments of the society. This in itself is exclusion.

The objectives of the paper

BRICS countries have been identified as sharing some common characteristics in areas that include innovation, development challenges and opportunities, and similarities in economic indicators. Our focus in the analysis

Table 1: Economic and social indicators on BRICS versus OECD.

	BRICS					OECD
	Brazil	Russia	India	China	South Africa	
Population in millions	201	144	1211	1357	52	N/A
GDP (in millions)	2246	2096	1871	9185	382	N/A
	(2013)	(2013)	(2013)	(2013)	(2012)	
Unemployment rate (% of labour force, annual average)	6.6	5.5	3.7	4.1	25.1	8.1
	(2013)	(2013)	(2012)	(2013)	(2012)	(2013)
Unemployment rate for population aged under 25 years (%)	14.6	13.8	28[a]	–	51.5	19[b]
	(2012)	(2013)	(2011)		(2012)	(2011)
Poverty: Headcount ratio at $1.90 a day (2011, % of population)	5.5	0.1	21.2	–	16.6	–
Income distribution (2009–2013)						
Income share held by highest 10% of population[c]	41.8	32.2	30.0	–	51.3	9.6
	(2013)	(2012)	(2011)		(2011)	(2012)
Income share held by lowest 20% of population[d]	3.3	5.9	8.2	-	2.5	-
	(2013)	(2012)	(2011)		(2011)	
Gini index[e]	0.50	0.42	0.37	0.47	0.64	0.31
	(2012)	(2012)	(2010)	(2013)	(2009)	(2011)
						0.32
						(2012)

[a]Age 15–29 years.
[b]Online OECD Employment database: http://www.oecd.org/employment/onlineoecdemploymentdatabase.htm, Accessed: 10 March, 2016.
[c]World Bank: http://data.worldbank.org/indicator/SI.DST.10TH.10/countries, Accessed: 10 March, 2016.
[d]World Bank: http://data.worldbank.org/indicator/SI.DST.FRST.20/countries, Accessed: 10 March, 2016.
[e]Higher value indicates a higher level of inequality in a country. 0 = complete equality; 1 = complete inequality.
Source: http://www.ies.gov.in/pdfs/sunita-sanghi-and-a-srija.pdf, Accessed: 10 March, 2016.
Source: Adapted from BRICS (2014) and IMF (2015), OECD (2014).

and discussions that follow is to (1) examine the five BRICS countries and identify some of the IID specificities, and (2) examine the role that public policies play in supporting IID in each of the countries. Using the TH, we attempt to outline the roles played by university, industry and government as innovation actors and provide generalizable recommendations across the board. We locate the discussions within the NSI framework as the guiding conceptual and analytic framework.

Some basic information on BRICS

In order to make a case for the need to stimulate IID and the relevant public policy support, it is important to demonstrate why BRICS countries need IID in the first place. In Table 1, we present brief data on BRICS poverty,[2] Gini Index/inequality,[3] and unemployment. We compare the data on BRICS with OECD averages. The aim is to highlight the potential roles that IID can play in addressing societal challenges, such as poverty, inequality and unemployment in BRICS, thereby ensuring that innovation contributes to socio-economic development.

'Reducing inequality, while at the same time promoting more and better jobs in the emerging economies (EEs), BRICS included, requires a multipronged approach' (OECD 2011a, 49). We argue in this paper for innovation practices and policies that foster inclusive developmentare imperative in this multipronged approach. In a 'Special Focus: Inequality in Emerging Economies (EEs)', OECD (2011a) records that 'inequality in Brazil, Russia and South Africa have reached high levels and remain high' (OECD 2011a, 49). Inequality is still a major challenge, Keeley (2015). As Figure 1 below illustrates, the Gini Index or Coefficient (a proxy for the measure of [income] inequality) for South Africa is more than twice the OECD average, while Brazil's is almost twice.

The data presented above help to justify the aim of this paper, that is, the importance of innovation as a mechanism for addressing socio-economic challenges and promoting inclusive development in BRICS. Although we do not analyze these development indicators in-depth, as this is outside the scope of this paper, we argue that they help to demonstrate the importance of IID and therefore deserve a brief mention. Furthermore, these basic data help to provide background information useful in situating the literature review and discussions that follow within the BRICS context.

The rest of the paper is organized as follows. In the next section, we review the relevant literature while in the section following that we briefly explain the methodology. In the next two sections, we present and discuss the findings and then conclude, respectively.

Review of relevant literature

Even in recent studies, literature and policy priority setting (see, for example, IMF 2014; IMF 2016), IID is still not sufficiently considered and recognized as one of the essential mechanisms for global growth and development. This represents a gap this paper seeks to address. Therefore, in order to cover the issues raised in this paper, the review of relevant literature attempts, in broad terms, to address some basic information (such as inequality levels) about each country. The aim is to help contextualize the literature review, analysis and discussions. We also briefly review literature on innovation and innovation system for each country, and highlight TH actors involved in IID support

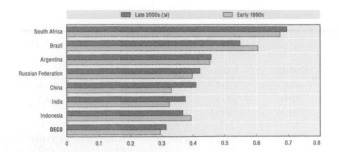

Figure 1: Change in inequality levels, early 1990s versus late 2000s.
Source: OECD 2011a

and related public policies. We provide brief statements, where applicable, on whether there are papers discussing IID and related public/innovation policies. In cases where literature on any of these elements is limited or non-existent for one or two countries,[4] we highlight it as a gap, which we hope this paper helps to fill.

Triple helix and national systems of innovation frameworks

The triple helix (TH) argues for a shift from bilateral interactions of double helixes to interactions of a triple helix consisting of university, industry and government relationships in innovation ecosystems. This makeup, according to the TH framework, fosters innovation, the creation of hybrid enterprises and joint innovative projects in the modern knowledge economy (Etzkowitz and Dzisah 2008, 10; Etzkowitz and Leydesdorff 2000).

As a framework, TH is widely used in the literature to analyze the relationships and interconnections of the main innovation stakeholders of university (or academia), industry and government. Using TH to analyze IID efforts, as we do in this paper, is novel. First, we do not intend to provide an in-depth analysis of TH in this paper. Second, we acknowledge that there is literature which focuses on discussions about the role of other innovation actors such as users, the informal sector and civil society, and the importance of public opinion (see for example, Etzkowitz and Zhou 2006) with respect to TH. However, by using TH, the framework helps to clarify that the paper focuses on the roles played by university, industry and government (U-I-G) in IID. In the discussions that follow, we also draw insights from the national systems of innovation (NSI) framework (Freeman 1987; Lundvall 1992; Nelson 1993) as it acknowledges the role of U-I-G as innovation actors within national systems.

BRICS – Brazil, Russia India, China and South Africa

Scerri and Lastres (2013) show that there are economic, innovation, policy and development complementarities, similarities and sufficient differences among BRICS countries justifying the study and analysis of these countries as a group of nations. The authors stress the need for further extensive studies on BRICS economies and innovation. In addition, they acknowledge that major challenges still remain in these countries in the areas of inequality, exclusion and policy support for innovation. Although the work of Scerri and Lastres (2013)

does not directly address IID, we posit that innovation, as a tool for inclusive development, can be an important mechanism in addressing some of the gaps identified by Scerri and Lastres (2013).

In addition, IID and public (particularly innovation) policies can play significant roles in efforts to lower socio-economic challenges such as inequality (the gap between the rich and poor) and exclusion (Galbraith, Krytynskaia, and Wang 2004; OECD 2011a; BRICS 2014; Keeley 2015), poverty (World Bank 2004; OECD 2011b; IMF 2014) and unemployment (World Bank 2005, 2010; NDP 2011; Gupta 2013; Marcelle 2014; UNCTAD 2014), all of which impact on development.

In the sections that follow, we review some of the relevant literature useful in analyzing IID and public policy, focusing on perspectives from BRICS countries and the roles that university, industry and government (i.e. triple helix actors, THA) play in the processes involved. We start with a review of literature on Brazil.

Brazil

In Brazil, although public policies have played a positive role in promoting innovation, differences in income distribution and inequality remain a source of concern. This strengthens the argument for a new approach to addressing inclusion, combating marginalization and promoting innovation that address societal needs (Cassiolato and Soares 2013). This point highlights the role of IID in public policies.

In a similar vein, Koeller and Gordon (2013) advocate an NSI that adequately addresses the very important connection between innovation policy and development policy, arguing that development policies underpin other policies, including innovation policies. The authors, in line with the characterization of development advanced by Cassiolato and Lastres (2008), stress that there is need to define NSI in ways that suit the contexts and perspectives of Brazil and that attention must be paid to structural implications that have bearings on innovation and development. In addition, they explain that policies have impacts on the roles that institutions play in development and innovation processes. Here, again, we find the need for active interactions and collaborations between TH actors, as NSI stakeholders, in fostering development and policies necessary for supporting IID in fast-developing countries.

Furthermore, the authors observe that in Brazil, structural barriers and ill-conceived intrinsic logic have in the past enabled the perpetuation of underdevelopment. In noting the challenges in the social and economic structures, Koeller and Gordon (2013) submit that these gaps need to be considered during policy formulation and addressed by appropriate development and targeted innovation policies. Here we see a clear case for THA working together in order to foster innovation, enhance policy and achieve development goals. A key role for government, the authors submit, is the formulation of specific policies targeted at IID. In this paper, we posit that for such development and innovation policies to be effective, it is important that they not only focus on mission-oriented innovation projects and activities, but also strive to

Table 2: Findings: triple helix actors roles in IID and policy support.

BRICS Country	Triple Helix Actors (THA)		
	University	Industry	Government
Brazil	• Establishment of incubator movements, science parks, and angel networks by universities, research institutions, firms and municipalities (Etzkowitz and Dzisah 2008) • Innovation research, knowledge production and dissemination with context specificity to Brazil (Cassiolato and Lastres 2008) • Departure from US and other imported capabilities towards the development and dependence on domestic capabilities (Scerri and Lastres 2013) • Greater research, innovation collaborations at national and international levels • Connection between innovation policy and development policy; development policies underpin innovation policies (Koeller and Gordon 2013)	• Commercialization of IID/grassroots innovation products • Firms creating incubators around universities • Influenced by government policy and although started in the software sector, the incubator movement in Brazil now permeates the entire innovation ecosystem, involving many academics teaching courses in entrepreneurship, various universities, departments and industrial sectors across Brazil (Araujo and Costa 2014).	• The establishment of infrastructure and institutions, e.g. BNDES (Scerri and Lastres 2013) • The setup of research centres and boosts in innovation, e.g. Brazilian Innovation Agency and CNPq, the National Council for Scientific and Tech. Development (insert refs). • Remarkable increases in innovation investment and new models of funding, e.g. the Newton Fund set up to bolster Brazil-UK collaboration in science and innovation • Finance and various other institutions as a way of promoting innovation and fostering development (Scerri and Lastres 2013) • Positive impacts from public policies and policy support for innovation in general (Cassiolato and Soares 2013)
Russia	• Putting IID into policymakers' agenda (Gokhberg and Kuznetsova 2011b)	• Lags behind, e.g. China, due to factors that include poor protection of IPR, limited finance, low investments in ICT, low/inefficient R&D and skills gaps (Gianella and Tompson 2007; EBRD 2011)	• As semantically IID is not in circulation in policy papers, there are no measures or interventions of government for IID support. The more or less close government initiatives are measures to support social entrepreneurship though providing grants for social entrepreneurs.
India	• Scouting, documenting of GRI; providing its infrastructure for testing and validation of innovation; providing incubation services for innovators in situ (training); incubator units • In general, university became a networking base for pushing GRI movement and involving government and business in its development (Ustyuzhantseva 2015)	• Start-up firms, industrial clusters (Basant and Chandra 2007) • Commercialization, innovation distribution, mass production (Ustyuzhantseva 2015)	• Setting up institutional infrastructure for IID development (NIF, GIAN, GTIAF, and MVIF).[a] Created institutions provide IID with micro venture finances, help in patenting, business incubation (NIF, 2006–2012). • Awards, competitions and exhibitions to scout and develop the best innovations for inclusive development
BRICS Country	University	Industry	Government

(Continued)

Table 2: Continued.

BRICS Country	Triple Helix Actors (THA)		
	University	Industry	Government
China	• Selection, improvement and diffusion of technology (Yao, Mosi, and Xiaodong 2015) • Dissemination and absorption of knowledge, access to science and technology knowledge (Yao, Mosi, and Xiaodong 2015) • Primary intermediary service, training skilled labour and educating high level talents (Yao, Mosi, and Xiaodong 2015)	• Builds cross-sectoral value network (Xing, Tong, and Chen 2010) • Integrates stakeholders' capability and resources, thus improving value chain imperfections and institutional void in BOP market (Xing, Tong, and Chen 2010) • 'Triggers' and leads inclusive innovation activities in NSI (Yao, Mosi, and Xiaodong 2015) • Provides insight into real or potential demand of low-income people, R&D activities, flexible and low cost manufacturing systems, customized distributing channels and market promotion (Yao, Mosi, and Xiaodong 2015)	• Provides rules, regulations makers and of innovation activities, act as enablers (Yao, Mosi, and Xiaodong 2015) • Policymaking, stimulate innovation, a conducive environment and risk protection mechanisms; builds relationships and trust among stakeholders (Xing, Tong, and Chen 2010) • Sets up institutions beneficial to inclusive innovation and eliminates barriers (Shao, Xiao-qiang, and Yun-huan 2011) • As an enabler, improves the efficiency and diffusion of regional inclusive innovation and helps the poor build capability (Shao, Xiao-qiang, and Yun-huan 2011) • Provides financial support and subsidies, help to set up production bases, to help expand the domestic and overseas markets (Xing, Tong, and Chen 2010)
South Africa	• Support for the capability of researchers within public research institutions. • 'Entrepreneurial universities' – manage and process contracts, patent applications, royalties and protecting IPRs, consultancy, university-owned companies (on innovation, not necessarily for IID in all cases) • Incubators; • Carry out, co-ordinate and support research activity in general, and contribute to firms' innovation (Kruss 2008) • Bringing updated knowledge from developed countries to existing local firms – relevant for IID (Albuquerque et al. 2015) • Fostering interactions, knowledge exchange, and learning processes with industry (Albuquerque et al. 2015)	• Production of 'consumer goods serving low income/BoP markets (through the BoP learning lab)' (COFISA 2009, v) • Harnessing existing 'points of interaction' between universities and private research institutes (Kruss 2008; Albuquerque et al. 2015) • Increasing use of external knowledge partners (universities, public research institutes and technology centres) • Taking advantage of innovation-related capabilities, specialized selection (in firms and SMEs alike), greater interaction (with suppliers and users), networks, strategic cooperation (Albuquerque et al. 2015) – mostly focused on innovation in general, not necessarily for IID in all cases.	• Funding R&D and innovation (not necessarily IID) through relevant agencies, e.g. Human Science Research Council (HSRC); 'Entrepreneurial State' initiatives – government entities – start entrepreneurial projects • Supportive government programmes, such as THRIP, Innovation Fund, sectoral and other Special Purpose Vehicles, as well as plans for a Provincial Innovation Council and a Regional Innovation Forum (Kruss 2008; COFISA 2009) • Knowledge production/circulation, e.g. IID seminars, workshops, stimulating collaborations (COFISA 2009) • Government initiatives and funding seen as promoting innovation networks, partnerships and interactions between universities, industry and government (Kruss 2008) • Policymaking, though not entirely yet clear how effective these are (Phiri et al. 2013)

ᵃNational Innovation Foundation (NIF), Grassroots Innovation Augmentation Network (GIAN), Grassroots Technological Innovation Acquisition Fund (GTIAF) and Micro Venture Innovation Fund (MVIF).

Source: Based on research data

address grassroots innovation[5], recognized as IID in this paper. This balance is essential for development. Nevertheless, significant gaps remain in the literature both on IID and policies.

Building on the notion of innovation as the engine of economic growth, Koeller and Gordon (2013), in analyzing Brazil, recommend that innovation policies should be thought of as development policies. This recommendation however carries with it some caveats, e.g. growth does not automatically imply development. In order to reach development, innovation policies must be integrated with development policies, and be specific to the context of underdevelopment, for instance in the areas outlined above – poverty reduction, inequality and employment. We argue that to help address these (under)development challenges, THA have important roles to play through the formulation of policies, scaling up (where applicable) and use of innovation projects, activities, products and services that are inclusive.

Russia

The concept of inclusive innovation emerged in developing countries, as a result of poverty, people's lack of access to their basic needs (food, clean water, housing) and because they were deprived of the benefits of scientific and technological progress (Mohnen and Stare 2013). However, inclusive development is considered as a process of socio-economic integration (Foster and Heeks 2013), aimed at providing access to basic goods and services and meeting the needs of marginalized groups of society, mainly at the bottom of the pyramid (World Bank 2013). We have discussed our preference for and use of IID in the first section of this paper. We retain the 'inclusive innovation' phrase in cases where authors have specifically used it or where it appears in the literature being reviewed to reduce the chances of misrepresentation, for example, as in this instance and in the discussions on India, below.

The analysis of Russian scientific literature reveals an obvious delineation of the study of socio-economic development of Russia and issues related to innovation development (Sokolov 2006; Gokhberg et al. 2009). The first time the question of inclusion in innovation policy was raised in expert discussions on Russian development strategy was in 2011. The strategy discussions examined this issue in terms of inclusion of Russia's wider population in the innovation process with the aim of achieving two goals, to: (1) overcome social disparities and (2) involve the wider society in innovation activities. Citizens have to see themselves as part of innovation and development processes (Foresight 2011).

The final results of the discussions were summarized and included in Russia's innovation development scenario. For the first time, the social function of innovation, including the integration of vulnerable groups in the innovation process, was included in a strategy developed by experts and proposed to the government for implementation (Стратегия 2020 2011). In support of the proposed strategy, various scholars published a number of papers devoted to the importance of integrating social needs into national innovation policy. Gokhberg and Kuznetsova

(2011a), for example, emphasized the need for integrating vulnerable population groups into innovation processes as one of the key directions of public regulation aimed at supporting innovation in Russia.

The authors consider inclusive innovation as part of the social function, which they argue, also includes the development of human capital in the innovation sphere and supporting the creative class. In another work, Gokhberg and Kuznetsova (2011b) identified three main sets of possible inclusive innovation support programmes: support for youth and children creativity; development of mass innovative entrepreneurship through building a system that promotes creative ideas; and inculcation of innovative products and practices into social services, healthcare and other public services which are provided for vulnerable groups of society. Thus, innovation for inclusive development entered policymakers' agenda. However, experts noted that the implementation of this strategy might be limited by political opportunities in Russia (Дмитриев 2011).

Various Russian scholars, mainly economists, who examine the Russian innovation system within the concept of the triple helix, reflect a general tendency to underestimate society (and its needs) as one of the main actors in innovation processes (Маховикова and Ефимова 2010; Smorodinskaya 2012). Triple helix in the Russian context is interpreted as the networking of academia, industry and government, with a predominance of pair communications where government is always one party in each communication (Smorodinskaya 2011; Dezhina 2014; Zaini, Lyan, and Rebentisch 2015). In this context, industry and academia build their relationship indirectly, through government agencies and officials. Issues of government innovation policymakers' accountability to society and innovation benefactors have also been raised (Dezhina and Kiseleva 2008). Despite the acknowledgement for the need for innovation in Russia, innovation is still considered in terms of providing cross-sectoral interaction at the level of the region or industry (Акбердина and Малышев 2011; Smorodinskaya 2011; Пахомова 2012), with far less attention paid to the inclusion of the wider society in innovation processes.

India

India demonstrates great potential in building an IID ecosystem, which could not happen without systematic support from the government through policy initiatives and measures. At policy level, the idea of inclusive development has been stated in various documents (11th Five-Year Plan 2012; Mid Term Appraisal 2012; Ustyuzhantseva 2014), although the issues of social development and poverty reduction through inclusive growth arose before. The Indian government stated inclusive growth as one of the basic goals of legislation and budgetary allocations (George, McGahan, and Prabhu 2012). The grassroots innovation movement started in India in the late 1980s as a means of scouting for and documentation of innovation developed by ordinary people, i.e. grassroots innovators. One of the most important findings of research devoted to this movement was reference to the base of pyramid (BoP) as a source of innovation, and not just as innovation recipients (Gupta 2006). Thus, by 2007, scholars had considered the development of inclusive innovation in two ways. The first was in increasing capacity of formal institutions to meet the needs of BoP (Utz and Dahlman 2007), while the second was in promoting and supporting grassroots innovation (GRI) through adding value to these innovations (Gupta 2007; Utz and Dahlman 2007).

In India, inclusive innovations are mostly considered as grassroots innovations (Gupta 2006), frugal innovations, which are often called 'jugaad' (Rajou et al. 2012), and innovation for the BoP (Prahalad 2004). Gupta argues that 'jugaad' innovation cannot be considered as inclusive innovation, as, in fact, it is a makeshift approach, not geared towards sustainable solutions (Gupta 2014). As we have argued earlier, a more appropriate approach to 'inclusive innovation' would be to conceptualize it as innovation – in all its forms – used for inclusive development that seeks to involve the entire society as much as possible. Nevertheless, GRI has become one of the main focuses of India's government policy measures and infrastructure to support and promote inclusive innovation.

India's innovation policy and innovation systems research ecosystem has evolved around the broad OECD framework, upon which a country's innovation system is analyzed from the perspective of technical innovation and the supporting system (Oslo Manual 2005; Sinha 2011). OECD's efforts at formulating an inclusive growth agenda began in 2012. The first evaluation of Indian innovation policy with regard to inclusive development was done at the OECD Conference in Paris in 2012. It was observed that despite the relatively early emergence of the 'inclusion'[6] concept in India, the top-down model of innovation development had prevailed. So, this model has not provided inclusive growth in India (Krishna 2014). The economic reforms of the 1990s resulted in rapid economic growth and the development of innovative industries. However, this growth also brought issues of growing inequality and the need for inclusive growth into the political discourse (SIP 2015).

Utz and Dahlman (2007) highlighted a number of IID challenges in India such as high transaction cost of documenting innovation, the need for systematic and extended value addition, commercialization through diffusion and dissemination, funding of industrial production and distribution of innovation. The authors proposed a number of initiatives and interventions for government, business and NGOs useful in overcoming the challenges. For example, they drew attention to the efforts of the informal sector to create and absorb knowledge and to the grassroots innovation initiatives in particular. Besides the need for an assessment of the impact of this innovation on improving the livelihoods of people in the informal sector, the authors recommend such specific measures as creating a national fund to acquire rights to grassroots innovation, developing common fabrication laboratories and testing centres for innovation validation, and developing a nationwide strategic plan to add value to local knowledge, innovation

and practices through collaboration with public and private R&D institutions.

The government applied some of the proposed initiatives. For instance, India has placed inclusive innovation as a main policy focus for the 2010 decade (OAPM 2011). It includes the creation of institutional and financial infrastructure for inclusive innovations (such as National Innovation Foundation, India Inclusive Innovation Fund and Micro Venture Innovation Fund) (Heeks et al. 2013; Ustyuzhantseva 2015). Despite such impressive steps, efforts to capture inclusive innovation at policy level are still evolving. The new science, technology and innovation policy, which was released in 2013, pronounced Indian society as the main NSI stakeholder (STIP 2013). This new policy, however, has been criticized for replicating old and ineffective 'linear model of innovation' (Krishna 2013).

China

Inclusive growth is an indispensable solution to China's future development. Taking full advantage of the enormous consumption, production and entrepreneurship potential contained among the poor, inclusive innovation not only provides enterprises with the competitive advantage difficult to imitate (Zhao et al. 2014), but also offers equal opportunities for the poor who survive at the social network margin to participate in economic growth, make a contribution and share reasonably in the fruits of growth (Tang 2013; Xing, Zhou, and Tong 2013).

The birth of inclusive innovation in Chinese scholarship was greeted with theoretical research. Wu and Jiang (2012) proposed three theoretical mechanisms on how inclusive innovation promotes inclusive development, namely, reducing barriers, upgrading ability and changing institutions through innovation in the perspective of social exclusion. Li (2013) submits that in China institutional arrangements based on cooperation is the way to realize inclusive growth.

Empirical studies, such as Gao, Zhou, and Cao (2013) conceptualize 'inclusive industrial innovation' and propose that it is necessary to establish and improve the public service system, improve the income distribution system, carry out market-oriented reform, and transform government's functions in order to achieve the goal of inclusive industrial innovation. Xing, Tong, and Chen (2010), in a multiple case study carried out on six business enterprise operations in the rural areas of China, found five key components of an enterprise's business model, namely: local capability, value proposition, value network, key activities and profit model. Tang (2013) explored five factors that influence inclusive innovation behaviour of Chinese small and micro technology-based firms. These are capital, allocation of resources, organization network, agency, partnerships and entrepreneurship of the CEO.

However, most research on inclusive innovation in China is conducted from the perspective of innovation systems – because inclusive innovation faces challenges such as lack of market information, deficiency in knowledge and skills, imperfect institutional systems, backward infrastructure and limited access to financial services.

These constraints cannot be solved by a single enterprise, but should be supported by other stakeholders, such as government, NGOs, local communities, research institutes, universities, financial institutes and intermediary institutions. Thus, regional inclusive innovation systems can be constructed by the three subsystems: 'innovation entity subsystem', 'innovation support subsystem', and 'innovation environment subsystem' (Shao, Xing, and Tong 2011).

Gao, Liu, and Zhou (2014) propose an evaluation system for Chinese inclusive innovation. The result shows that Chinese (regional) inclusive innovation performance declines from the east to the midwest and is closely linked to local economic development, openness and industrial structure. The implication for development is that more attention should be paid to rural areas and regions, underpinned by the national systems of innovation framework and innovation systems thinking, as opposed to economic development at national level. The recommendation is for China to maintain a reasonable industrial structure, build a balanced innovation system and avoid the polarization of resource allocation. It is hoped that this will enhance innovation efficiency and performance.

South Africa

Addressing poverty and inequality is South Africa's greatest challenge – Foreword, by Asad Alam, Country Director for South Africa, World Bank (World Bank 2014).

Evidence from South Africa demonstrate the existence of close cooperation between industry and the higher education sector, as well as between university, industry and government (Kruss 2008). The essence of these interactions is to help bridge identified gaps in education and skills, with the ultimate aim of knowledge generation and dissemination. Studies have also focused on the nature and forms that these interactions take. However, little is known about how these cooperations influence IID and related policy, which this paper investigates. Kruss (2008), in analysing these interactions, explored the shifting relationships and partnership between triple helix actors, and the impact this has on research in universities, and on knowledge production and dissemination in South Africa. In this analysis, Kruss found that old and new forms of organizational relationships co-exist, with old forms tending to prevail, arguing that these may have counterproductive implications, thus advocating a balance between old and new forms of partnership, each with its peculiar functions and the creation of new forms of 'knowledge-intensive networks' (2008, 3, 15).

Consequently, we observe that all three levels of government (national, provincial and local), as well as public research institutions, are involved in supporting innovation in South Africa (COFISA 2009, 24). The presence of science and technology parks, which act as vehicles for university-industry cooperation, further demonstrate South Africa's resolve to enhance economic development through science, technology and innovation (COFISA 2009). However, in spite of these efforts, challenges remain in areas that include measurement of innovation, capabilities (organizational and individual), limited

attention to innovation in most of academia, the role of THA on IID and public policy support for IID (COFISA 2009, 47; OECD 2012, 2013a, 2013b, 2015; Daniels forthcoming).

There are also concerns related to government initiatives seen as impediments to innovation networks, an example of which is the intellectual property act, although designed in good faith 'may stifle, rather than support innovation' (COFISA 2009, 56). As Phiri et al. (2013) maintain, policies adopted in South Africa's post-democratic era 'have produced and reproduced social and economic inequalities which have hampered inclusive innovation, development and the nascent burgeoning of innovation in the informal sector'. Some studies have observed that in spite of the acknowledged cooperation and interactions among triple helix actors, and the potentials of IID in South Africa, efforts aimed at addressing the needs of the poor and marginalized people are still at their rudimentary stage (Lorentzen and Mohamed 2009; Pogue and Abrahams 2012; see also, for example, Soares and Cassiolato 2013). In addition, the ability and efforts to measure IID's contributions to South Africa's national growth and socio-economic development is still weak (Daniels 2014; Daniels forthcoming). These gaps and weaknesses identified are not peculiar to South Africa or BRICS, but rather a situation currently prevailing in many other global South countries, resulting in part from the choice of measurement indicators. We provide more insights on these points in the discussion section.

Synthesizing the insights from the literature reviewed

In this paper, we have chosen to explore individual countries independently in the review of literature, while taking the opposite approach in the discussion, i.e. discuss the findings from the countries together as a whole. The rationale behind this stems from the reasoning that a cross-country review for the individual actors[7] involved in IID in each of the BRICS countries examined would be impractical to achieve within the limited length of a journal. This underpins the rationale for doing a country-level review as opposed to an actor-level review.

In examining the literature, we find gaps in the way that IID is conceptualized and theorized. We note that such gaps increase the difficulty of academic research and operationalization in industry, and limit our knowledge of IID ecosystems, institutions (formal, informal, intermediaries, hybrids), landscape and dynamics (Kruss 2008). We also observe the lack of suitable indicators and the inability to effectively measure the impact on and contributions of (inclusive) innovation on socio-economic development (Kraemer-Mbula and Wamae 2010; OECD 2012). Against this backdrop, we argue in this paper that there is a need to diversify IID literature from case studies on isolated products and services, towards providing a more macro approach, which we do in this paper. The insights from our analysis of the literature on IID reveal that IID can be a useful mechanism for ensuring that the benefits of innovation are optimized, while also remaining relevant to poor and rural communities.

Methodology

In synthesizing the insights from the literature review, we provided the rationale for doing a country-level review as opposed to an actor-level review. We draw from academic materials that support, disprove or justify the existence (or not) of IID, the roles of university, industry and government (as triple helix actors [THAs]) and appropriate public policies that support (or not) IID in BRICS. The sources of data utilized also include project and official government (e.g. strategy and policy) documents. Furthermore, we build on our knowledge and experience of this sector. While some documents draw extensively from primary data captured in the respective countries, others build on materials from international scholars, authors and organizations external to the respective countries.

Given that this paper is of exploratory nature and part of the initial stage of an IID research agenda we recently embarked on, we see the findings and research questions posed as starting points that should be followed by further empirical research to verify information and gain a greater depth of understanding of the issues addressed. The data analysis methodology follows the TH framework and focuses on the roles of three main innovation actors: university, industry and government (Table 2).

Discussion

Cassiolato and Lastres showed that development is neither linear nor sequential, but rather is a unique process that 'depends on several aspects related to political, economic, historic and cultural specificities that occur from long-run structural changes that generate ruptures with historically established patterns' (2008, 7). In this sense, we see the importance of context-specificity, learning, innovation, instability (ruptures and reassembling) and dynamism as exemplified by the triple helix framework, thus linking back to the active engagement and roles played by triple helix actors.

The findings highlight roles played by university, industry and government in innovation, with bearing on IID. With two billion BRICS citizens estimated to join the global middle class by 2030 (Cassiolato and Soares 2013), this potentially has a significant impact on global poverty levels, inequality and social exclusion. Brazil presents an example of THA role in innovation by the establishment of incubator movements, science parks, and angel networks by universities, research institutions, firms and municipalities. This has helped to advance knowledge capitalization and the 'realization that the incubator is fundamentally a teaching programme for individuals to learn to function as an organization', not simply an activity reserved for firms (Etzkowitz and Dzisah 2008, 5).

Other financing mechanisms, institutional settings, organizational interfaces and knowledge production and dissemination arrangements are being put in place. Brazil presents an interesting and important example of government support for innovation through efforts that focus on the creation and strengthening of science and technology (S&T), finance and various other institutions as a way of promoting innovation and fostering development (Scerri and Lastres 2013). Efforts in this direction

include the creation of National Bank for Social and Economic Development (BNDES).

At the 6th BRICS summit in July 2014, the creation of BRICS Bank, seen as a 'rival' or 'alternative' to the US-dominated World Bank and IMF, is another such step that echoes this strategy (Griffith-Jones 2014). These steps reflect the choice of development and innovation strategies designed to achieve self-sufficiency, or, at least, less dependence, on external capabilities and finance while Brazil charts its development course. The results have been astounding, with world-renowned scholars such as Nobel Economist Joseph Stiglitz in support of the paradigm shift and originality in approach.

In knowledge production and dissemination, we see Brazil involved in international research and innovation collaborations, financed by government agencies[8] (e.g. BNDES) with a view to improving understanding of how innovation can better serve humanity and respond to global challenges. How these collaborations between TH actors contribute to IID is yet to be measured in precise terms. Nevertheless, they signal interest towards advancing research and innovation and demonstrate commitments by the nation's THA focused development strategies.

In spite of these incubators, knowledge and finance initiatives, increases in capital, investment, growth and development recorded in Brazil, 'the Brazilian economy, despite the liberalization process in the 1990s, remains the most closed amongst the BRICS countries' (Scerri and Lastres 2013, 28). The focus on the creation of formal institutions means that significant attention is shifted from the informal economy and sectors, which as we know, accounts for a significant proportion of innovation in BRICS countries and the global South in general (ILO 2012). The implication is less support for innovation practitioners at grassroots level, thereby impacting negatively on IID and similar efforts with potentials to address the needs of the poor and marginalized segments of the economy. We submit that maintaining this balance is critical to inclusive development and that careful operationalization of the triple helix framework in this context presents an opportunity towards achieving this goal.

Russia has the weakest presentation of IID agenda among BRICS. In Russia, IID is not yet recognized at the level of national innovation policy, although progress is slowly being made in this regard in terms of supporting social entrepreneurship. The national strategy on innovation development is yet to address social innovation components and does not adequately consider society as a beneficiary of development arising from innovation. As pointed out in the previous section, the role of industry in innovation in Russia lags behind countries such as China. In addition to the reasons provided earlier, other factors include the low share of industry enterprises performing technological innovation, 9.4% of all industry enterprises, with more than 75% of these companies being large business. Small businesses involved in innovative activity account for 4.8% of small enterprises out of the total number of small-scale business (Малое и среднее предпринимательство в России 2014). Thus, the general level of innovative activity in Russia is considerably low. It is still too early to tell whether recent government initiatives (e.g. the Skolkova project) and policy changes aimed at encouraging innovation in academia and industry have been successful (EBRD 2011).

Why does academia not involve industry in IID discussions in Russia? Consideration of the main actors from the triple helix model perspective is helpful in answering this question. As Smorodinskaya (2011) explains, one-on-one interactions dominate in the Russian quasi-market economy and government is an indispensable participant in these interactions. This results in the exclusion of direct interaction between industry and academia, which can only occur through intermediary bureaucrats. But IID is mostly a product of joint involvement of all NIS actors, based on knowledge, innovation and consensus spaces (Etzkowitz and Ranga 2010). So, as long as there are no or weak knowledge, innovation and consensus spaces in Russia, the IID agenda will remain separate from industry and can only be transferred to industry by government.

Among the BRICS countries, India has one of the most developed IID support systems. The academic sector was the trigger that created the system first involving government agencies and then industry. In fact, academia initiated the knowledge and innovation spaces between industry and government to reveal and recognize grassroots people as a source of knowledge. Students and academic staff initiated scouting and documenting IID from the grassroots. If an innovation with market potential is identified, it is transferred to the incubation system, arranged by the government through various institutions, which provide a wide range of services such as testing, validating, patenting and business incubating. If the innovation is of social importance and aimed at improving citizens' livelihoods, but has weak or no market potential, it is eligible for other forms of institutional support, for example by the Grassroots Technological Innovation Acquisition Fund (GTIAF). The GTIAF acquires rights to technologies for generating public goods from the innovators and escalates the process of awarding the people who developed the GRI.

The other option for grassroots innovators is to become entrepreneurs under the support of the Micro Venture Innovation Fund (MVIF), which was set up with financial assistance from the Small Industry Development Bank of India (SIDBI). These organizations provide financing exclusively for risky innovations with limited or no commercial market. One of the main criteria for the selection of projects for funding is their social value and social benefit.

These measures and initiatives are supported in various policy documents released by the Indian government. The main message of the 11th Five-Year Plan (2007–2012) is a statement of the Indian specific way of innovation development, aimed at improving the quality of people's lives through innovation (11th Five Year Plan 2007). This approach was developed further in the 12th Five-Year Plan (2012–2017). According to the Plan, the Indian model of innovation should provide affordable innovations to meet people's needs (in transport, healthcare, water resources, etc.) and ensure inclusive growth of the country. The New Science, Technology and Innovation Policy released in 2013 declares society as the main goal of innovation development and the central stakeholder of the national innovation system that is supposed to ensure inclusive innovative development in India (STIP 2013).

In the case of China, the Chinese economy recorded a 7.4% annual growth rate in 2014 (IMF 2014) with growth of up to 10% reported in previous years. But this growth performance has come at a high cost, as the gap between rich and poor (i.e. inequality) has widened dramatically, and the environment has suffered immense damage. What is more, this growth has not brought citizens sufficient benefits, as it has been driven more by investment than consumption.

China's 12th Five Year Plan (2011–2015) places great emphasis on 'inclusive growth', promoted by the Asian Development Bank (ADB). The concept of inclusive growth and inclusive innovation is an attempt to transform China's economic and social development model. It seeks to address rising inequality and create an environment for more sustainable growth by prioritizing more equitable wealth distribution, increased domestic consumption, and improved social infrastructure and social safety nets. During 2009–2010, President Hu Jintao raised the concept of 'inclusive growth/innovation' in three important public speeches, appealing that the fundamental purpose of such growth and innovation should be to ensure that the fruits of economic globalization and economic development benefit every citizen.

In spite of its chequered history, recent evidence indicates that South Africa is making efforts at using innovation and public policies as mechanisms for inclusive development. For example, conceptually, South Africa prefers the notion of 'innovation for inclusive development' as opposed to 'inclusive innovation'. In addition, the country aims to have inclusion as part of all public policies, institutions and projects in contrast to approaching inclusion only in terms of innovation (OECD 2015a; 2015b). It is important to point out that this was not the case in the past. Phiri et al., for example, submit that policies adopted in South Africa's post-democratic era 'have produced and reproduced social and economic inequalities which have hampered inclusive innovation, development and the nascent burgeoning of innovation in the informal sector' (2013, 1). As Daniels (2014) and UNDP (2014) found, appropriate public policies required to promote and support the contributions of innovation to socio-economic development are either non-existent or weak across developing countries; not in South Africa alone. Therefore, the strategies being proposed and adopted (see for example, NDP 2011) attest to the country's commitment to IID and the role of public policies in this regards (UNDP 2014).

The IID agenda is viable in BRICS. The governments of these countries realize the imbalanced character of growth and the potential of IID to address these imbalances. The presence of the IID agenda at a policy level in Brazil, India, China and South Africa proves this. These governments take the role of facilitator of IID through the creation of institutional and financial infrastructure. Industry becomes a more active participant in these processes, not just through producing goods *for* BoP, but also through building capacities for adding value and commercialization of IID developed *by* BoP. The same trend is observed for academia. Besides 'traditional' functions of knowledge production and dissemination, academic institutions establish incubators, testing labs, angel networks and other items of infrastructure and become a networking base for industry and policymakers.

The case of Russia demonstrates the importance of such networking and horizontal interactions between actors. Russia is the only country among BRICS that does not have an IID agenda at policy level. In view of weak private finance and institutional infrastructure for innovation development, the government remains the dominant actor, framing the core and direction of innovation activity, and the main source of financial support for innovation both in academia and industry. Through programmes and strategies (social, economic, innovation), and the funding of certain activities, the state delineates the fields of activities for the actors in the national innovation system. The bilinear model of communication between these actors obstructs the exchange of ideas and information about societal demands. This explains the situation when the IID agenda, which was raised by the expert community, was not accepted by the government and developed at policy level.

In summary, we draw some important lessons, based on this analysis of BRICS. We find that: (1) BRICS countries mostly focus on innovation in general, with far less attention paid to IID, the focus of this paper. One reason for this gap may lie in the weak conceptualization and theorizing of IID or the lack of appreciation of the potential roles of IID; (2) absence of specific public policy support for IID in BRICS; (3) emphasis on a narrow view of innovation as opposed to a broader, macro and all-encompassing view, provided that the innovation addresses socio-economic societal challenges; (4) dearth of capabilities needed to support and promote innovation as a mechanism for inclusive development; and finally, (5) gaps in the literature that advances knowledge of innovation for inclusive development.

Conclusion

In this paper, we have drawn from various literature in order to improve our understanding of IID in BRICS. We explained some of the roles that TH actors, as a source of knowledge and innovation, play in the process. The insights are useful in addressing important developmental challenges facing BRICS. BRICS countries, like other 'late-developing societies' (Kruss 2008, 5) face the burden of inequality and the challenge of fostering socio-economic development with strong innovation capabilities while combating social exclusion (Arocena and Sutz 2003). However, drawing from the findings of this paper and the fresh insights that can be gleaned, we conclude that BRICS countries, except for Russia, are actively engaging in activities and public policies to both support and promote innovation in the broad sense,

but less so in the case of inclusive innovation as a mechanism for inclusive development.

As noted earlier, Russia remains the only country among BRICS countries that does not have an IID agenda at academic or policy level. This situation, we explained, is exacerbated by weak private finance and institutional infrastructure for innovation development, and the government's dominant role in framing the core and direction of innovation activity, while, at the same time, acting as the main source of financial support for innovation both in academia and industry.

In India, we find the active involvement of THA in IID support and development with the academic sector as the main driver of these processes. IID was 'discovered' in the informal sector of economy and identified as GRI. This determined the specifics of its support – mostly at institutional level. Despite that the need for inclusive growth is recognized at policy level, there are still no detailed programmes or plans for implementation of this growth. Instead, measures are taken by institutions and organizations of THA to support people, businesses and production cycles for innovative ideas at grassroots. There is a lack of evaluation and measurement of effectiveness of GRI, which might be an impediment in development of programmes at policy level.

China presents an interesting case in which although IID and policy support exist, some scholars argue that university and research institutes, which play a central role in innovation, are excluded from the inclusive innovation system (Xing, Tong, and Chen 2010; Gao, Zhou, and Cao 2013).

These findings highlight questions such as: Should academic actors be involved in, for instance, grassroots innovation and if yes, to what extent? Should firms (industry) 'interfere' in informal sector innovation and if yes, how, why and under what circumstances? This paper substantiates the perspective that THA should be more engaged in IID. This argument serves as a contribution to theory, thus adding theoretical value to the country-level descriptions we provide in this paper.

On the basis of these analyses and findings, we conclude that TH actors can play important roles in areas that include measurement and analysis of innovation/IID, promoting IID, initiating and participating in IID policymaking (formulation, implementation, reviews, monitoring and evaluation), linking formal with informal (sectors, actors, knowledge), development of capabilities, and strengthening innovation ecosystems in ways that support IID. Another area of importance is in ensuring that innovation is relevant to the wider society, and better targeted at excluded and marginalized sections of the population, particularly those at the bottom of the pyramid.

These recommendations we conclude are critical to ensuring that the gains of innovation are optimized and that innovation is applied in a more efficient manner to address societal challenges aimed at improving the quality of life. In the specific cases of BRICS, the use of IID approaches, projects and activities is necessary to achieve and maintain balanced growth and development strategies that are not only smart (i.e. innovation-led) but also do not exacerbate poverty, inequality and social exclusion. IID can be a useful tool in achieving the socio-economic and development goals of BRICS.

Further research

In summarizing the insights from this section, we note a mixed picture in BRICS THAs' roles in IID, with considerable impacts on development. Our findings reveal that India and China show the best levels of performance among BRICS countries, followed by Brazil and South Africa, while Russia lags significantly behind. Nonetheless, BRICS countries are all making an effort, but specific questions still remain, which further research could examine. Some of these questions are: How can or should a TH framework function in the context of IID and in the way it links to public policy for IID in BRICS? What (new or additional) roles are there for THA in IID activities across BRICS innovation ecosystems, capabilities, learning and measurement of innovation, and why? What (if any) changes occur in the role and functionality of THA that arises from the specifics of IID, how do these changes take place, and what shifts occur in the innovation system, e.g. in blending formal and informal sectors or actors? What IID-relevant public policies and policy support for IID are available in BRICS countries and what roles do (or can) THA play in operationalizing these policies? Are these policies needed and if yes, why? How do they support / promote IID and what are the challenges and limitations and how can these be mitigated or managed?

The extensive literature review in this paper resulted in the identification of roles played by THA in IID. The examination suggests that empirical data needed to effectively map IID activities and policy support in BRICS, and thereby address the research questions posed in this paper are, presently, insufficient, weak or non-existent. Further research, therefore, could focus on the generation and dissemination of empirical data on IID. This research also highlights new possibilities related to changing public (including science, technology and innovation) policies, triple helix (TH) and the national systems of innovation (NSI) frameworks with a view to further research that examines the possibilities of extending public policies and existing frameworks to ensure that they provide for IID.

Acknowledgements

An earlier draft of this paper was prepared for the XIII Triple Helix International Conference 2015 (THA 2015), held on 21 –23 August 2015, at Tsinghua University, China and presented at the conference. The conference theme was: 'Academic-Industry-Government Triple Helix Model for Fast-Developing Countries'. Many thanks to conference participants who provided useful comments.

Disclosure statement

No potential conflict of interest was reported by the authors.

Notes

1. Used here in line with the ILO definition, it refers to economic activities by workers and economic units that are –

in law or in practice – not covered or insufficiently covered by formal arrangements. This means that they are operating outside the formal reach of the law; or although they are operating within the formal reach of the law, the law is not applied or not enforced; or the law discourages compliance because it is inappropriate, burdensome, or imposes excessive costs. (Accessed 4 July 2016, http://libguides.ilo.org/informal-economy-en)
2. For instance, although recent data indicate that poverty levels in South Africa have dropped since 2002 and 2006, reaching a low of 45.5% in 2011 (SSA 2014), this remains a high figure.
3. We acknowledge that there are various forms of inequality, such as, income, wage, earnings, urban, spatial and social inequality. Nevertheless, we do not address these distinctions as they are beyond the scope of this paper.
4. Such as in Russia, where inclusive innovation or innovation for inclusive development still does not feature in the mainstream literature or public policies.
5. Defined in terms of innovation from/for/at/by grassroots, based on Daniels (forthcoming) framework.
6. The epistemological roots of 'inclusive innovation' concepts in India can be traced to the Gandhian model of agricultural technologies development.
7. In addition to the specific roles of each actor in the different countries.
8. An example of which is the Mission-Oriented Finance for Innovation Conference (http://missionorientedfinance.com/) co-financed by BNDES in partnership with SPRU, hosted in London, UK, and the subsequent book (http://www.policy-network.net/publications/4860/Mission-Oriented-Finance-for-Innovation) edited by Mariana Mazzucato and Caetano C. R. Penna.

ORCID

Chux U. Daniels ⓘ http://orcid.org/0000-0002-5179-4176
Olga Ustyuzhantseva ⓘ http://orcid.org/0000-0003-3023-5428
Wei Yao ⓘ http://orcid.org/0000-0001-9224-4413

References

11th Five Year Plan. 2007-2012. *Planning Commission. Government of India*. Accessed February 18, 2014. http://planningcommission.nic.in/plans/planrel/fiveyr/11th/11_v1/11v1_ch8.pdf.
Albuquerque, E., W. Suzigan, G. Kruss, and K. Lee, eds. 2015. *Developing National Systems of Innovation: University–Industry Interactions in the Global South*. Cheltenham, UK: Edward Elgar Publishing Limited.
Anderson, J., and C. Markides. 2007. "Strategic Innovation at the Base of the Pyramid." *MIT Sloan Management Review* 49 (49116): 83–88.
Акбердина, В., и Е. Малышев. 2011. "Возможности взаимодействия государства, бизнеса и сферы образования в рамках модели «тройной спирали» на примере агропромышленного комплекса забайкальского края." *Экономика региона* №4: С.269–С.274.
Araujo, J. F. M., and E. M. Costa. 2014. "Genesis of the Incubator Movement in Brazil: How the Need for New Software Companies Helped Foster the Development of Dozens of Incubators all Over the Country." *Hélice* 3 (4).
Arocena, R., and J. Sutz. 2003. "Knowledge, Innovation and Learning: Systems and Policies in the North and in the South." In *Systems of Innovation and Development: Evidence From Brazil*, edited by J. E. Cassiolato, H. M. M. Lastres, and M. L. Maciel, 291–310. Cheltenham: Edward Elgar.
Basant, R., and P. Chandra. 2007. "Role of Educational and R&D Institutions in City Clusters, an Exploratory Study of

Bangalore and Pune Regions." *World Development* 35 (6): 1037–1055.

Brem, A., and P. Wolfram. 2014. "Research and Development From the Bottom Up: Introduction of Terminologies for New Product Development in Emerging Markets." *Journal of Innovation and Entrepreneurship* 3 (9): 1–22.

BRICS (BRICS Joint Statistical Publication). 2014. Brasil: IBGE (Instituto Brasileiro de Geografia Ee Estatística).

Cassiolato, J., and H. M. M. Lastres. 2008. "Discussing Innovation and Development: Converging Points between the Latin American School and the Innovation Systems Perspective." *Globelics Working Paper Series, Working Paper no. 08-02*. Accessed March 20, 2015. http://www.globelics.org/wp-content/uploads/2012/11/wpg0802.pdf.

Cassiolato, J., and M. Soares. 2013. "Introduction: BRICS National Systems of Innovation." In *The Role of the State*, edited by J. Cassiolato and M. Soares. London: Routledge.

COFISA (Cooperation Framework on Innovation Systems between Finland and South Africa). 2009. "Mapping Triple Helix Innovation Networks in the Western Cape." *Kaiser Associates Economic Development Practice*. Accessed March 13, 2015. http://www.kaiseredp.com/wp-content/uploads/2014/11/finrep_mthin_wc_oct2009.pdf.

Cozzens, S. E. 2010. "Innovation and Inequality." In *The Theory and Practice of Innovation Policy: An International Research Handbook*, edited by Ruud E. Smits, Stefan Kuhlmann, and Philip Shapira, Chapter 15. Cheltenham (UK) and Northampton, MA (USA): Edward Elgar.

Daniels, C. U. 2014. "Measuring Innovation in the Global South: An Annotated Bibliography." *A report on the measurement and analysis of innovation programme*, submitted to the Centre for Science, Technology and Innovation Indicators (CeSTII), Human Sciences Research Council (HSRC), South Africa.

Daniels, C. U. Forthcoming. "Policy Support for Innovation at Grassroots in Developing Countries: Perspectives From Nigeria." Accessed April 20, 2014. http://www.merit.unu.edu/MEIDE/papers/2015/Daniels_CD_1423048542.pdf.

Dezhina, I. 2014. "Technology Platforms in Russia: a Catalyst for Connecting Government, Science, and Business?" *Triple Helix. A Journal of University-Industry-Government Innovation and Entrepreneurship* October 1: 6. http://link.springer.com/article/10.1186/s40604-014-0006-x.

Dezhina, I., and V. Kiseleva. 2008. "State, Science and Business in Innovation System of Russia." *Research Paper Series*, issue 115P.

Дмитриев, Л. 2011. "Главная проблема «Стратегии-2020» - отсутствие политических возможностей для реализации." Accessed February 23, 2015. http://2020strategy.ru/news/32664731.html.

EBRD (European Bank for Reconstruction and Development). 2011. Diversifying Russia: Harnessing Regional Diversity.

Etzkowitz, H., and J. Dzisah. 2008. "Rethinking Development: Circulation in the Triple Helix." *Technology Analysis & Strategic Management* 20 (6): 653–666. doi:10.1080/09537320802426309.

Etzkowitz, H., and L. Leydesdorff. 2000. "The Dynamics of Innovation: From National Systems and "Mode 2" to a Triple Helix of University–Industry–Government Relations." *Research Policy* 29: 109–123.

Etzkowitz, H., and M. Ranga. 2010. "A Triple Helix System for Knowledge-Based Regional Development: From "Spheres" to "Spaces"." In VIII triple Helix Conference, Madrid, October.

Etzkowitz, H., and C. Zhou. 2006. "Triple Helix Twins: Innovation and Sustainability." *Science and Public Policy* 33 (1): 77–83.

Малое и среднее предпринимательство в России. 2014. Статистический сборник Федеральной службы государственной статистики. http://www.gks.ru/wps/wcm/connect/rosstat_main/rosstat/ru/statistics/publications/catalog/doc_1139841601359

Forsight. 2011. *Transcript of the meeting of the expert group №5 at a scientific seminar "Foresight"*, 14 October, 2011. Accessed February 22, 2015. http://strategy2020.rian.ru/stenograms/20111028/366186836_10.html.

Foster, C., and R. Heeks. 2013. "Conceptualising Inclusive Innovation: Modifying Systems of Innovation Frameworks to Understand Diffusion of New Technology to Low-Income Consumers." *The European Journal of Development Research* 25 (3): 333–355.

Freeman, C. 1987. *Technology Policy and Economic Performance: Lessons From Japan*. London: Pinter.

Galbraith, J. K., L. Krytynskaia, and Q. Wang. 2004. "The Experience of Rising Inequality in Russia and China During the Transition." *The European Journal of Comparative Economics* 1 (1): 87–106.

Gao, T., X. Liu, and J. Zhou. 2014. "The Measurement of China's Regional Inclusive Innovation Performance: Theoretical Model and its Empirical Verification." *Studies in Science of Science* 32 (4): 613–622.

Gao, Y., K. Zhou, and D. Cao. 2013. "Striding Across the Middle-Income Trap and the Inclusive Industrial Innovation." *Journal of Sichuan University of Science & Engineering (Social Sciences Edition)* 28 (1): 20–27.

George, G., A. McGahan, and J. Prabhu. 2012. "Innovation for Inclusive Growth: Towards a Theoretical Framework and a Research Agenda." *Journal of Management Studies* 49 (4): 661–683.

Gianella, C., and W. Tompson. 2007. "Stimulating Innovation in Russia: The Role of Institutions and Policies." OECD Working Paper 539, ECO/WKP(2006)67.

Griffith-Jones, S. 2014. "A BRICS Development Bank: A Dream Coming True?" *UNCTAD Discussion Papers*, No. 215. Accessed: April 30, 2015. http://unctad.org/en/PublicationsLibrary/osgdp20141_en.pdf.

Gokhberg, L., and Tatiana Kuznetsova. 2011a. "S&T and Innovation in Russia: Key Challenges of the Post-Crisis Period." *Journal of East-West Business* 17 (2–3): 73–89.

Gokhberg, L., and Tatiana Kuznetsova. 2011b. "Strategy 2020: New Outlines of Russian Innovation Policy." *Foresight-Russia* 5 (4): 8–30.

Gokhberg, L., N. Gorodnikova, T. Kuznetsova, A. Sokolov, and S. Zaichenko. 2009. "Prospective Agenda for S&T and Innovation Policies in Russia." In *BRICS and Development Alternatives: Innovation Systems and Policies*, edited by J. Cassiolato and V. Vitorino, 73–100. London: Anthem Press.

Gupta, A. 2006. "From Sink to Source: The Honey Bee Network Documents Indigenous Knowledge and Innovations in India." *Innovations: Technology, Governance, Globalization* 1: 49–66.

Gupta, A. 2007. "Towards an Inclusive Innovation Model for Sustainable Development." In *Global Business Policy Council of A.T. Kearney – CEO's Retreat*. Dubai, December 9–11, 2007. http://anilg.sristi.org/wp-content/Papers/Towards%20an%20inclusive%20innovation%20model%20for%20sustainable%20development.pdf

Gupta, A. 2013. "Policy Gaps for Promoting Green Grassroots Innovations and Traditional Knowledge in Developing Countries: Learning from Indian Experience." *IIMA Working Paper* No. 2013-02-02.

Gupta, A. 2014. "This is Not Jugaad: A Misnomer for Majority of the Grassroots Innovations." Accessed May 14, 2015. https://www.academia.edu/4170842/This_is_not_Jugaad.

Heeks, R., M. Amalia, R. Kintu, and N. Shah. 2013. "Inclusive Innovation: Definition, Conceptualisation and Future Research Priorities." *Working Paper*, No. 53.

ILO Department of Statistics. 2012. Statistical update on employment in the informal economy. http://laborsta.ilo.org/applv8/data/INFORMAL_ECONOMY/2012-06-Statistical%20update%20-%20v2.pdf

IMF (International Monetary Fund). 2014. *"South Africa: Selected Issues Paper."* IMF Country Report No. 14/339. Washington, D.C.: IMF.

IMF (International Monetary Fund). 2015. "Emerging Powers and Global Governance: Whither the IMF?" IMF WP/15/219. http://www.imf.org/external/pubs/ft/wp/2015/wp15219.pdf

IMF (International Monetary Fund). 2016. "IMF Annual Report 2016: Finding Solutions Together." Washington: IMF. http://www.imf.org/external/pubs/ft/ar/2016/eng/pdf/ar16_eng.pdf

Keeley, B. 2015. *Income Inequality: The Gap Between Rich and Poor. OECD Insights.* Paris: OECD Publishing.

Koeller, P., and J. Gordon. 2013. "Brazil." In *The Role of the State,* edited by J. Cassiolato and M. Soares, 23–80. London: Routledge.

Kraemer-Mbula, E., and W. Wamae, eds. 2010. *Innovation and the Development Agenda.* Ottawa: OECD/IDRC.

Krishna, V. 2013. "Science, Technology and Innovation Policy 2013: High on Goals, Low on Commitment." *Economic & Political Weekly* 48 (16): 15–19.

Krishna, V. 2014. "Innovation and Inequality: Indian Experience." In *OECD Advisory Group Meeting. Knowledge and Innovation for Inclusive Growth.* Paris.

Kruss, G. 2008. "Balancing old and new Organisational Forms: Changing Dynamics of Government, Industry and University Interaction in South Africa." *Technology Analysis & Strategic Management* 20 (6): 667–682. doi:10.1080/09537320802426358.

Li, Dongfang. 2013. "Institutional Arrangement of Inclusive Growth in China." *Science and Technology Management Research* 10: 1–5.

Lorentzen, J., and R. Mohamed. 2009. "Where Are the Poor in Innovation Studies?" Paper presented at the *7th Annual Globelics Conference,* Dakar, Senegal, 6-9 October. Accessed November 21, 2014. https://smartech.gatech.edu/bitstream/handle/1853/36668/1238509827_TM.pdf?sequence=1.

Lundvall, B. A., ed. 1992. *National Systems of Innovation: Towards a Theory of Innovation and Interactive Learning.* London: Pinter.

Marcelle, G. 2014. "Science Technology and Innovation Policy That is Responsive to Innovation Performers." In *Invited Chapter for International Research Handbook on Science, Technology and Innovation Policy in Developing Countries: Rationales and Relevance,* edited by S. Kuhlmann and G. Ordonez-Matamoros. Edward Elgar.

Маховикова, Г., и Н. Ефимова. 2010. *Инновационный менеджмент.* Москва: Эксмо.

Mid Term Appraisal for 11th Five-Year Plan 2007–2012. *Planning Commission, Government of India.* Accessed February 18, 2014. http://planningcommission.nic.in/plans/mta/11th_mta/chapterwise/chap19_science.pdf.

Mohnen, P., and M. Stare. 2013. "The Notion of Inclusive Innovation. No. 15." *Innovation for Growth – i4g,* Policy Brief No. 15. Luxembourg: European Commission. http://ec.europa.eu/research/innovation-union/pdf/expert-groups/i4g-reports/i4g_policy_brief_15_-%20_inclusive_innovation.pdf

NDP (National Planning Commission). 2011. *National Development Plan 2030: Executive Summary.* Pretoria: National Development Commission.

Nelson, R. R. 1993. *National Innovation Systems: a Comparative Analysis.* New York: Oxford University Press.

NIF (National Innovation Foundation). 2013. *Annual Report 2012-13.* Accessed January 8, 2015. http://www.nif.org.in/dwn_files/English_Annual%20Report_2012_13_NIF.pdf.

OAPM (Office of the Advisor to the Prime Minister). 2011. "Towards a More Inclusive and Innovative India." In *Office of the Advisor to the Prime Minister, Government of India.* New Delhi. Accessed March 2, 2015. http://www.iii.gov.in/images/stories/innovation/Innovation_Strategy.pdf.

OECD. 2011a. *Divided We Stand: why Inequality Keeps Rising.* Paris: OECD Publishing.

OECD. 2011b. *OECD Reviews of Labour Market and Social Policies: Russian Federation.* Paris: OECD Publishing.

OECD. 2012. *Innovation for Development: A Discussion on the Issues and an Overview of Work of the OECD Directorate for Science, Technology & Industry.* Paris: OECD.

OECD. 2013a. *OECD Economic Surveys: South Africa 2013.* Paris: OECD Publishing.

OECD. 2013b. "Innovation and Inclusive Development." Conference discussion report, Cape Town, South Africa, 21 November, 2012. February 2013 Revision. Paris: OECD.

OECD. 2014. "Society at a Glance 2014: OECD Social Indicators." OECD Publishing. http://dx.doi.org/10.1787/soc_glance-2014-e

OECD. 2015a. *OECD Economic Surveys: South Africa 2015.* Paris: OECD Publishing.

OECD. 2015b. *Innovation Policies for Inclusive Development: Scaling Up Inclusive Innovations.* Paris: OECD Publishing.

Oslo Manual. 2005. *Oslo Manual: Guidelines for Collecting and Interpreting Innovation Data.* 3rd ed. Accessed October 18, 2014. http://www.oecd.org/innovation/inno/oslomanualguidelinesforcollectingandinterpretinginnovationdata3rdedition.htm.

Пахомова, И. 2012. "Модель «тройной спирали» как механизм инновационного развития региона." *Научные ведомости Белгородского государственного университета. Серия: История. Политология. Экономика. Информатика,* №7-1. С.50-55.

Phiri, M., H. Makelane, N. Molotja, and T. Kupamupindi. 2013. "Inclusive Innovation in South Africa: Entrepreneurship and Inequality in the post Democratic Era." Accessed December 17, 2014. www.merit.unu.edu/MEIDE/papers/2013/Phiri_Makelane_Molotja_Kupamupindi.pdf.

Pogue, T., and L. Abrahams. 2012. "South Africa's National System of Innovation and Knowledge Economy Evolution: Thinking About "Less Favoured Regions"." *International Journal of Technological Learning, Innovation and Development* 5 (1-2): 58–82.

Prahalad, C. 2004. Fortune at the Bottom of the Pyramid: Eradicating Poverty Through Profits. Pearson Education Inc. 435 p.

Prahalad, C. K. 2012. "Bottom of the Pyramid as a Source of Breakthrough Innovations." *Journal of Product Innovation Management* 29 (1): 6–12.

Rajou, N., J. Prabhu, S. Ahuja, and K. Robert. 2012. "Jugaad Innovation: Think Frugal, Be Flexible, Generate Breakthrough Growth." *Jossey-Bass* 288 pages.

Scerri, M., and H. Lastres. 2013. *The Role of the State.* New Delhi: Routledge.

Shao, Xi, Xiao-qiang Xing, and Yun-huan Tong. 2011. "Research on Inclusive Regional Innovation System." *China Population, Resources and Environment* 21 (6): 24–27.

Sinha, R. K., 2011. "India's National Innovation System: Roadmap to 2020." *ASCI Journal of Management* 41 (1): 65–74.

Skills & Innovation Policy. 2015. *World Bank.* Accessed February 10, 2015. http://web.worldbank.org/WBSITE/EXTERNAL/WBI/WBIPROGRAMS/KFDLP/0,,menuPK:461238~pagePK:64156143~piPK:64154155~theSitePK:461198,00.html.

Стратегия инновационного развития Российской Федерации на период до 2020 года. 2011. Правительство РФ. http://government.ru/docs/9282/

Смородинская, Н. 2012. «Смена парадигмы мирового развития и становление сетевой экономики.» Экономическая социология. №4. С.95-115.

Smorodinskaya, N. 2012. "Russian Framework Conditions for Innovation in View of the Triple Helix Concept." The X Triple Helix International Conference, July 2011, Stanford University, USA https://www.researchgate.net/publication/244988940_Russian_Framework_Conditions_for_Innovation_in_View_of_the_Triple_Helix_Concept.

Soares, M. C. C., and J. E. Cassiolato. 2013. "Innovation Systems and Inclusive Development: Some evidence based on empirical work." Accessed February 13, 2015. http://www.cdi.manchester.ac.uk/newsandevents/documents/SoaresCassiolatoPreWorkshopPaper.pdf.

Sokolov, A. 2006. "Identification of National S&T Priority Areas with Respect to the Promotion of Innovation and Economic Growth: The Case of Russia." In *Bulgarian Integration Into Europe and NATO*, 92–109. Amsterdam: IOS Press.

SSA (Statistics South Africa). 2014. *Poverty Trends in South Africa – An Examination of Absolute Poverty Between 2006 and 2011*. Pretoria: SSA.

STIP (Science, technology and innovation policy of India). 2013. *Department of Science and Technology, Government of India*. Accessed September 10, 2014. http://www.dst.gov.in/sti-policy-eng.pdf.

Tang, P. 2013. "Inclusive Innovation Research of Technology-Based Small and Micro Enterprise." *Science & Technology Progress and Policy* 30 (18): 29–32.

UNCTAD (United Nations Conference on Trade and Development). 2014. "Innovation Policy Tools for Inclusive Development."

UNDP. 2014. *The Impacts of Social and Economic Inequality on Economic Development in South Africa*. New York: UNDP.

Ustyuzhantseva, O. 2014. "From Policy Statements to Real Policy." *Current Science* 106 (11): 1472–1474.

Ustyuzhantseva, O. 2015. "Institutionalisation of Grassroots Innovation in India." *Current Science* 108: 2015, 1476–1482.

Utz, A., and C. Dahlman. 2007. *Unleashing India's Innovation: Toward Sustainable and Inclusive Growth*. Ed. Mark A. Dutz. Washington, D.C.: World Bank.

World Bank. 2004. *Inequality and Economic Development in Brazil*. Washington, D.C.: A World Bank Country Study.

World Bank. 2005. *Russian Federation: Reducing Poverty Through Growth and Social Policy Reform*. Report No. 28923-RU.

World Bank. 2010. "Innovation Policy: A Guide for Developing Countries." In *International Bank for Reconstruction and Development/World Bank. No. 54893*, Washington, D.C.

World Bank. 2013. "China: Inclusive Innovation for Sustainable Inclusive Growth." Document of the *World Bank No. 82519*. TA-P128575-TAS-BB, Washington, D.C, October.

World Bank. 2014. "South Africa Economic Update: Fiscal Policy and Redistribution in an Unequal Society." *South Africa Economic Update, No.6*. Washington, DC: World Bank Group.

Wu, Xiao-bo, and Yan-bin Jiang. 2012. "Where Inclusive Innovation Emerges: Conceptual Framework and Research Agenda." *Journal of Systems & Management* 21 (6): 736–747.

Xing, Xiao-qiang, Jiang-hua Zhou, and Yun-huan Tong. 2013. "Inclusive Innovation: Concept, Characters and key Successful Factors." *Studies in Science of Science* 31 (6): 923–931.

Xing, Xiaoqiang, Yun-huan Tong, and Xiaopeng Chen. 2010. "The Business Model at the Base of the Pyramid Markets: a Multi-Case Study." *Management World* 10: 108–125.

Yao, Wei, Weng Mosi, and Xiaodong Zou. 2015. "The Role of Entrepreneur University in Rational Inclusive Innovation System: Evidence from China." *Working Paper*.

Zaini, R. M., D. E. Lyan, and E. Rebentisch. 2015. "Start-up Research Universities, High Aspirations in a Complex Reality: a Russian Start-up University Case Analysis Using Stakeholder Value Analysis and System Dynamics Modelling." *Triple Helix* 2 (4).

Zhao, Wu, Sun Yongkang, Zhu Mingxuan, and Gao Ying. 2014. "Inclusive Innovation: Evolution, Mechanism and Path Selection." *Science & Technology Progress and Policy* 31 (6): 6–10.

Design patents: A snapshot study of the BRICS economies

Kashmiri Lal

This study is an attempt to examine the profile of design patents among the BRICS economies by examining design patents issued by the United States Patents and Trademark Office (USPTO), for the period 2002–2011. During this period, China was granted the most patents followed by Brazil, India, South Africa and Russia. In terms of the proportion of design to the utility patents, three countries, viz. China, Brazil, and South Africa dominated. An analysis of the results showed that China had produced two-third of the total patents in the design category. For all the aforementioned countries, most of the design patents were granted within a period of one year from the date of filing. For India, one-third of the design patents were issued in the jewellery and ornaments category followed by the apparatus tools and equipment category. In the case of Brazil, Russia, China and South Africa, the largest segment for which design patents were issued was the apparatus tools and equipment, followed by goods packaging. Most of the design patents in the BRICS region were issued to individual inventors, and two inventor teams received the second highest number of patents. As the largest percentage of inventors belonged to the BRICS countries, the United States was observed as the most affiliate country, closely followed by the other European countries. However, some prominent Asian countries were also noted for their close ties with the region.

Introduction

Design activity[1] is an integral component of contemporary society. Design has played a key role in popularizing products and making brands successful. Products with appealing designs can be more popular than those with high functionality. For the success of any product or brand, there seems to be a strong relationship between its functionality and design. The shapes and designs of products have attained wide attention in the commercially driven lifestyle of the current era. Basically, at the heart of every good or service produced, sold and consumed is design (Bryson and Rusten 2011). Designs are applied to a wide variety of products, ranging from traditional handicrafts to modern consumer durables, automobiles, architectural structure and textiles (Export-Import Bank of India 2014). Furniture, kitchenware, and, more recently, computing seem to be driven as much by the changes in design as in the function (Risch 2013; Chan, Mihum, and Sosa 2014). Designs have been highlighted through many high-profile legal cases[2] (Atkins 2013); consequently, firms are now realizing the importance of design patents. Samsung obtained nearly 6159 design patents between 2001 and 2015, compared to fewer than 50 during 1998–2004. Overall, USPTO issued 14,766 design patents in 1998, increasing to 21,356 in 2011. During the limelight of Egyptian Goddess case, USPTO issued more than 25,000 design patents. The latest trends in design patents issued include a computer Icon (Microsoft), Tablet (LG), Ear buds (Sony), and mobile connection terminals (Samsung) (Gaff and Cuomo 2013). Design patents have found a wide variety of applications in the modern IT industry, for example, the shape of the computer, CPU, peripherals, mobiles, modems, etc. In addition to software displays, icons, graphical user interfaces, and computer-generated icons are other areas in which society has witnessed the wide application of design patents (Chen, Sung, and Su 2014). Growth of

design activity and its protection through patents help firms or owners to exploit their designs for third party uses via the exchange of royalty fees. This process on a large scale inculcates competition amongst firms, thus encouraging creativity and promoting aesthetically attractive products. To judge the 'design competitiveness' across nations, the Korean Institute of Design Promotion (2008) has developed a framework to evaluate 17 major countries on three pillars: 'public-, industrial- and civilian- design'. It was observed from this framework that developed countries like the USA, Italy, France and Germany lead in design application and sophistication. Countries from BRICS were also mentioned in the same framework. While China and India are at the bottom, Brazil is at the median. The Design Ladder (Danish Design Centre 2016) is another tool for rating a company's use of design. Developed in 2001, it comprises four stages that are vital for a company's design policy. The first stage (non-design) is when the firm employs no additional designer(s), and gives the responsibility of design to general employees. In the second stage (design as form giving), designers are employed by the firms to give form and shape to their products. In the third (design as process) and fourth (design as strategy) stages of the ladder, the firms try to give birth to specialized designs by incorporating design in their core strategy.

According to USPTO, there are six types of patent applications (shown in Table 1): utility; design; plant; reissue patents; defensive publication; and statutory invention registration. Among them, patents on utility, design, and plants are commonly used. Design patents are issued for a new, original and ornamental design of an article to be manufactured. The major difference between utility and design patents is that the former generally permits its owner to exclude others from making, using or selling the invention for a period of up to 20 years from the date of filing, while the latter permits its owner to

Table 1: Types of patents according to USPTO.

Utility patents	Design patents	Plant patents
Issued for the invention of a new and useful process, machine, manufacture, or composition of a matter or a new and useful improvement. Protection: 20 years	Issued for a new, original and ornamental design embodied in or applied to an article of manufacture. Protection: 14 years	Issued for a new and distinct invented or discovered asexually reproduced plant including cultivated sports mutants, hybrids and newly found seedlings, other than a tuber propagated plant, or a plant found in an uncultivated state. Protection: 20 years
Reissue patents	Defensive publication	Statutory invention registration
Issued to correct an error in an already issued utility, design or plant patent. Protection: Remains the same, but scope can be changed as result of the reissue of patent.	Issued instead of regular utility, design, or plant patents. Protection: It offers limited protection, defensive in nature to prevent others from patenting an invention, design, or plant patents.	This invention replaced the earlier defensive publication in 1985–86 and offers similar protection.

Source: Compiled from source: USPTO (2016d).

exclusively use the design for a period of 14 years. Contrasting features of protection of design patents among the BRICS economies are outlined in Table 2. Design patents are limited to a single claim, while utility patents have multiple claims. Design patent applications are processed more quickly than utility patents. Approximately 90% of the patents issued by the USPTO in recent years have been in the category of utility patents, followed by design and plants patents, etc. In the context of design, other forms of protection are through *copyrights, trademarks, geographical indicators (GIs)* and *trade secrets*. The purpose of *copyrights* is to protect original works of authorship, including production, distribution, translation, etc. (Davidsson 2004). *Copyrights* protect the expression, not the idea. The validity of a copyright under international law is the leftover life of the creator plus 50 years. *Trademarks* are used to represent a particular good and/or service provided by a particular person or business entity. They can be described in the form of shapes, letters, words, numerals, drawings and symbols. Vocal sounds, music and fragrance can also be trademarked. Each and every country has its own trademark office where application for trademark registration can be made. Trademarks are used by consumers to identify the source of originality of the goods, in the form of logos, names, designs or other indicia conveying the trustworthiness of their services. *Trade dr*ess is an associated term commonly used with trademarks. To better understand trademark and trade dress, one can relate these to a

beverage and its refilled bottle. The shape of the bottle is an example of trade dress, while names, such as *Thumps-Up^{TM}*, *Coca-Cola^{TM}*, or *Pepsi^{TM}* are trademarks. The shape of the bottle can also be protected under design patents for a stipulated period, whereas trade dress protection last indefinitely, as long as the owner of the make is using it for commercial gains. *Geographical indicators* (GIs) are signs used on products that have specific geographical origins and possess qualities or a reputation that are due to that origin. *Cognac, Scotch, Porto, Havana, Tequila* and *Darjeeling Tea* are some of the world-famous examples of GIs (World Intellectual Property Organization 2016b). GIs have wide application, from traditional handicraft and agriculture sectors to the modern manufacturing sector of items including food and natural products. *Trade secrets* are the other specialized form of intellectual property consisting of information in the form of shapes, formula, recipes, programmes, devices, techniques and processes. In the modern era of technology, firms use confidential business information such as customer list, sales tactics and corporate/marketing strategy as a vital arm of trade secrets.

Design protection varies from country to country. For example, the USA protects designs through patents granted by USPTO. Further, design patents in the USA can be used either independently or in conjunction with utility patents, trade dress, trademarks, copyrights and trade secrets to protect the novel attributes of products. In many countries, design protection is implemented

Table 2: Features of BRICS countries' design patent acts.

Brazil	Russia	India
Protection of design patents under the Brazilian Act is for 10 years from the date of filing; it can be extended for three successive five-year periods.	The term of protection for design patents under the Russian Act is for 15 years, starting from the date the application is submitted.	A registered design is protected for 10 years under the Indian Design Act 2000, from the date of registration or from the priority date, and is renewable for five years upon application prior to expiry.
China	South Africa	
Protection of design patents under China's Patent Act is for 10 years from the date of filing. Most of the foreign application filed were for utility patents, while domestic applications were mostly for industrial design.	Designs are protected in South Africa under Design Act 1993. Design is categorized into two: aesthetic and functional. Aesthetic designs are required to be new and original, while functional designs are subject to a novelty examinations. Protection is granted for 15 years for aesthetic designs and 10 years for functional designs from the date of registration.	

Source: Derived from Centre for WTO Studies (2013).

through 'industrial design' registrations conducted by the patent office. According to the World Intellectual Property Organization (2016a), 44% of design registrations are accounted for by the BRICS economies, for which the largest share (95%) is accounted for by China.

In the case of India, not much growth is seen in design patents, as its share of total patents is nominal. China, conversely, has a considerable share of design patents in the total output of patents. India has been patenting in the soft field of drugs, pharmaceuticals, and chemicals, etc., which is generally not suited for design patents. China's patent output has been in the field of electronics, telecommunication and advertising which is most suited for design patents. According to Thomson's Trademark Report (2012), China has established its trademark supremacy among the top 10 countries in the world by publishing the highest number of trademarks in 2012. Second behind China is the USA, where design activities are pursued vigorously. China's 1.4 million trademark applications in 2011 almost equal the sum of the USA, France, Russia and European Union trademark offices. According to a report by the World Intellectual Property Organization (2015), about 3.3 million industrial design registrations were in force worldwide, with nearly 1.15 million active industrial design registrations; China accounted for about one-third of the world total. An estimated 601,100 industrial designs were registered in 2014.

In view of the rising graph of design patents and design registrations at the global level, this study delves into the interesting stream of design patent literature and statistics to illuminate the status of design patents for the BRICS economies. This is a snap shot kind of study that presents various trends and profiles of design patents in the form of graphs, figures and tables. Divided into five parts, the first part of the study has laid the foundation by introducing the theme to the audience, while the next part describes the methodology used for the study. The part thereafter contextualizes the relevance of the study, while the penultimate part contains the analysis and discussions, which flows into various subsections. A summary and conclusions are presented in the final part.

Methodology
Patents categorized as 'design patents' for the five BRICS economies were looked up using the USPTO database for the period between 2002 and 2011. The results included 271 records for India, 362 for Brazil, 51 for Russia, 5,580 for China and 215 for South Arica. To make the study more interesting and multi-sourced, data from the World Intellectual Property Organization have also been incorporated. The profile of retrieved records was analyzed according to subfields like document number, publication date, title, inventor name, assignee, application number, filing date, primary class, etc. Accordingly, various activities and trends were analyzed and are presented in the form of graphs, figures and tabulations.

Relevance of the study
This study is an attempt to examine the profile of design patents among the BRICS economies issued by the United States Patents and Trademark Office (USPTO). The main motivations for the study are: first, design patents remain a relatively less studied[3] branch of intellectual property as compared to their senior sibling, utility patents, not only in the BRICS economies but also on a global level (Lee and Sunder 2013); second, over the past years the BRICS economies have rapidly increased their global share of design registrations from 26.5% in 2003 to 40% in 2012. In competitiveness and innovation, design patents help economic development by encouraging creativity through the latest type of innovation called 'form innovation' (Chen and Chang 2014). A product has two important features, i.e. *function* – how it works and *form* – how it looks. It is often design, appearance or form that first catches the attention of the consumer. Many scholastic writings and discourses have categorized this innovation as 'soft-, aesthetic-, design-, innovation' (Satish 2011; Marzal and Esparza 2007). In conjunction with 'functional innovation', form innovation protects design patents through form, shape and appearance. Together, they spur innovations, technological advancements and economic growth.

BRICS is a conglomeration of five major emerging economies, i.e. Brazil, Russia, India, China and South Africa. The group was originally known as 'BRIC' before the inclusion of South Africa in 2010. Although, the BRICS nations are developing countries, they are known as fast growing industrialized economies that have wide impact on global affairs. The BRICS nations have about 40% of the world's population and more than quarter of the world's land area. Owning about 16% of global utility patents, the group has a combined GDP of US$16.039-trillion, equivalent to one-fifth of the world's GDP.

Politically, China is a one-party state and Russia has highly centralized state control. Brazil, India and South Africa are democracies, but with significant corruption and/or ethnic strife. China outpaces the group in economic size, growth and trade. Brazil has a predominantly urban population, while India is largely rural. Russia has been witnessing an aging population, while India's population is comparatively young. Russia and China are established global players with permanent seats on the United Nations Security Council (UNSC). India, Brazil and South Africa aspire to become global players but are currently relegated to the position of regional power houses (Observer Research Foundation 2013).

Trade among the BRICS nations can be highly complementary. Brazil and Russia are strong in commodity and natural resources while China and India are net importers of these resources. China dominates the manufacturing sector, while India's strength lies in the generic pharmaceuticals, software engineering, textiles and business process outsourcing. Value added by the BRICS economies in manufacturing has increased from 2.6% in 1971 to 16.5% in 2008. The high GDP growth of BRICS economies, combined with its large population base, provides BRICS with the potential to generate demand. Thus, the sustained GDP growth of the BRICS nations gives these economies acting collectively the potential to be drivers of the global economy (Centre for WTO Studies 2013).

Table 3: Cross-country comparison in design patents assigned, 2002–2011.

Years	Brazil	Russia	India	China	South Africa
2002	15	4	7	212	9
2003	49	–	12	271	15
2004	55	4	11	349	14
2005	23	2	18	278	16
2006	26	2	20	541	14
2007	22	4	26	626	30
2008	31	4	37	895	32
2009	49	9	42	776	40
2010	44	13	59	833	25
2011	48	9	39	799	20
Total	362	51	271	5580	215
Compound annual growth rate (%)	42	33	50	44	42

Source: USPTO database

In addition, the oldest, well-developed Free Trade Agreements (FTAs) exist between most of the BRICS economies. South Africa is a member of the Southern Africa Customs Union (SACU) which consists of five member nations from the southern Africa region. On the other hand, Brazil is a part of MERCOSUR which consists of five sovereign South American states. India and China are members of the Asia Pacific Trade Agreement (APTA) which consists of five Asian countries. India and Brazil represent developing countries of Asian, African and South American origin in the Global System Trade Preferences (GSTP). China and Russia, as Pacific Rim countries, are the representative members of the Asia Pacific Economic Cooperation (APEC). To strengthen intra-BRICS trade, the group is at an advanced stage of an idea to set up funds under a new Development Bank to undertake green projects and develop green technology towards sustainable development. The new Development Bank with an initial corpus fund of US$100-billion would be in the same line as the IMF (*The Hindu* 2015).

In view of the large population, fast growth and strong trade network of the BRICS economies, a large number of firms from the region have placed emphasis on design patents, not only to protect their products from unauthorized use or infringements, but also to make products attractive, innovative and competitive. Their knowledge of natural resources, the political situation and trade among the BRICS economies would not only help firms from the region to link up for intra-product development, but also to gear up for a greater share of the designing activity in the world's output. Design patents especially hold importance in the region due to the fact that the manufacturing sector of the BRICS economies forms a major part of their local economies. Firms from developed countries are well aware that the use of attractive or appealing designs makes a critical difference to the profitability of a product or service, potentially leading to increased demand and revenues. However, firms from developing countries, especially from the BRICS economies, need a consolidation of ideas so that design-intensive enterprises have the potential to become the engines of economic and social development (World Intellectual Property Organization 2016c). It was observed that there is not enough academic literature on design patents

devoted to the emerging BRICS economies; and hence, this study may prove helpful for future studies.

Analysis

As shown in Table 3, China was granted the most design patents during 2002–2011, followed by Brazil, India, South Africa and Russia. On comparing the five BRICS countries, publication of design patents grew with a compound annual growth rate of 50% for India, followed by 44% for China, 42% each for Brazil and South Africa, and 33% for Russia. China was granted a large number of design patents in 2008, while India had the most design patents granted in 2010. It can be observed that most of the BRICS economies have seen a rapid hike in the granting of design patents after 2008 and 2009. This may be due to an enhanced awareness of design application, the proliferation of design education and competitiveness amongst design firms.

Ratio of design patents to utility patents

From Figure 1, it is clear that three countries, viz. China, Brazil and South Africa, dominate in terms of a comparison of design patents vs. utility patents. This comparison shows how design activities are vigorously pursued in an individual country in comparison with utility patents. China, during the period 2002–2008, registered more than half of its patents in the field of design patents. It attained the pinnacle in 2003 accounting for about 60% of China's utility patents; however, this has declined gradually, becoming about 19% in 2011. India and Russia have a very small proportion of design patents in

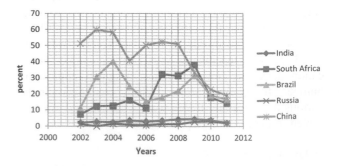

Figure 1: Ratio of design patents to utility patents, 2002–2011. **Source:** USPTO database

Table 4: Time taken to assign design patents, 2002–2011 (%).

Time in years	Brazil	Russia	India	China	South Africa
5 years and above	0.0	0.0	0.4	0.2	0.4
4 years	0.0	2.0	1.8	0.4	1.9
3 years	5.5	11.8	4.1	2.9	5.1
2 years	30.4	45.1	24.4	22.2	41.9
One year	57.7	41.2	59.4	62.3	46.0
Less than one year	9.1	0.0	10.0	12.0	5.1
Total	100	100.0	100.0	100.0	100.0

Source: USPTO database

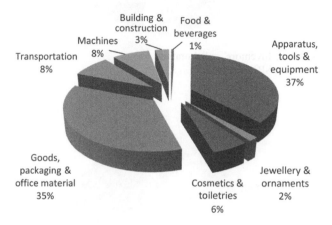

Figure 2: Subject focus, Brazil.
Source: USPTO database

comparison to utility patents. Thus, utility invention is the main focus of these two countries. Having a proportion of patents in the design category emphasizes the design competitiveness of a particular nation.

Time taken to assign design patents

From Table 4, a clear trend emerges that for all the countries in this study, most of the design patents were granted within a year from the date of filing, followed by two years, and less than a year (except Russia, where 11.8% design patents took three years to grant). Trends from the BRICS region almost coincides with the world trends, where most of the design patents were also granted within a year of filing. However, this is in contrast with utility patents, where time taken for granting a patent is generally longer than for design patents. This establishes

the fact already mentioned above that design patents are granted more quickly than invention patents.

Subject focus of design patents

Subject concentration of design patents for all five BRICS countries is shown in Figures 2–6. For India, 33% of design patents were issued in the subject area of jewellery and ornaments, followed by 27% for apparatus tools and equipment. Due to the large consumer base and higher demand, India remains the largest importer of gold metal in the world. Also, 15% of design patents were issued in the packaging of goods. However, according to another conservative estimate based on a survey of firms from the design sector by the Confederation of Indian Industry (2016), architecture, interior, animation, social media and fashion are the leading domains in the Indian design sector. With changes likely to happen in the economic geography in the future, design domains such as graphic design, industrial design and human-computer interaction (HCI) design are growing at a fast pace.

For the rest of the countries, viz. Brazil, Russia, China and South Africa, the largest segment for which design patents were issued is apparatus tools and equipment, followed by goods packaging. The entire BRICS region contrasts with the world trend where the subject's[4] focus is otherwise different.

Inventors' collaboration pattern

The inventors' collaboration pattern for the respective BRICS countries has been diagrammed in Figures 7–11. As has been evident from the graphs, most of the design patents in the BRICS region were issued to single inventors followed by two-inventor teams. Approximately

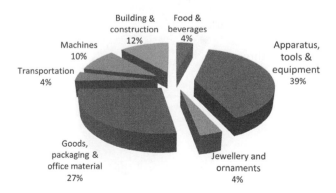

Figure 3: Subject focus, Russia.
Source: USPTO database

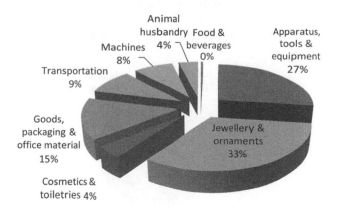

Figure 4: Subject focus, India.
Source: USPTO database

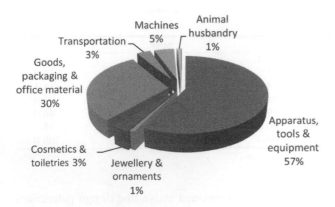

Figure 5: Subject focus, China.
Source: USPTO database

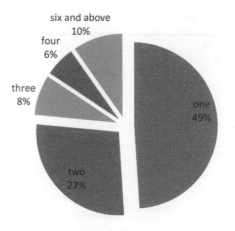

Figure 8: Inventors' collaboration, Russia.
Source: USPTO database

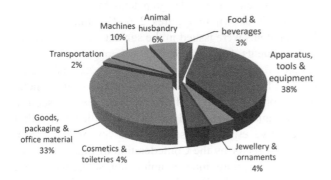

Figure 6: Subject focus, South Africa.
Source: USPTO database

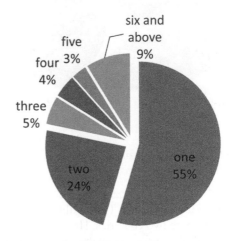

Figure 9: Inventors' collaboration, India.
Source: USPTO database

nine and ten per cent of design patents issued to multi-inventors (six and above) were in the case of India and Russia, respectively. However, similar collaborations were not observed in the other three nations in the region. The design patents which are granted as 'individual', however, top the list, followed by the ownership of proprietary firms like Samsung, LG, Microsoft, Apple, etc.

Inventors' affiliation pattern
Inventors' affiliations are shown in Figures 12–16. The largest percentage of inventors came from the BRICS countries, while the USA was observed as the most major affiliate country in the region, followed by other advanced countries like Britain, Germany, and Canada. However, some prominent Asian countries like Japan, South Korea, Singapore, Taiwan and Hong Kong were also noted for their close ties with the region, primarily

with China and Russia. For instance, Taiwan, in the case of China, is the third largest affiliate nation; similarly, South Korea and Singapore, in the case of Russia, are third largest affiliating nations after the US. This has been of particular importance since Japan, Taiwan, Germany, South Korea, Canada, the United Kingdom and Hong Kong have a sizeable[5] number of the global design patents, after the United States.

To summarize inventors' affiliations, we can observe that multinationals from developed countries (North America, Europe and Asia) have established some form

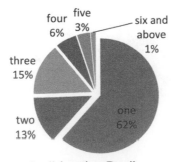

Figure 7: Inventors' collaboration, Brazil.
Source: USPTO database

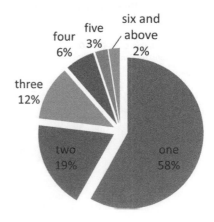

Figure 10: Inventors' collaboration, China.
Source: USPTO database

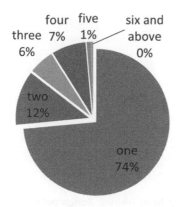

Figure 11: Inventors' collaboration, South Africa.
Source: USPTO database

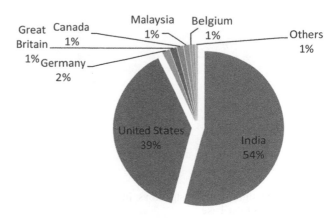

Figure 14: Inventors' affiliation, India.
Source: USPTO database

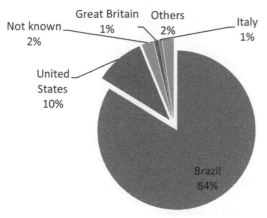

Figure 12: Inventors' affiliation, Brazil.
Source: USPTO database

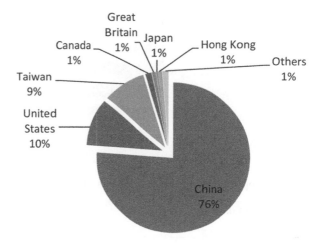

Figure 15: Inventors' affiliation, China.
Source: USPTO database

of tie-ups in the BRICS economies to start designing firms, both to cut costs and to customize their products to suit local needs.

Design patents assigned vs. filed

It is observed in Figure 17 that about 90% of design patents filed in the region were successfully granted. The success rate of granting design patents is as high as 97% in the case of South Africa. Thus, it means the rejection

rate of design patents is lower than that of utility patents. The latter category has more competition than the former and thus involves a more stringent patent-granting regime.

Top firms owing design patents

Consolidated data showing the ownership of design patents for the BRICS economies appear in Figures 18–22. For Brazil, the firm Grendene S.A. owned the most design patents (37), followed by Johnson & Johnson Industria E., Comercio Ltda (33) and Natura Cosmeticos

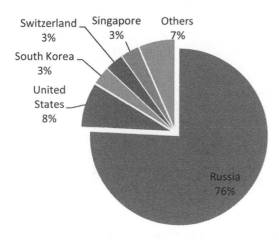

Figure 13: Inventors' affiliation, Russia.
Source: USPTO database

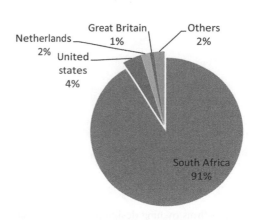

Figure 16: Inventors' affiliation, South Africa.
Source: USPTO database

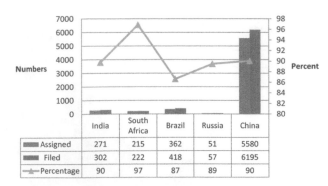

Figure 17: Design patents assigned vs. filed, 2002–2011.
Source: USPTO database

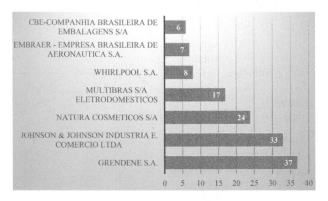

Figure 18: Top firms owning design patents for Brazil, 2002–2011.
Source: USPTO database

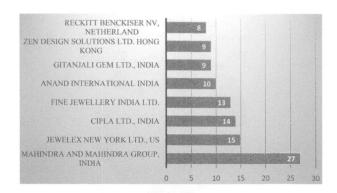

Figure 20: Top firms owning design patents for India, 2002–2011.
Source: USPTO database

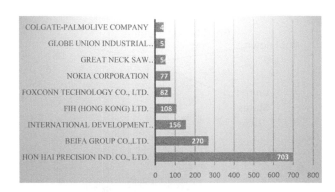

Figure 21: Top firms owning design patents for China, 2002–2011.
Source: USPTO database

S/A (24). In the case of Russia, the firm Obshtshestvos Ogranichennoy Otvetstvennostyu owned the most design patents. For India, the firm Mahindra and Mahindra held ownership of the highest number of design patents (27) followed by the US firm, Jewelex New York Ltd (14). Reckitt Benckiser, NV from the Netherlands and Len Design from Hong Kong owned the most design patents in the foreign nation category. Indigenous firms from India were in a good position, as five among the top eight firms were from India.

Among the largest haul of design patents for China, the Hon Hai Precision Ind. Co., Ltd. topped the list of Chinese firms, owning 703 design patents. Other firms recognized as owning design patents from China were

the Beifa Group Co., Ltd. (270), International Development Corporation (156) and Fih (Hong Kong) Ltd. (108). The two South African firms, Cochrane Steel Products (Proprietary) Ltd. and Zodiac Pool Care Inc. were recognized for owning the highest number of design patens, closely followed by the Coca-Cola and Bonfit America Inc. However, there are a large number of design patents owned by individuals in each country in the BRICS region. The Chinese firm Hon Hai Precision Ind. Co., Ltd. held the distinction of being the top firm from the BRICS region and the eighth largest in the world, accounting for 1762 design patents in the period between 1991 and 2015.

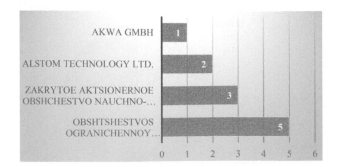

Figure 19: Top firms owning design patents for Russia, 2002–2011.
Source: USPTO database

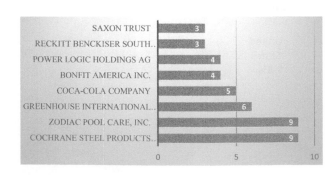

Figure 22: Top firms owning design patents for South Africa, 2002–2011.
Source: USPTO database

Findings and conclusions

- The highest number of design patents among the countries examined, for the period 2002 to 2011, was awarded to China (5580), followed by Brazil (362), India (271), South Africa (215) and Russia 51.

- The number of design patents awarded in the period 2002 to 2011, grew at a compound annual growth rate of about 50% for India, followed by 44% for China, 42% each for Brazil and South Africa, and 33% for Russia.

- Most of the economies in the study have registered rapid increases in the granting of design patents since 2009.

- For all the nations in the study, the largest number of design patents was awarded within a year or two from the date of filing. This is in contrast with utility patents, where an average pendency stands more than four years.

- Among all the BRICS countries, the largest number of design patents was awarded in the apparatus, tools and equipment category, followed by goods, packaging and office material. However, for India, the largest number of design patents was awarded for jewellery and ornaments design, followed by apparatus and equipment. India is a major exporter of jewellery, and design patents in this domain show Indian firms' increasing awareness of the importance of adding value to their products.

- 10,398 inventors collaborated in the case of China, followed by 674 for India, 658 for Brazil, 326 for South Africa, and 124 for Russia. Greater inventor collaboration was observed in the case of India, as there were 2.49 inventors per patent, closely followed by Russia with 2.43 inventors per patent. In the case of Brazil, this ratio was 1.82, while in China and South Africa it was 1.86 and 1.52, respectively.

- Most of the patents in the design category were awarded to single inventors, followed by inventor teams of two. As evidenced in the above point, there were very few patents in the multi-inventor category (six and above) for South Africa, Brazil and China. However, for India and Russia, there were encouraging trends for multi-inventor design patents.

- A large percentage of inventors belong to the BRICS region. The United States is observed as a major affiliate country, followed by Great Britain, Germany, the Netherlands and Canada. India and China have some network collaboration with their Asian counterparts. Taiwan, in the case of China, is the third largest collaborator. India is actively collaborating with Malaysia. Japan, South Korea, Singapore and Malaysia are also developing partnerships with the BRICS economies. This partnership can reap rich dividends as firms like Samsung, Sony, and LG are from Japan and South Korea. These firms also hold the distinction[6] of being among the world's top firms in the field of design patents.

- Most of design applications from South Africa (97%) were awarded patents, followed by 90% each for China and India, 89% for Russia and 87% for Brazil. These high percentage rates are good evidence showing that rejections rates for granting design patents are far lower than for utility patents.

- In 271 design patents for India, 78 firms were involved. Large numbers of firms were from the US (41) followed by India (25). However, fewer firms were observed from Great Britain, Hong Kong, Switzerland, Singapore, Germany, Netherland and Belgium.

- For Brazil, the firm Grendene S.A. owned the highest number of design patens, followed by Johnson & Johnson Industria E. Comercio Ltda, and Natura Cosmeticos S/A. While for Russia, the firm Obshtshestvos Ogranichennoy Otvetstvennostyu owned the most design patents. For India, the firm Mahindra and Mahindra held ownership of most design patents, followed by the US firm, Jewelex New York Ltd. Similarly, Reckitt Benckiser, NV from Netherlands and Len Design from Hong Kong held the highest number of design patents from foreign nations. Indian firms are in a good position, as among the top eight firms, five are from India.

- The Hon Hai Precision Ind. Co., Ltd. topped Chinese firms in terms of owning the most design patents. Other firms recognized as owning design patents from China are the Beifa Group Co. Ltd., International Development Corporation and Fih (Hong Kong) Limited. South African firms, Cochrane Steel Products (Proprietary) Limited and Zodiac Pool Care, Inc. are recognized for owning the most design patents, closely followed by Coca-Cola and Bonfit America Inc. However, there are a large number of design patents owned by individuals in each country in the BRICS region.

Notes

1. Design activity primarily covers the features, shapes, configuration, pattern, ornamental or composition, of lines or colours applied to any article in two-dimensional or three-dimensional form or in both forms. It can result from or be applied in industrial, mechanical, chemical or manual processes either separately or in any combinations.

2. Egyptian Goddess Inc. vs. Swisa Inc. and Apple Inc. vs. Samsung Inc. (Carani 2013).

3. Most of the studies have focused on issues of copyrights, trademarks and utility/invention patents.

4. For the period 2001–2015, the worldwide number of design patents issued in the top five subject categories are: recording, communication, or information retrieval (38,017); furnishing (23,772); transportation (21,122); tools and hardware (17,947); and equipment for food and drink preparation (17,494) (USPTO 2016a).

5. At present, worldwide, the United States holds 59% of the design patents and the non-US share is 41%. Countries like Japan, Taiwan, Germany, South Korea, Canada, the United Kingdom and Hong Kong together account for 28% of non-US design patents (USPTO 2016c).

6. For the period 2001–2015, the major South Korean firm, the Samsung Electronics Co. Ltd, held ownership of the world's largest number of design patents (6159) and stood at number one. A firm from Japan, Sony Corporation, stood in third place with 3182 design patents, followed by LG Electronics Inc., in fifth place with 2440 design patents (USPTO 2016b).

References

Atkins, E. B. 2013. "Unchecked Monopolies: The Questionable Constitutionality of Design Patent and Product Design Trade Dress Overlap in Light of Egyptian Goddess, Inc. v. Swisa, Inc." *Intellectual Property* 4 (3): 57–71.

"BRICS to set up Green Fund." 2015. *The Hindu*, February 6. http://www.thehindu.com/sci-tech/energy-and-environment/brics-to-set-up-green-fund/article6865844.ece.

Bryson, J. R., and G. Rusten. 2011. *Design Economies and the Changing World Economy.* New York, NY: Routledge. http://samples.sainsburysebooks.co.uk/9781136883620_ sample_844372.pdf.

Carani, C. V. 2013. "Apple v. Samsung Design Patents Take Center Stage." *Landslide* 5 (3), http://www.americanbar. org/content/dam/aba/administrative/litigation/materials/aba-annual-2013/written_materials/2_1_apple_vs_samsung_FN. authcheckdam.pdf.

Centre for WTO Studies. 2013. *BRICS-Trade Policies, Institutions and Areas for Deepening Cooperation.* New Delhi: Indian Institute of Foreign Trade.

Chan, T., J. Mihum, and M. Sosa. 2014. *The Evolution of Product form-identifying and Analyzing Styles in Design Patents.* INSEAD-The business School for the world. Ayer Rajah Avenue, Singapore.

Chen, R., and H. Y. Chang. 2014. "Patent Strategy and Layout Instrument–Derivative Design Patents." *Journal of Management and Strategy* 5 (3): 35–44. doi:10.5430/jms. v5n3p35.

Chen, R., Y. S. Sung, and P. T. Su. 2014. "Design Patents for Animated Images: Development Trends." *Journal of Intellectual Property Rights* 19 (9): 43–48.

Confederation of Indian Industry. 2016. "India Design Report." http://www.cii.in/webcms/Upload/a2.pdf.

Danish Design Centre. 2016. "The Design Ladder: Four Steps of Design Use." http://danskdesigncenter.dk/en/design-ladder-four-steps-design-use.

Davidsson, O. 2004. "Game Design Patents, Protecting the Internal Mechanism of Video Games." *Master Thesis.*, IT University of Goteborg, Sweden.

Export-Import Bank of India. 2014. "R&D Development in BRICS-an Insight." *Occasional Paper No. 168.* Mumbai.

Gaff, B. M., and P. T. Cuomo. 2013, March. "Design Patents." *Computer* 46, 8–10.

Korean Institute of Design Promotion. 2008. "National Design Competitive Power." https://www.dexigner.com/images/article/17907/KIDP_National_Design_Competitiveness. pdf.

Lee, P., and M. Sunder. 2013. "Design Patents-law Without Design." *Stanford Technology Law Review* 17: 277–304.

Marzal, A. J., and E. T. Esparza. 2007. "Innovation Assessment in Traditional Industries. A Proposal of Aesthetic Innovation indicators." *Scientometrics* 72 (1): 33–57. http://link. springer.com/article/10.1007/s11192-007-1708-x.

Observer Research Foundation. 2013. *A Long Term Vision for BRICS-submission to the BRICS.* New Delhi: Academic Forum.

Risch, M. 2013. "Functionality and Graphical User Interface Design Patents." *Stanford Technology Law Review* 17: 53–105.

Satish, N. G. 2011. "Soft Innovations-A Case of Chinese Design Patents on Pens." *ASCI Journal of Management* 41 (1): 123–30. http://aspire-dst.org.in/documents/soft-innovations-a-case-of-chinese-design-patents-on-pens.pdf.

Thomson Reuters. 2012. "Thomson's Trademark Report". http:// trademarks.thomsonreuters.com/sites/default/files/rsrc_assets/ docs/State_of_TrademarkReport_FINAL.pdf.

USPTO. 2016a. "Design Report 1991–2015." http://www.uspto. gov/web/offices/ac/ido/oeip/taf/design.pdf.

USPTO. 2016b. "Design Patenting by Organizations 2015." http://www.uspto.gov/web/offices/ac/ido/oeip/taf/topod_15. pdf.

USPTO. 2016c. "Design Patents, 1991–2015." A Patent Technology Monitoring Team Report.

USPTO. 2016d. "Types of Patents." https://www.uspto.gov/web/ offices/ac/ido/oeip/taf/patdesc.htm.

World Intellectual Property Organization. 2015. "World Intellectual Property Indicators." http://www.wipo.int/ edocs/pubdocs/en/wipo_pub_941_2015.pdf.

World Intellectual Property Organization. 2016a. "Industrial Designs Highlights." http://www.wipo.int/export/sites/ www/ipstats/en/wipi/2015/pdf/wipi_2015_designs.pdf.

World Intellectual Property Organization. 2016b. "Geographical Indications-an Introduction." http://www.wipo.int/edocs/ pubdocs/en/geographical/952/wipo_pub_952.pdf.

World Intellectual Property Organization. 2016c. "Unlocking Design Potential in Developing Countries." http://www. wipo.int/wipo_magazine/en/2016/03/article_0002.html.

Outward foreign direct investment (OFDI) and knowledge flow in the context of emerging MNEs: Cases from China, India and South Africa

Angathevar Baskaran Ⓘ, Ju Liu Ⓘ, Hui Yan and Mammo Muchie Ⓘ

The paper explores the factors driving Outward Foreign Direct Investment (OFDI) by Emerging multinational enterprisess (EMNEs) and the patterns of knowledge transfer in six cases of EMNEs from three BRICS' economies (India, China and South Africa). It found that there are significant differences between the OFDI from EMNEs and Developed multinational enterprise (DMNEs), which cannot be explained by using traditional FDI models. The way that EMNEs enter and operate in developed and developing countries are different. Knowledge transfers between EMNEs and developing host economies are predominantly one way and the former transfers more technology and knowledge than they gain. In the case of EMNEs and developed host economies, the knowledge and technology transfers appears to be more evenly matched, a two-way street benefitting both parties. The paper makes two major contributions: (i) it attempts to identify and distinguish the factors driving OFDI and patterns of knowledge transfer of OFDI from EMNEs and shows how they differ from DMNEs; (ii) it highlights aspects of OFDI by EMNEs such as expansion into countries outside their respective regions, and different patterns of technology and knowledge transfer in the South and North respectively.

List of abbreviations

BRICS	Brazil, Russia, India, China and South Africa
CNPC	China National Petroleum Corporation
DMNEs	Developed multinational enterprises
EMNEs	Emerging multinational enterprises
FDI	Foreign Direct Investment
LLL	Linkage, leverage and learning model
OFDI	Outward Foreign Direct Investment
OLI	Ownership, Locational, and Internalization Advantage Framework
R&D	Research and Development
TWMNE	Third world multinational enterprise
UNCTAD	United Nations Conference on Trade and Development
ZTE	Global telecommunications equipment and network solution provider, China

Introduction

Until the year 2000, only a small number of studies (e.g. Lall et al. 1983; Agarwal 1985; Tolentino 1993; Cai 1999) focused on the outward FDI from developing economies. Since the mid-1990s, OFDI from these countries has grown significantly (UNCTAD 2006; The Financial Times 2014) due to increases in foreign exchange reserves and growth of MNEs, which are capable of investing overseas, along with changes in the global economy. OFDI figures for 2013 across the world confirm this trend (see Tables 1 and 2). As traditional FDI theories are based on MNEs from developed countries, a number of studies have examined whether the OFDI by the EMNEs are same as or different from that of DMNEs: a) in terms of motivating factors, technology and knowledge flows

(e.g. Witt and Lewin 2007; Fleury and Fleury 2011; Mani 2013; Jeenanunta et al. 2013; Aminullah et al. 2013; Norasingh 2013); b) whether existing theoretical concepts can be applied similarly to these firms, or if there is a need to develop new conceptual and analytical tools (e.g. Liu, Buck, and Shu 2005; Mathews 2006; Luo and Tung 2007).

Ramamurti (2012, 41) identified two extreme views: one arguing that EMNEs can be understood only with new theory (e.g. Mathews 2002) and the other asserting that existing theory is quite adequate to explain EMNEs (e.g. Narula 2006), and suggested: 'truth is somewhere in between and that [the] real challenge is to discover which aspects of existing theory are universally valid, which aspects are not, and what to do about the latter'. An overview of the literature shows that there is a majority view that the traditional FDI theories do not fully help explain various aspects of OFDI from EMNEs. However, there is no consensus on a new dominant theoretical framework, which can help explain all aspects OFDI from EMNEs, including the factors driving OFDI by EMNEs and the patterns of knowledge transfer (see e.g. Mathews 2006; Li 2007; Yamakawa, Peng, and Deeds 2008; Kim and Rhe 2009; Ramamurti 2012).

The paper explores the factors driving OFDI by EMNEs and the patterns of knowledge transfer, as gaps remain in the understanding of these two aspects, by examining six cases of MNEs from three BRICS' economies (India, China and South Africa). It employs descriptive data for a period of about 10 years in each case which

Table 1: OFDI from different regions across the world.

Region	OFDI (US$ in billion) – 2013	Trend over 2012 (%)
Asia Pacific	177.91	−1.94
Europe	246.99	+16.79
North America	127.26	+9.5
Latin America & Caribbean	17.53	Almost double
Middle East & Africa	48.02	+21.81

Source: The fDi Report 2014 (The Financial Times Limited)

were gathered from secondary sources including EMNEs' annual financial reports, press releases, websites and other sources. Relying on secondary data does not affect the robustness of the findings and results from the case studies, as the analysis of developments in each case covers the entire 10-year period. This period is sufficient for in-depth developments to emerge and to trace how things evolved in each case. It also allows for distinguishing differences among the cases. Furthermore, a good number of interviews and comments from senior managers of selected EMNEs (except one South African case) specific to this research were included from the secondary sources.

The contribution of this paper is threefold: (i) based on the empirical evidence from six detailed case studies of EMNEs, it attempts to identify and distinguish the factors driving OFDI and patterns of knowledge transfer of OFDI from EMNEs and show how they differ from DMNEs; (ii) it also highlights aspects of OFDI by EMNEs such as expansion into countries outside their regions, non-infrastructure/natural resources sectors, different patterns of technology and knowledge transfer in the South and North respectively. This differs from the existing OFDI literature (iii) it applies grounded and appreciative approaches to the six case studies of EMNEs from three important emerging economies. It also draws lessons on how best to undertake further research on the difference of OFDI between EMNEs and DMNEs.

Literature review and conceptual framework
The literature on OFDI can be categorized as region specific, country specific, and comparative studies across

countries and regions. Studies that focus on OFDI from particular countries such as China, India, Brazil, Malaysia and South Africa have been growing in recent years. For example, an increasing number of studies are focusing on China (e.g. Kiggundu 2008; Zhang 2009); India (e.g. Rajan 2009; Athukorala 2009), South Africa (e.g. UNCTAD 2005; Draper, Kiratu, and Samuel 2010), and other developing and newly industrialized countries such as Singapore, Thailand and Malaysia (Ellingsen, Likuma-huwa, and Nunnenkamp 2006; Wee 2007). However, there are still a number of areas of OFDI where we need to gain a clearer understanding of several issues. These include motivating factors, the differences in the EMNEs oper-ations in developing and developed host countries, and the process of intra- and inter-firm knowledge transfer across borders and within host countries. This study attempts to contribute to this gap in OFDI literature by analyzing six cases, two each from China, India and South Africa. Furthermore, we contribute to comparative studies in this area, as there are few such studies (e.g. Kumar and Chadha 2009).

Traditional FDI theories have been developed from the experience of developed countries' MNEs. Table 3 shows some of the major traditional FDI theories and models including the classification of FDI by UNCTAD. As tra-ditional FDI theories are based on the behaviour and experience of DMNEs, attempts have been made over the years to apply or justify the application of traditional FDI theories to developing countries' MNEs. With more and more empirical studies on EMNEs over the years, there is a general agreement that not all aspects of EMNEs can be captured and explained by traditional FDI theories. Yet, there is no agreed unified new theory or model which can be applicable to both DMNEs and EMNEs, or to EMNEs alone.

Lecraw (1993, 589) studied the OFDI by Indonesian MNEs and found that they 'have gone abroad not only to exploit their ownership advantages but also to access and develop ownership advantages they did not previously possess'. Dunning, van Hoesel, and Narula (1996) attempted to explain the OFDI from the developing countries using data from Taiwan and Korea. They

Table 2: OFDI from countries of Asia Pacific, Latin America & Caribbean, and Middle East & Africa.

Asia Pacific	Capital investment (US$ in billion)	Market share (%)	Latin America & Caribbean	Capital investment (US$ in billion)	Market share (%)	Middle East & Africa	Capital investment (US$ in billion)	Market share (%)
Japan	50.04	28.13	Brazil	6.38	36.41	UAE	14.68	30.56
Hong Kong	48.18	27.08	Mexico	4.29	24.47	Kuwait	10.73	22.35
China	18.97	10.67	Bermuda	1.94	11.08	South Africa	5.45	11.36
India	13.52	7.60	Argentina	1.37	7.84	Mauritius	3.25	6.77
Singapore	12.48	7.01	Chile	1.09	6.21	Israel	3.12	6.49
South Korea	8.59	4.83	Colombia	1.09	6.19	Nigeria	3.06	6.37
Australia	8.35	4.69	Jamaica	0.44	2.54	Saudi Arabia	1.50	3.13
Taiwan	5.14	2.89	Honduras	0.37	2.10	Qatar	1.46	3.04
Thailand	4.27	2.40	Guatemala	0.19	1.10	Egypt	1.12	2.33
Malaysia	2.56	1.44	Bahamas	0.10	0.55	Bahrain	0.56	1.17
Other	5.82	3.26	Other	0.26	1.51	Other	3.09	6.43
TOTAL	**177.91**	**100**	**TOTAL**	**17.53**	**100**	**TOTAL**	**48.02**	**100**

Source: The fDi Report 2014 (The Financial Times Limited), p.5, 11, and 13

Table 3: Some major traditional FDI & internationalization theories/ models.

FDI theories	Author(s)	Highlights
Industrial organization/monopolistic advantage theory	Hymer (1960) Published in 1976	Identified two major determinants of FDI: removal of competition and firm-specific advantages and FDI takes place only if the benefits of exploiting firm specific advantages outweigh the relative costs of operation overseas.
Product cycle theory	Vernon (1966)	Identifies four stages of production cycle: innovation, growth, maturity and decline to explain some FDI by the US companies in Western Europe.
Internalization theory	Buckley and Casson (1976) Hennart (1982) Rugman (1982)	Buckley and Casson founded internalization theory: MNCs organize their internal activities in order to develop specific advantages and to exploit them. Hennart developed two models of internalization – vertical and horizontal. Rugman argues for internalization as general theory of FDI and why the MNEs exist.
The eclectic/ OLI paradigm	Dunning (1973, 1980, 1988)	According to the framework, foreign investment occurs when three factors act simultaneously in a country: (i) ownership advantage; (ii) locational advantage; and (iii) internalization advantage of a firm. Precise configuration of the OLI parameters facing any particular firm, and the response of the firm to that configuration, is strongly contextual (i.e. they are different from company to company and also depend on specific socio, political, and economic conditions of the host country.
Uppsala internationalization model	Johanson and Vahlne (1977)	How firms gradually intensify their activities in foreign markets through continuous adjustments and learning by doing in ever-changing international markets.
POM model	Luostarinen (1979)	How companies from small and open markets internationalize.
Host country FDI policy/ choice of FDI location by MNEs	Dunning (1998)	The choice of location is an important factor in the FDI by transnational firms, and countries that want to attract FDI could influence their decision by setting in place an attractive FDI policy regime.
Host country's locational advantage	UNCTAD (1998)	A host country can influence directly/indirectly the locational advantage as it relates to the host country's own decision. It has three determinants: FDI policy framework; economic determinants; and business facilitation. These differ from country to country and from time to time.
Typology of FDI	UNCTAD (1998)	Classified FDI into four types: (i) Natural resource seeking; (ii) Market seeking; (iii) Efficiency seeking or export oriented; and (iv) Strategic asset seeking.

called these OFDIs the second wave third world multinational (TWMNE) activities (aided by government policies) which represented an intermediate stage between first wave TWMNE activities and a 'conventional' industrialized country's MNE activities, thereby implying the OFDI from emerging economies are different from the FDI by DMNEs. Mathews (2006) proposed a new 'linkage, leverage and learning' (LLL) model which argues that OFDI from ENMEs aims to achieve competitive advantages via external linkage, leverage and learning rather than exploiting existing internal advantages through internal control. Li (2007, 300) argued that neither the (Ownership, Locational, and Internalization advantage framework) OLI nor the LLL model alone is sufficient (but they are complementary) and attempted to integrate them 'into a holistic, dynamic and dialectical framework' to make it applicable to all types of MNEs in future.

Yamakawa, Peng, and Deeds (2008) developed the 'strategy tripod' framework which integrates resource-based, industry-based, and institution-based views to analyze OFDI from EMNEs. Extending and using this framework, Lu et al. (2010) studied OFDI by Chinese private MNEs and found that a high-technology and R&D base tends to motivate strategic asset seeking OFDI and export experience and high domestic competition tend to

induce market seeking OFDI. Also, supportive government policies motivate both types of OFDIs. Kim and Rhe (2009) examined the trends and determinants of South Korean OFDI and found that dynamic effects of economic development have influenced the changing character of OFDI. They also concluded 'the behaviour of South Korean firms does not completely comply with the traditional theories of FDI. Thus, we may argue that this applies not only to South Korean firms but also firms from other economically evolving countries' (Kim and Rhe, 2009, 132).

There is no agreement that a single theory or model can satisfactorily explain the complex nature of OFDI from EMNEs. Taking this into account, Figure 1 presents a conceptual framework of OFDI from EMNEs, which integrates different FDI models to understand and explain the factors driving the OFDI (institutional and business), mode of OFDI flow, destination of OFDI and knowledge flow between the EMNEs and host economies. As this is a qualitative study, it uses broad indicators for institutional and business factors and knowledge flows (see Figure 1). It also recognizes that there are certain aspects of EMNEs which cannot be explained by the traditional FDI theories and various EMNE models proposed in recent years.

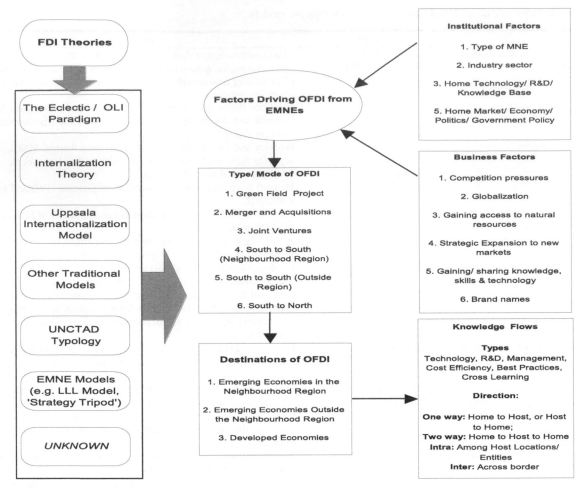

Figure 1: OFDI from emerging economy MNEs (EMNEs) – Drivers and knowledge flows.

Research methodology

We employ an interpretative research approach using the case study method and qualitative secondary data. A case study is more suitable than other methods when we are not clear about certain phenomenon or new research areas (Eisenhardt 1989; Eisenhardt and Graebner 2007). As highlighted in Figure 1, certain aspects of EMNEs cannot be explained by the traditional FDI theories and models. Therefore, the case study method is likely to help explore these aspects in greater depth. Even a single case study can help to undertake thorough analysis (McAdam and Marlow 2007) by answering questions such as 'what' and 'why' (Yin 1990). However, multiple case studies provide more external validity compared with a single case study. Therefore, we use multiple case studies. The data were gathered from secondary sources such as company annual reports, news/press releases, company websites and other published sources including interviews with senior executives of the case companies (except one South African case). Relying on secondary data does not affect the robustness of findings and results, as the analysis of developments in each case covers the entire 10-year period. This period is sufficient for all developments to emerge in detail and to trace how things evolved in each case and distinguish differences among the cases.

Using the conceptual framework presented in Figure 1, we have drawn an analytical framework to map and analyze data for the EMNEs with respect to: (i) factors driving the OFDI (institutional and business); (ii) type/ mode of OFDI; (iii) destination; and (iv) knowledge flow (see Tables 6 and 7).

Case studies

Using UNCTAD/Erasmus University database, six cases were selected from the top 100 non-financial EMNEs ranked by their foreign assets (US$ million and No. of employees, 2008). These are China National Petroleum Corporation (CNPC) (27th) and ZTE (79th) from China, Tata Steel (15th) and Hindalco Industries (29th) from India and MTN Group Limited (21st) and Sasol Limited (44th) from South Africa.[1]

This paper uses the qualitative interpretative research approach rather than quantified measurements to understand how OFDI from EMNEs evolved between 1999 and 2013, what are the motivating factors, and what are the patterns of technology/ knowledge transfer/ flow between the EMNEs and host economies.

China National Petroleum Corporation (CNPC, China)

CNPC, founded in 1988, is a state-owned MNE and one of the largest oil and gas producers in the world. CNPC's outward investment started from the early 1990s. Since then, its overseas oil and gas field production has increased tenfold as shown in Figure 2. By 2010, it had entered into

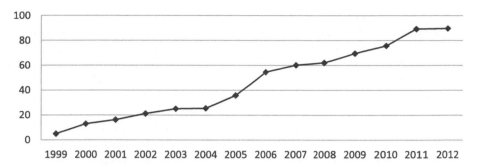

Figure 2: CNPC's annual field production of overseas crude oil over the last 10 years (million tons).
Source: Annual Reports of CNPC 1999–2000 to 2011–2012, Available at: http://www.cnpc.com.cn/cn/gywm/qywh/cbw/lncbw/

more than 30 overseas joint projects involving an investment of over US$5 billion (Finance and Economy 2010). By 2012, its operations had expanded to 63 countries and regions.

Since 2001, CNPC has recorded continuous growth by developing a number of large oil and gas projects across different regions on its own, through partnership with national oil companies (e.g. Malaysia, Algeria, Peru, Ecuador), as well as with other international oil companies (e.g. Iraq, Iran, Brazil, Mozambique, Australia). It has also used mergers and acquisitions. For example, in 2004 it purchased 100% of stocks from the Ay-Dan Petroleum Joint-stock Company (Kazakhstan) and in 2009 it acquired the Keppel Oil and Gas Services (Singapore) and the Mangistaumunaigas (MMG) from Central Asia Petroleum Ltd. at a cost of US$3.3 billion (CNPC News Release, 24 April 2009). In Canada, it acquired 60% of MacKay River and Dover oil sands assets of Athabasca Oil Sands Corp, which was 'a major breakthrough for CNPC in the overseas unconventional energy sector' (CNPC Annual Report 2009, 53). In 2013, it purchased all the shares of Petrobras Energia Peru S.A. for over US$ 2.6 billion (CNPC News Release, 15 November 2013).

According to CNPC's President, Jiang Jiemin, the global financial crisis and its increasing competitiveness have helped the company to expand its global business (CNPC Annual Report 2009, 9). It was ranked 83rd in 2001 when it first appeared in Fortune Global 500 and by 2014 it ranked 4th with an annual revenue of US$ 432.01 billion (CNPC News Release, 08 July 2014). Its global growth is also reflected in its technological and innovation growth in terms of registered patents. In 2010, it was granted 1701 patents (out of 2178 applications) including 300 invention patents (out of 841 applications). In 2013, it applied for 4481 patents and was granted 3639 (CNPC Annual Report, 2010, 14; 2013, 16). According to Jiang Jiemin, CNPC followed a strategy of making the human resources in its overseas operations 'more international, professional and local' (CNPC Annual Report 2011, 3), and its effort to 'promote indigenous innovation and deepen global cooperation and international exchange of technology' produced significant results. For example, its OFDI expansion enabled it to forge a partnership with Shell for establishing a shale oil joint research centre.

Investment in developing world
Since 1997, CNPC has been investing in Sudan providing finance, technology and training and

> helped Sudan become [a] petroleum exporter almost from scratch in just a few years, establishing a relatively sophisticated, state-of-the-art, large-scale and integrated petroleum industry system covering oil exploration and development, ground facilities construction, long-distance pipeline operation, refining and petrochemical production. (CNPC Annual Report 2007, 46–47)

In Kazakhstan and Venezuela, it invested in mature oil fields and used its 'proven technologies' as well as 'excellent management' to achieve significant annual growth in output. It followed a strategy of employing locals and providing them training. For example, in Aktobe city, Kazakhstan, with a population of 300,000 and a workforce of around 150,000, it employed about 20,000 locals, nearly 15% of the city's workforce (CNPC Annual Report 2007, 47–48).

Investment in developed economies
CNPC has forged alliances in the emerging and developed countries based on more evenly balanced two-way traffic of technology and knowledge flow. For example, its collaboration with Transneft (Russia) to jointly construct a 1000 km long pipeline between Russia and China has helped its engineers to master innovative construction techniques such as 'pipeline construction in multi-year frozen soil areas', 'construction on forest wetlands', and 'construction in swamps of permanently frozen soil' (CNPC Annual Report 2010, 36). Similarly, its joint venture with the US-based ION Geophysical Corporation helped provide global seismic contractors with 'brand new portfolios of first-class onshore geophysical products and premium services' such as enhanced drilling fluid and liquid cement systems, integrated matching technologies, and horizontal, underbalanced, and gas drilling (CNPC Annual Report 2009, 58). In 2010, it acquired oil sands assets from Canada's Athabasca Oil Sands Corp. and agreed to jointly develop them. In 2013, it acquired SPEC Engineering Inc., which gave access to the latter's technological expertise in land-based operations and troubleshooting experience in offshore production platforms. In 2011, CNPC bought 19.9% of Australia's LNG, which specializes in technology development and application of natural gas liquefaction and possesses natural gas liquefaction OSMR technology. This gave CNPC

preferential rights in using LNG's patented OSMR technology (CNPC News Release, 12 February 2010; 27 January 2011; 26 September 2013). In 2011, CNPC and INEOS Group Holdings PLC signed a 'strategic cooperation agreement to share refining and petrochemical technology and expertise between their respective businesses', which helped CNPC enter the high-end European market (CNPC News Release, 05 July 2011).

Technology and knowledge sharing in host economies
It appears that the OFDI by CNPC has resulted in uneven knowledge flow between CNPC and the developing host economies. That is to say that more technology and knowledge flowed from CNPCs to the host economies in terms of equipment, machinery, advanced technology, specialist expertize and training to local employees than the other way round. On the other hand, CNPC gained knowledge about local conditions, specific national managerial cultures, and international experience transferrable to its operations in other host economies and training for Chinese employees, and so on.

CNPC provided a range of technical services and expertize such as seismic data acquisition, processing and interpretation, apart from technology and products. For example, in 2004, a total of 43 seismic crews provided seismic data acquisition, processing and interpretation services in 22 countries. There were 38 well logging, 30 geologic logging and 29 testing crews in operation in Sudan, Iran, Kazakhstan, Venezuela, Pakistan, Azerbaijan, Syria, Libya and Algeria (CNPC Annual Report 2004, 28). By 2012, it was operating in 63 countries 'providing technical services in geophysical prospecting, well drilling, well logging and mud logging, as well as engineering and construction services for oil/gas field production capacity building projects, large refining and chemical installations, and pipelines and storage facilities' (CNPC Annual Report 2012, 38).

CNPC fostered international talents among its Chinese employees. For example, in 2007 it trained 370 senior management teams, employed 237 certified project management professionals and sent 128 professionals to study abroad. In 2011, it sent employees to Russia and the United States for financial and legal training, executive MBA courses and project management professional training. In 2012, 227 high-level executives received training at home and abroad. It sent Chinese employees for advanced management and technology training programmes in Canada, quality management training at Siemens, an internship programme at Baker Hughes, visiting scholar programmes at Stanford University and the University of Texas and the EMBA courses at the University of Houston (CNPC Annual Report 2007, 23–24, 2011, 16; 2012, 13–14; 2013, 15).

CNPC 'proactively promoted the local hiring of employees and created a communicative, coordinated and harmonious working environment' and provided 'well-tailored management knowledge, job-related skills and HSE training to foreign employees' (CNPC Annual Report, 2010, 10; 2011, 16). In 2010, it employed over 80,000 foreign employees, 90% of whom were locals in Sudan, Kazakhstan, Peru, Venezuela and Indonesia. By 2011, the percentage of foreign employees in overseas projects had reached 91% (CNPC Annual Report, 2010, 10; 2011, 16).

In Turkmenistan, CNPC trained 120 local employees as welders and worked with local schools to train more than 600 locals. It established the Surface Engineering Skills Training Centre to provide training courses on welding, oxy-fuel cutting and engineering machinery. It has trained 3850 local employees since 2010 and most of them 'have become skilled workers, with some standing out as technical managers' (CNPC Annual Report, 2012, 13–14). Similarly, in Iraq, the company has set up an integrated training centre and trained more than 110 local workers in welding skills, and has been collaborating with the University of Basrah (CNPC Annual Report, 2010, 10; 2013, 15).

In Myanmar, CNPC ran a three-phase training programme: (i) learning professional skills at the University of Yangon; (ii) Chinese and English language and pipeline management courses at Southwest Petroleum University, China; and (iii) hands-on experience at CNPC's pipeline transport stations. Similar three-stage training was provided in Venezuela: first stage induction training; second stage on-site training; and third stage advanced training for outstanding employees in China. An employee who received the three-stage training would be able to train other local employees. In Kazakhstan, CNPC set up training centres in Shymkent and Zhanazhol to provide simultaneous training to 25 welders, 30 pipeline workers and 30 riveters. It also worked with the local universities to run skill training. Furthermore, CNPC has been funding students in host countries to study Master's degree programmes in Chinese institutions such as China University of Petroleum (CNPC Annual Report, 2011, 16; 2012, 14; 2013, 15).

Strategic alliances with Western oil companies
OFDI in the developing countries appears to have helped CNPC to forge strategic alliances with other global oil companies based on two-way technology and knowledge flow both in developed host economies and at home in China. For example, its engineers were part of more than 10,000 managers, technicians and construction workers from over 20 countries involved in the implementation of the Amu Darya and Turkmenistan-China natural gas project in 2009. The 50/50 joint venture with Shell in Australia (2010) helped combine their 'strengths in technology, capital, experience on project management and marketing ability' and the CSG production in Australia. The relationship helped CNPC acquire 35% of Shell in Syria, and to forge R&D cooperation on unconventional oil and gas development (shale oil and gas, CBM). They jointly set up the Shale Oil Joint Research Centre in Beijing (in 2013) and agreed to 'strengthen long-term worldwide cooperation in unconventional resources, deep water, LNG, and upstream and downstream businesses' (CNPC News Release, 8 November 2013, 09 April 2014). They also collaborated with Exxon Mobile on joint technology R&D related to exploration and development, refining and chemicals and oilfield services. In 2012, CNPC and Siemens (Germany) agreed on procurement and supply of merchandise and service

sharing best practices and experiences, and cooperating in equipment manufacturing (CNPC Annual Report, 2009, 55; 2013, 18; CNPC News Release, 29 September 2012). In 2013, CNPC and Celanese (Dallas, Texas) forged a technology partnership to 'upgrade the quality of oil products, strengthen technological innovation capability and contribute more to air quality improvement' (CNPC News Release, 28 August 2013).

CNPC's strategic alliances not only helped it expand into the overseas market but also its own domestic oil and gas resources. In 2013, it entered into 37 joint exploration and development projects (16 conventional crude oil, 10 conventional natural gas, 10 CBM and 1 shale gas), which included partnerships with Eni S.p.A. (Italy), Chevron (US), Dart (Australia), Shell (Anglo-Dutch), Roc Oil (Australia), and Total (France) (CNPC Annual Report, 2010, 22; 2013, 24).

Zte (China)
Growth
ZTE is a global telecommunications equipment and network solution provider, founded in 1985 with its headquarters in Shenzhen. It is listed on both the Shenzhen and Hong Kong stock exchanges.[2] ZTE's strategy of globalization began in 1995 when it invested in Indonesia. In 2004, ZTE made rapid progress and expansion into Asian, African, South American and Russian markets, and in 2005 it set up 15 branches offering marketing, technological support, service and maintenance in Europe (Hou 2006).

The OFDI by ZTE followed two different modes: (i) building up functional facilities for marketing, production, and R&D as a telecommunication equipment supplier; and (ii) joining the telecommunication networks of different countries as an operator. Before 2006, ZTE mainly adopted the telecommunication supplier mode. In 2004, it invested RMB2.1 billion in overseas markets out of the RMB3.5 raised from its IPO on the Hong Kong Exchange. Its factories, research, training, service and engineering centres provide customized and innovative products and services to the local telecommunication operators and end customers. Since 2005, ZTE has also adopted a 'band-together' strategy to go global. It followed the Chinese companies abroad to help them build IT infrastructure

and provide telecommunication service. Having been regarded as a 'low-end mobile phone vendor', in 2009 ZTE overtook Nortel and Motorola to become the world's fifth largest telecoms infrastructure vendor, and became the sixth largest handset vendor by exporting over 60 million terminals. In 2011, ZTE replaced Apple as the world's 4th largest handset vendor, and in 2013, it became the fifth largest smartphone company in the world (ZTE 2004 to 2013).

Currently, ZTE delivers products and services to over 500 operators in more than 140 countries and has 107 subsidiaries and eight delivery centres abroad. It also has 18 overseas R&D centres including five in the US and two in Europe. Apart from its cooperation with top Chinese telecommunication firms such as China Mobile, China Telecom and China Unicom, it has also forged long-term partnerships with industry leaders globally such as France Telecom, UK's Vodafone, Australia's Telstra, Canadian Telus, MTN of South Africa, and America Movil of Mexico.

Over the years, ZTE's percentage of annual overseas business income to annual total income has increased continuously (see Figure 3), which shows the importance of its OFDI. By 2011, its overseas sales amounted to RMB20.8 billion (US$3.27 billion) accounting for 55.7% of its total sales. In 2011 and 2012, ZTE was ranked first in international patent applications by the World Intellectual Property Organization. It filed applications for 3906 patents in 2012, 37% more than a year earlier. ZTE's patent applications in 2012 greatly exceeded other companies in the telecommunications equipment industry such as Panasonic (2951), Ericsson (1197), Nokia Siemens (326), and Alcatel-Lucent (346). Between 2009 and 2012, ZTE invested more than RMB30 billion in R&D (ZTE Press News, 14 March 2013).

Skills and knowledge flow
ZTE has 3500 R&D staff and eight R&D centres including international facilities in India, Pakistan, and France. More than 5000 engineers specialize in 4G LTE technology development, and it operates eight R&D centres in China, the US and Europe (ZTE Press clipping, Maistre, 02 June 2014; Luk, 05 December 2013). ZTE actively

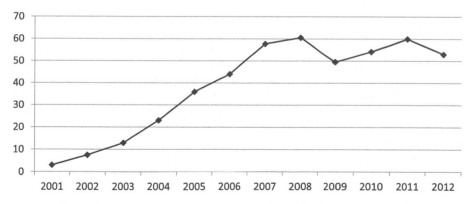

Figure 3: ZTE's percentage of annual overseas business income to annual total income.
Source: Annual Reports of ZTE 2001–2012, Available at: http://www.zte.com.cn/cn/about/investor_relations/corporate_report/

set up research centres overseas including in North America and Europe, and employed local talent. For example, according to Denson Xu, President and CEO of ZTE Canada, 'ZTE is aware of the needs to employ locals if it wants to develop and grow in Canada, so as to better understand the local culture and business environment', and '90% of its Canadian hires are local' (ZTE Press clipping, Lok and Na, 08 November 2013). ZTE has set up five R&D centres and one logistics centre in the US employing around 300 people. According to Lixin Cheng, CEO of ZTE USA, in 2012 ZTE invested US$30 million more in the US for 'leveraging local talent to bring new innovations to consumers' (ZTE Press clipping, Fiercewireless, 12 December 2012).

ZTE also invested significantly in training of local staff and forging links with local academic and research institutions. For example, in Romania it targeted training of 200 students in Bucharest initially and then planned to expand to other university centres. In Germany, it joined with Technical University Dresden to set up an R&D centre for innovation and patent development of long-term evolution technologies. ZTE linked with the Etisalat Academy, Dubai, to provide technical expertise to operators across the Middle East, Asia and Africa (ZTE Press clipping, Zawya, 19 March 2010). In 2011, it announced setting up a high-tech industrial park in Sao Paulo, Brazil which would employ about 2500 people by 2014 (ZTE Press clipping, Gomonews, 04 July 2011; Romania Insider, 12 December 2012).

In 2011, ZTE opened 'the UK Innovation Centre' (London), the first of 10 international innovation centres that ZTE was building to forge collaborative R&D programmes with major operators in different countries including Spain, France and Germany. According to Xie Daxiong, Executive Vice President of ZTE, 'The innovation centres under development by ZTE and its partners are a collective, proactive response to ... a transformative shift in the industry' (ZTE Press clipping, Smart Gorillas, 03 August 2011). Because of this transformation in the global industry, ZTE was able to forge a number of knowledge-sharing partnerships in developed countries (e.g. Dusseldorf Test & Innovation Centre with Vodafone, Poitiers in France, Kista in Sweden). Michael Stückmann, Chief Operating Officer of ZTE Vodafone Business Unit (Germany) said, 'ZTE will benefit from the excellent test conditions, outstanding infrastructure and the good support which the Centre has to offer' (ZTE Press clipping, Total Telecom, 08 March 2010).

The most interesting development was in India. ZTE made the India branch a global experts resource hub to support its operations and subsidiaries across the world. ZTE India employed over 90% Indian staff and they were sent to ZTE offices across the world. Jie Li, Director of Human Resources of ZTE announced: 'We are delighted to name ZTE India as our global expert resource hub. With its vast pool of skilled talent and resources and excellent language skills, India is the instinctive choice' and Indian 'employees will have the opportunity to work with and lead global teams on various innovative projects and expand their expertise in the latest technological

developments in the telecommunications arena' (ZTE Press clipping, Times of India, 19 August 2011).

Furthermore, in 2009 ZTE launched a global recruitment drive in universities in the US, France, Mexico, Colombia and Ethiopia to 'help inject fresh thinking and talent into ZTE from major regions in the world'. Also, ZTE was able to actively recruit talents from its competitors such as BlackBerry and Motorola Mobility. Adam Zeng, head of ZTE's mobile device business said: 'We hope the talent from BlackBerry can enhance our product security and design capability' and 'the company will continue to actively recruit more talent from BlackBerry and other competitors' (ZTE Press clipping, Luk and Osawa, 10 June 2014).

Strategic alliances

Like CNPC, the OFDI by ZTE in developing economies and second tier European countries appears to have helped it expand into developed countries and forge knowledge sharing alliances. By 2013, ZTE had reached cross-licensing agreements with Qualcomm, Siemens and Ericsson, Jazztel (Spain), Duxbury Networking (South Africa), and Telekom Austria. According to Wang Haibo, ZTE's Director of Legal Affairs, 'ZTE advocates an open and win-win model based on sharing' of knowledge (ZTE Press News, 14 March 2013).

In 2011, ZTE and British Telecom (BT) forged a partnership to develop next generation technologies. Clive Selley, CIO of BT Group said:

> We've been very impressed by the knowledge and sophistication of the ZTE's team and by combining our complementary strengths and expertise on a range of research projects, we expect this partnership to be very fruitful for both ourselves and our respective customers (ZTE Press clipping, Total Telecom, 02 June 2011).

ZTE also joined with Intel focusing on the new Intel® Atom™ Processor Z2580 platform to enhance its next generation smartphones' performance.

In summary, OFDI by ZTE shows that two-way knowledge flow occurred through technology transfer, recruitment and training of local talent, technical service, local R&D in the host economies (more in developed host countries than developing countries) and strategic alliances with other industry leaders. ZTE has learned and gained knowledge from emerging economies in different regions and shared the knowledge among its operations in different host countries.

Tata Steel (India)

Tata Steel Group has a presence in over 50 countries with manufacturing operations in 26 countries. In 2007, it acquired Corus (UK), later named Tata Steel Europe (TSE), which is Europe's second-largest steel producer with main operations in the UK and the Netherlands, and a global network employing 37,000 people. South East Asian operations comprise Tata Steel Thailand and NatSteel Holdings headquartered in Singapore, one of the largest steel producers in the Asia Pacific region. From the acquisition of NatSteel, Tata Steel gained markets in Vietnam, Thailand, Australia, China, Malaysia, Philippines and Singapore. Now, it is the tenth largest steel

Table 4: Tata Steel – Geographical distribution of revenues (% of total revenues).

Financial year	India	Asia (excluding India)	UK	EU (excluding UK)	Rest of the world
2007–08	15	12	37	32	5
2008–09	16.87	11.80	34.25	30.14	6.94
2009–10	–	–	–	–	–
2010–11	26	13	27	29	5
2011–12	27	13	26	29	5
2012–13	29	13	24	28	6

Source: Tata Steel Group, *Annual Report* (2007–2008 to 2012–2013)
Note: Data for 2009–2010 are not available.

producer in the world with 81,000 employees across five continents. Table 4 shows the importance of OFDI at Tata Steel and demonstrates that about two-thirds of its revenue is from outside India and over 50% from the UK and other EU countries.

Table 5 shows that before the onset of the global financial crisis in 2008–2009, the company invested heavily in Asian and EU countries, and then investment declined but somewhat stabilized in 2010–2011.

Tata Steel has adopted different integration strategies in Asia and Europe. In Asia, the focus is on sharing the know-how for achieving best operating practices and enhancing product quality among plants in different countries, while the focus is on quality and technology integration and sharing two-way best practices related to productivity, process improvement and cost efficiency between Corus (Europe) and Tata Steel in India.

The R&D activity within the Tata Steel Group takes place in five major centres: (i) IJmuiden Technology Centre (Netherlands); (ii) the Swinden Technology Centre (UK); (iii) Teesside Technology Centre (UK); (iv) Automotive Engineering Group (UK); and (v) Jamshedpur R&D Centre (India), which employs over 1000 researchers. The fact that four of them are located in Europe shows that the company has gained significant R&D capability by acquiring the Corus. Investment in Europe appears to have paid off and enhanced Tata Steel's R&D, product and process technology development capabilities, and also enabled it to forge strategic collaborations with others. For example, the European operations launched 14 major new products mainly in the automotive sector in 2012–2013 such as Bake Hardening 260 or 'BH260' which helps make lighter doors and bonnets, safety-critical automotive sheet steel called 'DP800GI' (Tata Steel 2012–2013, 23; Tata Steel Press Release, 11 July 2013), perforated armour steel-PAVISE™ SBS 600P, and special railway track, called SilentTrack®. Its UK operations helped penetrate the US

and European markets. For example, it won the bid to supply pipes for Enterprise Products Partners' new crude oil export pipeline in the Gulf of Mexico's Keathley Canyon, and also won a contract in 2009, worth €350 million to supply rails to the French railway operator SNCF (Tata Steel Press Release, 19 April 2012, and 29 September 2011).

Tata Steel has forged international R&D partnerships such as partnering in the multi-partner ULCOS Project aimed at developing technology to achieve a 50% reduction of CO_2 emissions per ton of steel produced, development of high strength and ductility steels in collaboration with Salzgitter, Germany, and physical vapour deposition based coating process in collaboration with POSCO, South Korea (Tata Steel 2008–2009, 59). Tata Steel Europe (TSE) has been working with the UK Ministry of Defence (MoD) to develop a new high strength armour plate at a lower cost, which was originally developed by Cambridge University, QinetiQ and the MoD Science and Technology Laboratory (Tata Steel 2008–2009, 71). Its Long Products Division was awarded the prestigious certificate by CARES, UK, for superior quality rebars, the only rebar manufacturer in India to be awarded this accreditation (Tata Steel 2008–2009, 51)

The interesting fact is that the knowledge, technology and experience gained in its Europe operations have been transferred to the whole group. For example:

> Building on the experience gained with Process Improvement Groups in Corus, the Tata Steel Group Process Improvement Teams have been set up with the aim to ensure that best practice in process technology is applied throughout the Group. This involves among other things, transferring technology that has been proven in one plant to similar other installations (Tata Steel 2008–2009, 72).

Also 'as a result of shared best practice from Corus' Scunthorpe and Teesside plants, the number of coke ovens in Batteries 5, 6 and 7 at Jamshedpur [in India]

Table 5: Tata Steel – Geographical distribution of capital employed (% of total capital).

Financial year	India	Asia (excluding India)	UK	EU (excluding UK)	Rest of the world
2007–08	20	7	43	27	3
2008–09	27.98	11.6	31.83	25.95	2.64
2009–10	–	–	–	–	–
2010–11	43	21	14	20	2
2011–12	33	35	8	20	4
2012–13	46	11	17	20	6

Source: Tata Steel Group, *Annual Report* (2007–2008 to 2012–2013)
Note: Data for 2009–2010 are not available.

that were subject to down-time during the year fell to zero from 48 in 2008' (Tata Steel 2009–2010, 24). Furthermore, investment in Europe enabled Tata Steel not only to expand upstream steel capacity expansion in India but also move into downstream value-added products and market globally in the automotive sector, aerospace products, high speed rails, special rails for metro and tramways, and speciality pipes and plates for the energy and power sector, and consumer goods (Tata Steel 2010 –2011, 20).

On the other hand, TSE gained managerial knowledge and experience from its home company in India, particularly to implement the 'One Company' operating model by setting up a single sales and marketing team, a consolidated supply chain organization with three steelmaking hubs, specialty businesses and pan-European support functions.

Hindalco industries (India)

Hindalco (Aditya Birla Group) is one of the largest producers of primary aluminium in Asia. In 2003, it acquired Nifty copper mine and Mt. Gorden copper mine in Australia, and Novelis in Canada in 2007, which helped it become one of the top five aluminium leaders in the world. Novelis has a large global presence and operates from 12 countries with 31 manufacturing plants, and employs 12,000 people. Novelis is the global leader in aluminium rolled products and aluminium can recycling. It produces about 19% of the world's flat-rolled aluminium products and is the number one producer in Europe and South America, and the second largest in North America and Asia (Hindalco 2004 to 2013). In addition to its aluminium rolling activities, Novelis operates bauxite mining, primary aluminium smelting and power generation facilities in Brazil. After the acquisition of Novelis, the high-quality assets of the closed Rogerstone plant in UK were moved to Hirakud in India (Hindalco Press Release, 20 June 2008).

According to Debnarayan Bhattacharya, managing director of Hindalco, the acquisition of Novelis helped Hindalco in many ways:

> Novelis is a strategic fit for Hindalco. We wanted to grow upstream (with value added products) to ensure sustainable profits … For value added products like cans, we needed to have technology and customer acceptance. Neither can be purchased from the market. Even if we invest time and develop technology, there is always a fear that it may not succeed. We have learnt many things from Novelis. We began with cultural integration, followed by finance and technology, and now marketing. For example, the energy efficiency of their plants was far better than Hindalco's … We can bring Novelis' technology into India and make cans and sheets for Indian consumers (Kalesh 2008; Business Standard, 7 March 2011).

Novelis helped Hindalco gain a major share in the automotive sheet market with customers including Audi, BMW, Chrysler, Ferrari, Ford, GM, Hyundai, Jaguar, Land Rover, Mercedes-Benz, Porsche and Volvo. By moving the 'high-quality assets of the closed Rogerstone plant in the UK' to Hirakud in India, it created a hub for can

body stock to meet the increasing demands of the beverage can market in India and neighbouring countries. Like Tata Steel, the acquisition of Novelis has helped Hindalco forge international research partnerships with global leaders. For example, it joined hands with Thyssen-Krupp to develop a joining technology that could help carmakers to dramatically reduce the weight of vehicles (Hindalco Press Release, 02 September 2010; 10 April 2012; 04 June 2013).

When Hindalco acquired Novelis for US$6 billion, it was financed by US$3.5 billion of its own cash and US $2.5 billion debt. But unlike many similar buyouts, Hindalco avoided raising finance for acquisition by leveraging Novelis' balance sheet and instead, paid US$3.1 billion through recourse financing on Hindalco's own corporate guarantee and US$450 million by liquidating some treasury stocks. No new debt was added to the Novelis' books so that Novelis could avoid facing liquidity problems and recover quickly from inherited financial problems (Business Standard, 7 March 2011). In 2010, Novelis experienced 'a remarkable turnaround in the midst of extremely challenging circumstances. In the economy that was still emerging from recession, Novelis reported record results in terms of record adjusted EBITDA,[3] liquidity and free cash flow' (Hindalco Press Release, 03 September 2010). Novelis posted an EBIDTA of US $754 million and its liquidity improved by US$640 million to US$1 billion and net profit stood at US$400 million and sales at US$8.7 billion (Datta and Anand 2010). With increasing profitability, Hindalco was able to move its acquisition debt to Novelis' books. The financial integration of Novelis and Hindalco indeed enabled Hindalco to fund its US$4.6 billion expansion plan in India and 'within four years, half of the US$3.5 billion that Hindalco spent to buy Novelis has come back into its fold' (Business Standard, 7 March 2011). Novelis accounted for nearly 68% of Hindalco's consolidated revenues and around a third of its profit in 2011–2012.

Apart from gaining high technology from Novelis, Hindalco was able to create body making capacity in India by relocating its (non-performing) Rogerstone plant from the UK to Hirakud in India. This helped Hindalco gain competitive advantages both in India and overseas. Bhattacharya, managing director of Hindalco stated:

> In the can body plant that we are putting up in Hirakud, we have made sure that even with the highest subsidy that a Chinese manufacturer can enjoy, we will not only be competitive in India but also take on the Chinese in China (Business Standard, 7 March 2011).

While Hindalco gained Novelis' high technology, Hindalco's cost management expertise helped Novelis to turn around its financial downtrend quickly. The Australian subsidiary has also made a quick turnaround due to sustained cost management processes learned from Indian operations (Hindalco Press Release, 02 September 2010). In the words of Bhattacharya: 'Indian companies are very cost conscious, and we wanted to impart that in Novelis. That's what our effort was, and I will say that we are reasonably successful in that'. It is evident that there has been a two-way transfer of technology and knowledge between Hindalco and Novelis, with particularly the home company benefitting from the knowledge and technology transfers. As Bhattacharya stated: 'There is still significant disparity in skill levels. We have a lot to learn [in India] and have to complete our learning before we try to replicate what they [Novelis] do (Business Standard, 7 March 2011).

MTN group (South Africa)

Between 1994 and 2001, MTN was known as M-Cell Limited. In 2002, it was renamed 'MTN Group Limited'. MTN is a multinational telecommunications group, listed on the Johannesburg Stock Exchange. 'MTN's vision is to be the leader in telecommunications in emerging markets' (MTN Annual Report 2008, 11). It started investing in regional African market to diversify its revenue sources and has emerged as the industry leader in Africa. Today MTN operates in 22 countries in Africa and the Middle East (with clear market leadership in 15 countries). By 2009, total subscribers had increased to 116 million and 70% of the earnings of MTN came from outside South Africa from OFDI. This shows that MTN's revenue diversification strategy through mainly acquisitions in emerging markets proved effective. It helped MTN develop competitive advantage by adding 'more depth' to its 'management teams as well as professional and telecommunication-specific skills' (MTN Annual Report 2006, 22).

OFDI by MTN across Africa and other countries helped the company foster knowledge sharing across entities in these countries and facilitated sharing of innovation and best practices. For example, in 2012 the company launched numerous new services and products and 'many operations benefited from the experience of others, with regard to, for example, subscriber registration, subscriber acquisition, the execution of handset or device strategies and cost management' (MTN Annual Report 2012, 35). Also, they enabled MTN to forge technology alliances or knowledge partnerships with leading companies in different sectors which complemented MTN's core operations. For example, MTN Uganda and Fundamo, a leader of mobile financial services, jointly launched in 2009 MTN MobileMoney in Uganda. Similarly, MTN and MFS Africa partnered to launch the online money transfer service MTNMMO.COM. In 2011, MFS Africa pioneered the fully mobile-based life insurance service Mi-Life in Ghana partnering with MTN, Hollard Insurance and Microensure.

The MTN Group and TRACE (leading international brand and media platform focused on music and sports celebrities with over 60 million subscribers in 151 countries) have joined forces to offer innovative entertainment services to the fast-growing youth segment within the African mobile market, the first of its kind in Africa. MTN also partnered with Microsoft to provide Windows 8 and Windows Phone 8 operating systems to its customers. American Tower Corporation and MTN formed a joint venture in Uganda after working together in Ghana where the former's 'tower expertise, operational excellence and a focus on delivering growth and value from

the asset portfolio, are highly complemented by MTN's regional operational experience' (MTN Group Media Release, 09 December 2011). MTN has also formed Africa Internet Holding (AIH) by partnering with Rocket Internet and Millicom International Cellular (each becoming 33.3% shareholders) to develop Internet businesses in Africa.

For MTN 'relationships with third party developers were important to enable and increase access to innovative financial services such as merchandise payments, online payments and insurance solutions'. Its leading presence across Africa helped forge such partnerships (MTN Group Media Release, 13 November 2012).

Sasol Limited (South Africa)

Sasol is an integrated energy and chemicals company with operations in South Africa, Europe, the Middle East, Asia and the Americas. Its OFDI is directed towards either new investments in developing economies in Africa and Asia to secure oil, gas and coal supplies for its global operations or to consolidating existing investments in the developed economies such as the US and European countries to improve performance. Sasol made major investments in gas-to-liquids (GTL) projects in Nigeria and Qatar. The acquisition of Exxon Mobil's European Wax emulsion business helped its operations expand in Europe. Sasol Polymers has made a significant investment in Malaysia. Because of the global expansion of its operations, by the end of 2006, 33% of the Group's turnover was contributed by operations outside South Africa (Sasol Annual Report 2006, 7).

Sasol is a technology leader in the industry with expertise in coal and gas processing, Fischer-Tropsch catalysis and engineering research, refinery and fuels technologies, and chemical technologies. Particularly, it is one of the world's largest producers of synthetic fuels and a leader in GTL technology. Its current combined capital and operational expenditure for R&D is over R1 billion a year (90% invested in South Africa and about 10% overseas). This includes an investment of £15 million in UK-based OXIS Energy to develop next generation battery technology. It set up a new state-of-the-art research and technology (R&T) facility (Sasol One Site) in South Africa in 2012 to drive the company's global expansion. This includes 14 laboratories, a number of piloting facilities, 150 PhD students, about 100 engineers, 200 scientists, and 100 chemists and technologists. Flip de Wet, Managing Director of Sasol Technology stated: 'This facility will enable us to be more competitive and further push boundaries when it comes to our ambitious global growth program' (Sasol Media Release, 9 November 2012; 10 September 2012).

Sasol demonstrated its global technological leadership through expansion in the US and Canadian markets. The shale gas revolution created further growth and investment in the US market for Sasol, employing its 'transformational technologies in unlocking the value of these resources'. According to Sasol Senior Group Executive for Global Chemicals and North American Operations André de Ruyter's testimony to the US Congress: 'Sasol's gas-to-liquids (GTL) facility, the first of its kind in the U.S., will be a game-changer for America's energy future'. He further asserted: 'While natural gas is a major energy source for global power generation, it has lacked the versatility to address transportation needs. Now, with our proven GTL technology, natural gas can be transformed into a range of high-quality fuels and chemical products, maximizing in-country value'. In 2012, Sasol decided to set up 'a world-scale ethane cracker and an integrated GTL facility', near Westlake, Louisiana, to help 'further strengthening Sasol's position in the global chemicals market' (Sasol Media Release, 20 June 2013). It forged technological partnerships with leading companies such as Technip Stone & Webster Process Technology, ExxonMobil Chemical Technology Licensing, Univation Technologies, and Scientific Design Company to execute this project. Its technological leadership has led to a joint venture by Sasol and INEOS to manufacture high density polyethylene. It also partnered with General Electric to develop new water technology, which 'will further entrench' Sasol's 'position as a world-leader in gas-to-liquids technology and synthetic fuels production' (Sasol Media Release, 06 November 2013).

Similarly, Sasol's technological capabilities helped the company join with Talisman Energy in Canada to develop its shale gas operations. On the other hand, new technology products developed overseas have helped support Sasol's South African operations. For example, new ALCAT®TEAL highly purified tri-ethyl aluminium product developed by TEAL plant in Brunsbüttel, Germany, helped Sasol Polymers in South Africa (Sasol Media Release, 15 July 2013).

Synthesis of findings from the six cases

Tables 6 and 7 map the major drivers and type/mode of OFDI and the nature of knowledge flows between home and host economies, respectively.

We identify key factors such as R&D capability, technological and knowledge level, skills and products (mostly institutional factors) to trace the differences between the OFDI by EMNE and DMNE in terms of location decisions. The cases demonstrate that EMNEs tended to start OFDI in developing countries (either within their region or outside), as their R&D, technological and knowledge level, skills, and product range are limited. However, with increasing R&D, technological capability, product range and experience they subsequently entered the developed countries, and eventually emerged as technology and knowledge partners to become global leaders in their industry (as well as strong competitors). The experience of Tata Steel, CNPC, ZTE and, to a lesser extent, Hindalco, broadly demonstrate that their OFDI trends started with South to South (within region), then expanded to South to South (outside region), and finally entered into South to North. DMNEs do not face constraints which drive them to follow such a growth trajectory. However, this internationalization process of EMNEs cannot be

Table 6: Drivers of OFDI & Type/ mode of OFDI.

	CNPC	ZTE	Tata Steel	Hindalco	MTN	Sasol
Institutional factors behind OFDI from emerging MNEs						
Type of MNE (ownership)	State-owned	Public listed	PLC	PLC	PLC	PLC
Nature of industry sector	√	√	√	√	√	√
Home tech./ R&D/ Knowledge	√	√	√	√	√	√
Home market/ economy/ politics/govt. policy	Home market/ economy/ politics/ govt. policy	Home market/ economy/ govt. policy	Home market/ economy	Home market/ economy	Home market/ economy	Home market/ economy
Business factors behind OFDI from emerging MNEs						
Competition pressures		√	√	√	√	√
Globalization	√	√	√	√	√	√
Gaining access to natural resources	√		√	√		√
Strategic expansion to new markets	Expansion to new sources of oil supply	√	√	√	√	√
Gaining/ sharing knowledge, skills & technology	Intra/inter learning	Intra/inter learning	√	√	√	√
Brand names			√	√		√
Type/ Mode of OFDI from emerging MNEs						
Greenfield project		100% owned subsidiaries	100% subsidiary		√	√
Mergers and acquisitions	√		√	√	√	√
Joint ventures	√	Long-term partnerships	√		√	√
South to South (Neighbourhood Region)	√	√	√	√	√	√
South to South (Outside Neighbourhood Region)	√	√	√		√	√
South to North (Developed Economies)	Strategic partnerships	√	√	√	Strategic partnerships	√

Table 7: OFDI – Nature of knowledge flows: Evidence from the cases.

	CNPC	ZTE	Tata Steel	Hindalco	MTN	Sasol
Types of knowledge flow from OFDI						
Technology	√ Mainly Outflow	√ Mainly Outflow	√ Inflow	√ Inflow	√ Outflow	√ In/Out flow
R&D		√ In/Out flow	√ In/Out flow	√ Inflow		√ In/Out flow
Management		√ Inflow	√ Outflow	√ Outflow	√ Outflow	
Cost efficiency			√ Outflow	√ Outflow		
Best practices & cross learning	√ Outflow	√ Inflow	√ In/Out flow	√ In/Out flow	√ In/Outflow	√ In/Out flow
Direction of knowledge flow from OFDI						
Two way: Home to Host to Home	√ Uneven in developing countries	√ Strong but unequal among developed & developing hosts	√ Strong/ Equal	√ Strong/ Equal	√ Uneven	√ Uneven
Home to Host	√ Strong	√ Strong	√ Strong	√ Strong	√ Strong but through third parties	√ Strong
Host to Home	√ Weak but strong indirect through strategic partnerships	√ Strong in some cases and less strong in others	√ Strong	√ Strong	√ Strong but through third parties	√ Less strong
Intra: Among Host Locations/ Entities	√ Strong	√ Strong	√ Strong	√	√ Strong	√
Inter: Cross border	√	√	√	√	√	√

fully explained by the traditional MNE models of internationalization and FDI, as there are significant differences among the ways Chinese, Indian and South African EMNEs' internationalized.

The Chinese cases demonstrate that their home technology/knowledge base and competitive advantages and their foreign expansion coevolved. Indian MNEs had a weak knowledge base when they entered the developed countries. Both South African cases had a strong home technology/knowledge base from an early stage, but their internationalization process evolved differently. While, MTN expanded its operations mainly into African countries, Sasol simultaneously expanded into African, and other developing and developed countries. They demonstrate complex patterns of latecomer strategies towards internationalization. Similar complexity was highlighted by others. Jung and Rhe (2009) found the opposite trajectory where the OFDI was first concentrated in developed countries and then moved to developing countries. They argued that 'South Korean outward FDI possesses unique characteristics different from those of developed countries and developing countries' and explain this as the result of rapid development of home economy and 'market seeking' motive due to lack of capital and technology skills (Jung and Rhe 2009, 137, 139). Also, there are other aspects that need to be taken into account. For example, Indian software EMNEs tend to directly export and expand to developed countries

(US and Europe) while Chinese EMNEs face major entry barriers to the US and need to come up with strategies to overcome them (e.g. US telecommunications operators were asked by the government not to do business with Huawei and ZTE because of perceived security threats). Therefore, one can argue that internationalization of DMNEs and EMNEs and their OFDI operations are not similar.

The case studies demonstrate that the trajectory of OFDI by EMNE starts with large differences with DMNEs in the early phases, which later converge as its operations become mature and global. At the mature stage, when they enter the developed countries, their locational motive is not only 'market seeking', but also 'strategic asset seeking' and 'strategic alliance seeking'. This helps them enhance their technology and knowledge assets and to forge strategic partnerships with the global leaders in their industry. In other words, they emerge as strong competitors as well as knowledge partners in their industries. The 'technology/ knowledge alliance seeking' seems to be in response to market changes brought about by globalization and rapid technological changes, and increasing capital intensiveness of R&D. This appears to be an important aspect of OFDI by EMNEs which is different from DMNEs. However, the experience of South Korean MNEs shows similarity only with respect to 'market seeking' and 'strategic asset seeking' in developed countries, not 'strategic alliance seeking' (Jung and

Rhe 2009, 137, 139). This shows the complex nature of OFDI from EMNEs mainly driven by the impact of globalization and different strategies employed by them to internationalize.

Dunning (2000) presented an updated eclectic paradigm of international production to 'still claim to be the dominant paradigm explaining the extent and pattern of the foreign value added activities of firms in a globalising, knowledge intensive and alliance based market economy' (Dunning 2000, 163). However, the complexities of EMNE operations suggest that we need to look beyond traditional FDI models to explain their activities. For example, according to CNPC's President, Jiang Jiemin, the global financial crisis is one of the main factors that helped the company to expand its global business. Therefore, it is evident that there is a need for an alternative theoretical model which can help develop a better understanding of EMNE activities. There are alternative 'latecomer catch-up' and 'internationalization' models proposed by different scholars. For example, Meyer and Thaijongrak (2013) presented the usefulness of the extended internationalization process model (IPM) (originally known as Uppsala model) to explain the evolution of EMNEs, specifically focusing on the role of acquisitions in the internationalization process. They used six Thai case studies to illustrate this. Mathews (2006) in his 'linkage, leverage and learning' (LLL) model highlighted the important role of global value chains in creating opportunities for latecomer firms in emerging economies to forge linkages and leverage that to acquire technology, knowledge and market access and accumulate capabilities through sustained and repeated learning process. There are other models such as Li's (2007) 'holistic, dynamic and dialectical framework' and Yamakawa et al.'s (2008) 'strategy tripod' framework that attempt to present alternative models to traditional FDI models. These attempts suggest that it is necessary to combine the traditional FDI models like 'OLI' with 'latecomer catch-up strategies' and come up with an integrated model that can help understand the dynamics of EMNEs.

Furthermore, there appear to be significant differences between EMNEs and DMNEs in their modes of OFDI and the way they operate in both developed and developing host countries. The modes of OFDI by EMNEs in developing countries are mainly joint ventures and 100% owned subsidiaries, while in developed countries they are joint ventures and mergers and acquisitions (often of failing or underperforming companies, as they offer fewer barriers to entry). There are no such constraints that force DMNEs to choose specific modes of FDI, particularly in host developed countries.

EMNEs appear to follow the business philosophy of 'mutual benefit' and 'an open and win-win model based on sharing' of knowledge (as in the words of Wang Haibo, ZTE's Director of Legal Affairs). This is evident from the cases from China and India (e.g. Hindalco's financing of Novelis) and to a lesser extent South Africa. It suggests significant business cultural differences between EMNEs and DMNEs that warrant further research.

Among the institutional factors, the key motivational factors which drive OFDI are strengths and weaknesses of home technology and R&D base, home market and economic conditions. For example, the weaknesses of home technology and R&D base of Indian EMNEs were the major drivers behind their OFDI in developed countries. In contrast, the strength of technological and R&D base at home drove OFDI by the Chinese and South African EMNEs. Again, dominance and securing of the long-term future of the home base first played a major role behind OFDI by the Indian EMNEs and, to some extent, the Chinese EMNEs. The South African EMNEs were not driven by concerns about securing their home base, but mainly by market seeking in regional and global markets. Political factors appear to play a significant role (the home government) in determining the nature of OFDI when the EMNE is state-owned and state holding as in the case of China. Previous studies also suggest that government policy plays an important role in OFDI by EMNEs, particularly in countries like South Korea and China (e.g. Lu et al., 2010). Although this study did not focus on this aspect, there appears to be significant differences in the way government policy impacts on OFDI in different emerging economies. The impact appears stronger on Chinese MNEs than in India and South Africa. These differences among the institutional motivational factors across different case countries are difficult to explain using traditional FDI theories alone.

In relation to business factors, the key factors which drive OFDI are globalization, strategic expansion to new markets, and gaining/sharing knowledge and technology. Certainly, the globalization process has created necessary conditions for OFDI by EMNEs and forced them to think globally, and intelligently and selectively seek markets, raw materials, skills and knowledge, and technology, both in the South and North. For example, MTN's OFDI is mainly driven by dominating the regional market in Africa and less by 'skills and knowledge' seeking. ZTE's OFDI in selected Southern countries like India is driven more by seeking skills and knowledge, than by market. However, its OFDI in developed economies is driven by both 'skills, knowledge and technology' and 'market' seeking. Similarly, in the two Indian cases, their OFDI in the developed countries is driven more by technology and R&D than just market advantages. On the other hand, Sasol's OFDI in developed countries is mainly driven by market seeking and to a lesser extent by knowledge and technology partnerships seeking. In the case of Tata Steel, its OFDI in developing countries is driven mainly by market and efficiency seeking. OFDI by CNPC is largely driven by strategic expansion into new markets, particularly in developing countries, so as to gain leverage with global oil companies towards forging strategic partnerships in gaining/sharing knowledge, skills and technology. What we see from these case studies is that business factors capturing the motivations and drivers of OFDI are interlinked, presenting complex dynamics that require explanation by going beyond existing traditional FDI theories.

Similarly, the knowledge and technology flow from OFDI can be traced in two patterns: (i) in the case of developing countries hosts, it is mainly one way, i.e. from the home to host economy; (ii) in the case of developed countries hosts, it is two-way and more evenly matched between host and home economies; however, in some cases of developing countries hosts and home economies, the knowledge flow is two-way (e.g. China's ZTE's operations in India).

Conclusions

The findings from the six case studies suggest that there are significant differences between the OFDIs of EMNEs and DMNEs. The cases show that they cannot be explained by using traditional FDI models. Institutional factors which drive OFDI such as home technology and R&D base as well as home market and economic conditions are assumed to play similar roles in both DMNEs and EMNEs. However, a closer analysis reveals that there are complex differences. It is not always the strength of home technology and R&D base which drives OFDI by EMNEs, but also their weaknesses. Again, dominance and securing of long-term future of the home base can be the main drivers as shown by Indian EMNEs. Their long-term strategic objectives are different. They use their OFDI in developing markets as leverage to expand into developed markets and forge strategic technology and knowledge-sharing partnerships with established global leaders in the industry. This is evident from the cases from China and India and, though to a lesser extent, from the South African cases.

The investigation shows that the internationalization pattern of EMNES is complex, driven by business factors such as competition pressure, globalization, strategic expansion into new markets, and gaining/sharing knowledge and technology. EMNEs are intelligently and selectively seeking markets, raw materials, skills and knowledge, and technology, both in the South and North. Furthermore, within the same country, OFDI from one EMNE can be driven more by dominating regional markets and less by seeking skills and knowledge, whereas OFDI by another EMNE can be the opposite.

There are also significant differences in the way that EMNEs enter and operate in developed and developing countries. OFDI by EMNEs in developing countries are mainly through joint ventures and 100% owned subsidiaries; however, in developed countries, it is through joint ventures, mergers and acquisitions (often takeover of failing or underperforming companies), as this helps overcome entry barriers.

Knowledge transfers between EMNEs and developing host economies are predominantly one way, with the former transferring more technology and knowledge than they gain. In the case of EMNEs and developed host economies, the knowledge and technology transfers appears to be more evenly matched, a two-way street benefitting both parties almost equally. Increasingly, as shown from all the cases, EMNEs have become global players and have successfully forged global partnerships and alliances with other industry leaders, which sometimes extend to active participation in EMNEs' home economies (e.g. CNPC). It is clear that increasingly EMNEs have become sources of cutting edge technology and

knowledge that has helped to forge strategic partnerships with global industry leaders for sharing technology and knowledge.

Another interesting finding is that EMNEs' style of operation, both in developing and developed markets, appears to be significantly different from that of DMNEs. They are mainly driven by the business philosophy of 'mutual benefit' and appear to be more willing than DMNEs to share knowledge. This may be due to cultural business differences, which merits further research.

To conclude, the case studies have shown that there are complex aspects of OFDI from EMNEs which cannot be explained by existing FDI theories. These appear to be due to the impact of globalization, rapid technological change, increasing capital intensity of R&D and varied and dynamic strategic responses from EMNEs towards internationalization and catch-up. A theoretical model that integrates both 'latecomer strategies for catch-up' and 'the traditional FDI' models is necessary to understand fully the dynamics of EMNEs, including their OFDI patterns.

One of the limitations of this study is that it does not analyze the role of home government policies and strategies towards supporting and creating favourable environment for the OFDI activities of their local champions. Indeed, it is widely reported in the literature that the majority of emerging markets do not provide a conducive environment for the OFDI activities of their firms, placing them at a competitive disadvantage compared with their developed country counterparts. However, some emerging countries, particularly China, have introduced a number of initiatives to promote OFDI by domestic firms. Indeed, as a consequence of these initiatives, China's OFDI flows have grown rapidly over the years. The role of home government policies towards supporting EMNEs is a separate research topic that needs further research.

Although the limited number of cases studied is a constraint towards making generalizations for theory building, the evidence from the cases shows that new, unified conceptual frameworks and theories can provide the basis for furthering both empirical insights and varied appreciative conceptual frames in OFDI research.

Based on the evidence from this study, some policy implications have been drawn. Firstly, South-South OFDI appears to play an important role in significant technology and knowledge transfers, particularly to the host economies. It is thus very important the policymakers in the South actively seek and promote OFDI from the South.

Secondly, it is evident that OFDI from EMNEs in developed economies often involves acquisition of failing or underperforming high technology companies. This can help their survival and retain technological capabilities, knowledge and talents in the host economy. It is also evident that OFDI from EMNEs in developed economies results in an evenly matched two-way flow of technology and knowledge which is beneficial mutually. It is important that policymakers in developed countries encourage OFDI from the South in the targeted areas.

Thirdly, it is clear that EMNEs have increasingly become sources of cutting edge technology and knowledge, which has helped forge partnerships with other global industry leaders (DMNEs). Policymakers in both emerging and developed economies should actively support and encourage such partnerships.

As the emerging economies of the world continue to grow (China now is the biggest economy in the world, measured on purchasing power parity (PPP)), EMNEs are likely to play a significant role in the world economy. It is, therefore, an important task to undertake further and more comprehensive research on how OFDI from EMNEs is likely to impact on economic growth and development.

Acknowledgement

This paper largely draws on our following working papers and conference papers: Baskaran, Liu, and Muchie (2010), Baskaran, Liu, and Muchie (2011), Baskaran, Liu, and Muchie (2012), Baskaran, Liu, and Muchie (2014), and Baskaran et al. (2016).

Disclosure statement

No potential conflict of interest was reported by the authors.

Notes

1. One natural resource based and one technology intensive company were selected from China and South Africa out of total nine and eight available from UNCTAD/Erasmus University database respectively. But both cases selected from India are natural resources based, as there were only five companies in total.
2. Although it is a public listed company, it is considered a state holding company (while CNPC is a fully state-owned company), as the government has significant control over the company. The other category of state ownership is 'state invested company' where the government control is less or none.
3. Earnings before interest, taxes, depreciation and amortization.

ORCID

Angathevar Baskaran http://orcid.org/0000-0002-5723-8795
Ju Liu http://orcid.org/0000-0002-1352-1986
Mammo Muchie http://orcid.org/0000-0003-4831-3113

References

Agarwal, J. P. 1985. *Pros and Cons of Third World Multinationals: A Case Study of India*, J.C.B. Mohr (Paul Siebeck) Tubingen, Kieler Studien, 195.
Aminullah, E., T. Fizzanty, K. Kusnandar, and R. Wijayanti. 2013. "Technology Transfer Through OFDI: the Case of Indonesian Natural Resource-Based MNEs." *Asian Journal of Technology Innovation* 21 (sup1): 104–118.
Athukorala, P. C. 2009. "Outward Foreign Direct Investment From India." *Asian Development Review: Studies of Asian and Pacific Economic Issues* 26 (2): 131–153.
Baskaran, A., J. Liu, Y. Hui, and M. Muchie. 2016. "Outward Foreign Direct Investment (OFDI) and Knowledge Flow in the Context of Emerging MNEs: Cases from China, India and South Africa". *International Conference on Innovations, Trade and Development*, November 25-26, 2016, Centre for Development Economics and Innovation Studies and Department of Economics, Punjabi University, India.
Baskaran, A., J. Liu, and M. Muchie. 2010. "Exploring the Outflow of FDI from the Developing Economies: Case Studies from China, India and South Africa". *IERI Working Paper 2010-007*, Institute for Economic Research on Innovation, Tshwane University of Technology, Pretoria, South Africa.
Baskaran, A., J. Liu, and M. Muchie. 2011. "Exploring the Outflow of FDI from the Developing Economies: Selected Case Studies". *DIR Research Series Working Paper no. 149*, DIR Research Center on Development and International Relations, Aalborg University, Denmark.
Baskaran, A., J. Liu, and M. Muchie. 2012. "Exploring the Outflow of FDI from the Emerging Economies: Case Studies from China, India and South Africa." *Third Copenhagen Conference on 'Emerging Multinationals': Outward Investment from Emerging Economies*, Copenhagen, Denmark, 25-26 October 2012.
Baskaran, A., J. Liu, and M. Muchie. 2014. "Exploring the Outflow of FDI and Knowledge Flows in the context of Emerging Economy MNEs: Cases from India, South Africa and China". *The 12th GLOBELICS International Conference*, Addis Ababa, Ethiopia, 29-31, October 2014.
Buckley, P. J., and M. C. Casson. 1976. *The Future of Multinational Enterprise*. London: Macmillan.
Business Standard. 2011. "The financial integration of Novelis with Hindalco has given the latter the heft to push ahead with its ambitious growth plans in India." 07 March.
Cai, K. G. 1999. "Outward Foreign Direct Investment: A Novel Dimension of China's Integration Into the Regional and Global Economy." *The China Quarterly* 160: 856–880.
CNPC. 2003 to 2013. "Annual Report." CNPC, Beijing, China. Available at: http://classic.cnpc.com.cn/en/press/publications/.
CNPC. Various Dates, 2006 to 2014. "News Release". Available at: http://classic.cnpc.com.cn/en/press/newsreleases/.
Datta, K., and A. Anand. 2010. "Inside the Novelis Turnaround." *Economic Times*, 04 October.
Draper, P., S. Kiratu, and C. Samuel. 2010. "The Role of South African FDI in Southern Africa." *Discussion Paper 8/2010*, German Development Institute, Bonn.
Dunning, J. H. 1973. "The Determinants of International Production." *Oxford Economic Papers* 25 (3): 289–336.
Dunning, J. H. 1980. "Toward an Eclectic Theory of International Production: Some Empirical Tests." *Journal of International Business Studies* 11 (1): 9–31.
Dunning, J. H. 1988. "The Eclectic Paradigm of International Production: A Restatement and Some Possible Extensions." *Journal of International Business Studies* 19 (1): 1–31.
Dunning, J. H. 1998. "Location and the Multinational Enterprise: A Neglected Factor?" *Journal of International Business Studies* 29 (1): 45–66.
Dunning, J. H. 2000. "The Eclectic Paradigm as an Envelope for Economic and Business Theories of MNE Activity." *International Business Review* 9 (1): 163–190.
Dunning, J. H., R. van Hoesel, and R. Narula. 1996. "Explaining the 'new' wave of outward FDI from developing coountries : the case of Taiwan and Korea." *Research Memorandum 009*, Maastricht University, Maastricht Economic Research Institute on Innovation and Technology (MERIT).
Eisenhardt, K. M. 1989. "Building Theories From Case Study Research." *The Academy of Management Review* 14 (4): 532–550.
Eisenhardt, K. M., and M. E. Graebner. 2007. "Theory Building From Cases: Opportunities and Challenges." *Academy of Management Journal* 50 (1): 25–32.
Ellingsen, G., W. Likumahuwa, and P. Nunnenkamp. 2006. "Outward FDI by Singapore: a Different Animal?" *Transnational Corporations* 15 (2): 1–40.
Finance and Economy. 2010. "Oil Production of CNPC Kazakhstan Project Expected to Exceed 10 Million Tons." 18, p.
Fleury, A., and M. T. L. Fleury. 2011. *Brazilian Multinationals: Competences for Internationalization*. Cambridge: Cambridge University Press.

Hennart, J. F. 1982. *A Theory of Multinational Enterprise.* Ann Arbor, MI: University of Michigan Press.

Hindalco. 2004 to 2013. "Annual Report". Available at: http://www.hindalco.com/reports-presentations.

Hindalco. Various Dates, 2011 to 2014. "Press Release". Available at: http://www.hindalco.com/press-releases.

Hou, W. 2006. "Firm's Internationalisation: the ZTE Model." *Entrepreneur Information* 5: 117–118.

Hymer, S. H. 1960. "The International Operations of National Firms: A Study of Direct Foreign Investment." *PhD Dissertation (Published posthumously)*, Cambridge, Mass: The MIT Press, 1976.

Jeenanunta, C., N. Rittippant, P. Chongphaisal, A. Thumsamisorn, and T. Visanvetchakij. 2013. "Knowledge Transfer of Outward Foreign Direct Investment by Thai Multinational Enterprises." *Asian Journal of Technology Innovation* 21 (Sup.1): 64–81.

Johanson, J., and J. E. Vahlne. 1977. "The Internationalization Process of the Firm—A Model of Knowledge Development and Increasing Foreign Market Commitments." *Journal of International Business Studies* 8 (1): 23–32.

Kalesh, B. 2008. "Towards Total Integration." *Mint*, 16 June.

Kiggundu, M. N. 2008. "A Profile of China's Outward Foreign Direct Investment to Africa." *Proceedings of the American Society of Business and Behavioral Sciences* 15 (1): 130–144.

Kim, J. M., and D. K. Rhe. 2009. "Trends and Determinants of South Korean Outward Foreign Direct Investment." *Copenhagen Journal of Asian Studies* 27 (1): 126–154.

Kumar, N., and A. Chadha. 2009. "India's Outward Foreign Direct Investments in Steel Industry in a Chinese Comparative Perspective." *Industrial and Corporate Change* 18 (2): 249–267.

Lall, S., E. Chen, J. Katz, B. Kosacoff, and A. Villela. 1983. *The new Multinationals: The Spread of Third World Enterprises.* Chichester: Wiley.

Lecraw, D. J. 1993. "Outward Direct Investment by Indonesian Firms: Motivation and Effects." *Journal of International Business Studies* 24 (3): 589–600.

Li, P. P. 2007. "Toward an Integrated Theory of Multinational Evolution: The evidence of Chinese multinational enterprises as latecomers." *Journal of International Management* 13 (3): 296–318.

Liu, X., T. Buck, and C. Shu. 2005. "Chinese Economic Development, the Next Stage: Outward FDI?" *International Business Review* 14 (1): 97–115.

Lu, J., X. Liu, and H. Wang. 2010. "Motives for Outward FDI of Chinese Private Firms: Firm Resources, Industry Dynamics, and Government Policies." *Management and Organization Review* 7 (2): 223–248.

Luo, Y., and R. L. Tung. 2007. "International Expansion of Emerging Market Enterprises: A Springboard Perspective." *Journal of International Business Studies* 38 (4): 481–498.

Luostarinen, R. 1979. *Internationalization of the Firm.* Helsinki: Acta Acadamie Oeconomicae, Helsinki School of Economics.

McAdam, M., and S. Marlow. 2007. "Building Futures or Stealing Secrets?: Entrepreneurial Cooperation and Conflict Within Business Incubators." *International Small Business Journal* 25 (4): 361–382.

Mani, S. 2013. "Outward Foreign Direct Investment From India and Knowledge Flows, the Case of Three Automotive Firms." *Asian Journal of Technology Innovation* 21 (Sup 1): 25–38.

Mathews, J. A. 2002. "Dragon Multinational: A New Model for Global Growth." Macquarie University ResearchOnline.

Mathews, J. A. 2006. "Dragon Multinationals: New Players in 21st Century Globalization." *Asia Pacific Journal of Management* 23 (1): 5–27.

Meyer, K. E., and O. Thaijongrak. 2013. "The Dynamics of Emerging Economy MNEs: How the Internationalization Process Model can Guide Future Research." *Asia Pacific Journal of Management* 30 (4): 1125–1153.

MTN Group. 2004 to 2013. "Annual Report". Available at: https://www.mtn.com/investors/financials/pages/annualreports.aspx.

MTN Group. Various Dates, 2001 to 2014. "Media Release". Available at: https://www.mtn.com/PressOffice/Pages/pressrelease.aspx.

Narula, R. 2006. "Globalization, new Ecologies, new Zoologies, and the Purported Death of the Eclectic Paradigm." *Asia Pacific Journal of Management* 23 (2): 143–151.

Norasingh, X. 2013. "Foreign Direct Investment and Knowledge Transfer in Laos." *Asian Journal of Technology Innovation* 21 (Sup 1): 139–156.

Rajan, R. 2009. "Outward Foreign Direct Investment from India: Trends, Determinants, and Implications." *ISAS Working Paper No. 66*, Institute of South Asian Studies, National University of Singapore.

Ramamurti, R. 2012. "What is Really Different About Emerging Market Multinationals?" *Global Strategy Journal* 2 (1): 41–47.

Rugman, A. M. (ed.). 1982. *New Theories of the Multinational Enterprise.* London: Croom Helm.

Sasol Group. 2004 to 2013. "Annual Report". Available at: http://www.sasol.com/investor-centre/publications/integrated-report-1.

Sasol Group. Various Dates, 2004 to 2014. "Media Release". Available at: http://www.sasol.co.za/media-centre/media-releases/latest-media-releases.

Tata Steel. 2004-2005 to 2012-2013. "Annual Reports." Tata Steel Limited, Mumbai, India. Available at: http://www.tatasteel.com/investors/performance/annual-report.asp.

Tata Steel. Various Dates, 2008 to 2014. "Press Release". Available at: http://www.tatasteel.com/media/press-release.asp.

The Financial Times Limited. 2014. "The FDI Report 2014: Global Greenfield investments trends." London: FT.

Tolentino, P. E. 1993. *Technological Innovation and Third World Multinationals.* London and New York: Routledge.

UNCTAD. 1998. "World Investment Report." Geneva: United Nations.

UNCTAD. 2005. "Case study on Outward Foreign Direct Investment by South African Enterprises." Geneva: United Nations.

UNCTAD. 2006. "World Investment Report 2006: FDI from Developing and Transition Economies, Implications for Development." Geneva: United Nations.

Vernon, R. 1966. "International Investment and International Trade in the Product Cycle." *Quarterly Journal of Economics* 80: 190–207.

Wee, K. H. 2007. "Outward Foreign Direct Investment by Enterprises From Thailand." *Transnational Corporations* 16 (1): 89–116.

Witt, M. A., and A. Y. Lewin. 2007. "Outward Foreign Direct Investment as escape Response to Home Country Institutional Constraints." *Journal of International Business Studies* 38 (4): 579–594.

Yamakawa, Y., M. W. Peng, and D. L. Deeds. 2008. "What Drives new Ventures to Internationalize From Emerging to Developed Economies?" *Entrepreneurship Theory and Practice* 32 (1): 59–82.

Yin, R. K. 1990. Case study research: design and methods. Applied Social Research Methods Series, Beverly Hills, CA, USA: Sage Publications.

Zhang, N. 2009. "Strategic Reflection of "Going Global" From Chinese Petroleum Firms." *China Business Update* 6 (1): 56–59.

ZTE. 2004 to 2013. "Annual Reports." ZTN, Shenzhen, China. Available at: http://wwwen.zte.com.cn/en/about/investor_relations/corporate_report/.

ZTE. Various Dates, 2007 to 2014. "Press Clipping". Available at: http://wwwen.zte.com.cn/en/press_center/press_clipping/.

ZTE. Various Dates, 2007 to 2014. "Press News". Available at: http://wwwen.zte.com.cn/en/press_center/news/.

Foreign R&D units in India and China: An empirical exploration

Swapan Kumar Patra ⓘ

Among the many corporate functions of multinational enterprises (MNEs), their foreign R&D is considered to be the least mobilized. Firms usually keep their crucial R&D activities close to their home base. However, since the 1990s MNEs from developing countries are offshoring their R&D activities to developing Asian countries, particularly in India and China. With this recent trend, the foreign R&D by MNEs is becoming important and has attracted attention all over the globe. This study is an attempt to map this recent trend from an in-house developed database on the foreign R&D units in India and China. It also investigates the major motives of firms for choosing India and China as favourable R&D destinations. It is observed from this study that both 'market-driven' and 'technology-driven' factors are the major motives for MNEs to invest in R&D in these two emerging economies. Firms prefer R&D locations in India and China, where there are knowledge centres with an abundant supply of qualified and highly skilled human resources available at comparatively lower cost. As a result, MNEs' foreign-based subsidiaries are now increasingly playing a greater role in the generation, use and transmission of knowledge.

Introduction

The theories of the multinational enterprises (MNEs) contend that technological innovations at home or abroad are the main source of a firm's competitive advantage. Many scholarly works show that the major reason for the dispersion of MNEs is to secure new technological competencies distributed globally. A firm's global growth can be considered as a consequence of home based 'ownership advantages' and 'competitive advantage' to be exploited in foreign markets (Kuemmerle 1999; Dunning and Lundan 2009; Narula and Zanfei 2005). Usually, among many other corporate functions, offshore R&D by MNEs are generally considered less mobile and one of the last corporate functions to internationalize in the value chain (Mansfield 1975; Mansfield, Teece, and Romeo 1979). Till the end of last century, internationalization of R&D was mainly distributed in and restricted to the 'triad' (Europe, North America and Japan) region (Rugman and Verbeke 2003). Various scholarly works showed that up to the mid-1990s the main aim or motive of MNEs to enter emerging economies was to explore new markets through the adaptation of products for these markets according to local customers' needs (Reddy 1997). However, present-day internationalization strategies of firms are significantly different from their earlier ones of the previous century. A major shift has taken place and a new mode of MNE's R&D and innovation activities is gradually emerging. MNEs and their R&D units are being established in emerging economies like India and China due to several knowledge-based factors. In addition, MNEs have started establishing their networks of subsidiaries for technology transfer, skills and asset creation across national borders between their headquarters and their globally distributed subsidiaries. Much empirical evidence suggests that there appears to be a two-way transfer of knowledge between MNEs and their R&D units in India, China and other emerging economies (Cantwell and Piscitello 2000, 2005; Zanfei 2000; UNCTAD 2005). Now, MNE's foreign-based subsidiaries are increasingly playing a greater role in the generation, use and transmission of knowledge. Also, MNEs are developing external networks of relationships with the local universities, R&D institutions and other actors to acquire external knowledge through foreign subsidiaries. As a result of this evolutionary process, the organization of MNEs is subject to both centripetal and centrifugal forces (Zanfei 2000).

In the light of the above discussion, this paper explores the following research questions in the broader analytical framework of globalization of R&D and the motivations of firms recently starting their R&D operations away from home bases, particularly in India and China: How many R&D units have foreign firms opened in India and China in recent years? What are the sector-wise numbers of foreign firms opening up their R&D units in India and China? What are the countries of origin of firms? What are the motivating factors to open up their R&D operations in India and China? Based on these research questions, the study investigates the number of foreign R&D units, their locations and the motivating factors for offshoring R&D operations in India and China.

The paper is divided into seven sections. The next two sections deal with the latest trends in foreign R&D in India and China. The section after these deals with the methodology and the limitations of the study. The section following presents the findings, which include the number of firms and their R&D units, country of origin of firms, locations of their R&D units and their motives for establishing R&D units in India and China. The final section comprises the concluding remarks, including policy recommendations and future research avenues.

Table 1: Sector-wise distribution of foreign firms and their R&D units in India.

Sectors	Sector code*	Number of firms	Percentage of total firms	Number of R&D units	Percentage of total R&D unit
Consumer discretionary	25	41	7.96	53	7.23
Consumer staples	30	20	3.88	30	4.09
Energy	10	3	0.58	3	0.40
Financials	40	2	0.38	2	0.27
Healthcare	35	72	13.98	86	11.73
Industrials	20	64	12.42	82	11.18
Information Technology	45	277	53.78	433	59.07
Materials	15	29	5.63	36	4.91
Telecommunications	50	7**	1.35	8	1.09
Total		515		733	

*Global Industry Classification Standard code

Foreign R&D in India

India and China have witnessed a major surge in Foreign Direct Investment (FDI) over the last two decades. India received US$2633 million in 1998 and US$34,417 million in 2014, while China received US$45 463 million in 1998 and US$128,500 million in 2014 (World Investment Reports). This increase in FDI has happened due to the respective governments gradually opening up their economies and putting in place FDI-favourable policies. However, in India, many sectors like retail, defence and print media are still lagging behind in FDI. The Indian government has put restrictions on FDI in these sectors implementing investment caps (Nagpal 2010). However, India is gradually becoming a favourable destination for many information and communication technology (ICT) and business process outsourcing (BPO) firms. It is one of the most popular offshoring R&D destinations because it has a huge reservoir of high-quality, low-cost manpower (Mrinalini and Wakdikar 2008; Mrinalini, Nath, and Sandhya 2013). To take advantage of this, an increasing number of foreign MNEs are offshoring their R&D in India (Krishna, Patra, and Bhattacharya 2012; Patra 2014; Patra and Krishna 2015). This trend has continued since the mid-1980s, when Texas Instruments first set up their R&D unit in Bangalore (Satyanand 2007). Since then, there has been a major surge in the establishment of new foreign R&D units in India. Many foreign firms have entered into the Indian market as wholly owned subsidiaries or other forms of R&D alliances, like contract research and so on (Balakrishna and Balakrishna 2003). The R&D projects conducted by firms are in the Indian market and there are also many projects for high-end product generation for the global market. For example, Intel has about 1000 engineers working on software and hardware designs for communication and semiconductor products. Many foreign firms are designing everything from auto parts to consumer electronics in India through outsourcing or setting up their own R&D facilities in different locations. Although, foreign firms are investing massively in India, MNEs' R&D investments in India are mainly concentrated in high-technology sectors, particularly in automotive, pharmaceutical, biotechnology IT and telecommunications industries. In terms of nationality, US, European and South Korean firms are the main R&D investors in India (Asakawa and Som 2008).

Foreign R&D in China

With the inflow of FDI into China, MNEs are increasingly involved with local R&D activities (Gassmann and Keupp 2008). This phenomenon has been observed since the second half of the 1990s (Boutellier, Gassmann, and von Zedtwitz 2008). In the 1990s, many high-technology firms explored China as a favourable destination for their R&D activities (Gassmann and Keupp 2008). MNEs from developed countries ventured into China as a source of technology for new product development. This phenomenon was largely unnoticed in the last decade of the twentieth century and has only started receiving serious worldwide attention since the beginning of twenty-first century (Boutellier, Gassmann, and von Zedtwitz 2008). China's economic census data for different years has confirmed that foreign R&D in China has increased significantly (Sun 2011). During the second half of the 1990s, technology-intensive MNEs explored China as a location for their R&D activities. Since 2000, the number of foreign R&D operations has increased dramatically. Initially, companies in the ICT sector, such as *Microsoft, Nortel, Ericsson* and *Nokia* started opening their R&D units. Many firms from the biomedical and automotive industries have now also opened their R&D operations in China.

However, there was very little systematic research on foreign R&D in China during the 1990s, as this was a relatively new and recent phenomenon. The major reasons for this limited research may be because of difficulties in doing research in China and because MNEs' R&D in China was limited in size and scope. The major works on foreign R&D in China were published by Greatwall (2002) and Walsh (2003). Since then, however, many research works on foreign R&D in China have followed.

Nortel Network established the first foreign R&D centre in China in 1994. Nortel Networks Corporation and Beijing University of Posts and Telecommunications jointly set up a R&D unit in 1994 (Wen and Lin 2005; Li and Li 2010). With China's announcement towards the end of 2001 that it would join the WTO, between November 2001 and May 2002 many MNEs opened

Table 2: Country of origin of foreign firms in India.

	Consumer discretionary	Consumer staples	Energy	Financials	Healthcare	Industrials	Information technology	Materials	Telecommunications Service	Total	Percentage
United States	12	8	1	1	35	23	219	12	1	312	60.58
Germany	5	1			7	5	3	8		29	5.63
Japan	9				3	5	7		1	25	4.85
United Kingdom	1	2	1	1	6	4	7	2	1	25	4.85
France	2	3			4	7	4		1	21	4.08
Sweden	1					8	2	1		12	2.33
Switzerland		1			4	2	4	1		12	2.33
China	1						5		1	7	1.36
Denmark					2	3		1	1	7	1.36
Netherlands			1		1	2	2	1		7	1.36
Korea	2				1		2		1	6	1.17
Canada					1		4			5	0.97
Finland						2	3			5	0.97
India	1	3				1				5	0.97
Belgium					1	1	2			4	0.78
Israel					1		3			4	0.78
Italy	3	1								4	0.78
Taiwan	1						3			4	0.78
Austria	1					1				2	0.39
Bermuda						2				2	0.39
Luxembourg							1	1		2	0.39
Malaysia	1				1					2	0.39
Singapore							2			2	0.39
Spain					1	1				2	0.39
Australia					1					1	0.19
Croatia					1					1	0.19
Czech	1									1	0.19
Iceland					1					1	0.19
Ireland					1					1	0.19
Norway							1			1	0.19
Not known							1			1	0.19
Saudi Arabia								1		1	0.19
South Africa								1		1	0.19

Figure 1: Locations of foreign R&D units in India.

R&D units in China announced a substantial increase in investment or manpower (Prater and Jiang 2008). Scholarly works and popular media like press releases by firms and newspaper, report from time to time that many of the 'Fortune 500' firms have opened their R&D units in major Chinese cities like Beijing, Shanghai and so on (Motohashi 2005, 2010).

With this trend, China has advanced from the '*design it abroad, make it in China, sell it in abroad*' cycle to being a major, globally recognized R&D player (Prater and Jiang 2008). This is confirmed by various press announcements in newspapers like the *People's Daily*, the *International Trade Daily*, the *China Daily*, and so on. An increase in the R&D investment trend is being observed, particularly in the computer and telecommunications industries, because both these industries are highly dependent on R&D and have fast production cycles (Walsh 2003). For example, Microsoft established its first Asia Research Institute (its second overseas base) in Beijing, in 1998. By 2010, Microsoft had doubled its R&D employees up to 3000, with about 1500 project-based researchers in China. Motorola had about 25 R&D centres in China and has invested about US$500 million in R&D (Li and Li 2010). Beside in the ICT industry, major investments in R&D have also been made in high-technology industries like biotechnology, pharmaceuticals and automotive. Although there is a substantial amount of investments in R&D from Hong Kong and Taiwan, the majority of MNEs are from developed countries, particularly from triad regions (Gassmann and Han 2004). Above all, foreign R&D has become a significant phenomenon in China. Thousands of foreign companies are engaged in R&D in China and, according to Sun (2011), based on 2004 China Economic Census data, published in 2006, foreign R&D centres have generated about 150,000 high-skilled jobs in China. Regarding the activity of foreign firms, most scholars argue that the majority of them are 'adaptive R&D' and are geared more towards product development than the actual research. This argument is supported by the fact that most of the R&D employees from these foreign R&D units have been technical persons without any higher level degrees (Walsh 2003; von Zedtwitz 2004; Sun, Du, and Huang 2006; Sun 2009).

However, some empirical evidence indicates that foreign R&D in India and China is still limited in sectorial and regional scope. Further, Bruche (2009) argues that investment initially is focused more towards *resource-seeking* in India and *market seeking* in China. The main reason for this is that India, on the one hand, has a huge pool of low-cost, high-skilled manpower, whereas China, on the other, has 1.3 billion potential customers and an increasing middle-class population (Bruche 2009).

Methodology and limitations of the study
Methodology
The Technology Information Forecasting and Assessment Council (TIFAC) of the Government of India has conducted a survey of foreign MNEs and their R&D units in India between 1998 and 2003. The TIFAC survey (2005) was based on the survey of about 100 foreign firms with R&D centres in India. Zinnov, a management consultant firm located in Bangalore published a report called *Impact of MNC R&D centres in India* estimating that there were about 671 MNEs with R&D centres in India in 2010 (*Silicon India*, 31st March 2010). This study was based on the 2005 TIFAC report as well as other firms' related R&D information, mostly from business newspapers, firms' annual report and different web resources. All foreign firms' R&D-related information was collected and stored in a relational database for this study. The database was developed and regularly updated based on different types of news sources. The following newspapers and other sources from India were mainly consulted: the *Hindu Business Line*, the *Economic Times*, the *Financial Express*, the *Business Standard*, the *Times of India*, the *Hindustan Times*, *Live Mint* and so on. From China, the following two news sources were consulted: the *China Daily* and *Xinhua* (English). News information from the LEXIS-NEXIS database, firms' annual reports and different newswires compensated for the gaps.

This paper is an exploration of foreign firms (MNEs) and their R&D units in India and China. On the basis of available data published in different newspapers and secondary sources, the study identified a sample of 515

Table 3: Motives of firms for establishing R&D units in India.

Motives	Keywords	Frequency	Percentage
Market-driven	Market size	290	87.35
	Market proximity	5	1.51
	Local sales	36	10.84
	Customer	45	13.55
Production-driven	Manufacturing	45	13.55
	Localization	4	1.20
Technology-driven	Talent	112	33.73
	Qualified human resources	12	3.61
	Technology	138	41.57
Innovation-driven (push factors)	Assets buildings	2	0.60
	Ideas	4	1.20
	Products	82	24.70
	Process	13	3.92
	Location	24	7.23
Cost-driven	Cost difference	62	18.67
Policy-driven	Standards	7	2.11
	Tax Incentives	7	2.11
	IP	10	3.01
	Intellectual property	17	5.12

foreign firms and 733 R&D units in India for the period between 1985 and 2014. About 323 firms and 589 R&D units were identified in China for the period between 1990 and 2014. It is important here to note that a firm may have many R&D units in different cities. These R&D units or centres carry out R&D in the respective countries. (Note that the terms 'units' and 'centres' mean the same and are used interchangeably in this study.)

The sampled firms were classified based on the Global Industry Classification Standard (GICS). The GICS system was developed by Standard & Poor (S&P) and MSCI. It is a widely used industrial classification system. It divided all the industries into 11 sectors, 24 industry groups, 68 industries and 157 sub-industries.

The motivations of firms were extracted from the database and the motivating factors examined using interpretive content analysis and qualitative content analysis (Drisko and Maschi 2016). *Content analysis* is 'a research technique for making replicable and valid inferences from texts (or other meaningful matter) to the contexts of their use' (Krippendorff 2013, 24). Basic content analyses use word counts and other quantitative analytic methods to analyze data. For the content analysis in this this study we used freely available text processing software called TextStat 2 available at http://neon.niederlandistik.fu-berlin.de/textstat/. This open source software program analyzes the text corpora and displays word occurrences and concordances to search terms.

Limitations of the study

The area is very dynamic because firms quite frequently announce the establishment of their new R&D units in India and China. Many such announcements may go unnoticed. So, the actual estimate of foreign R&D units may vary. The database significantly covers foreign R&D units in India. However, in the case of China, this paper is based on the database of 323 firms with 589 R&D units which have Greenfield investment. Although popular newspapers, scholarly articles and reports have observed that China has more R&D units than India, this study found the opposite because of a number of limitations. Firstly, this study only took into consideration newspapers and reports published in the English language. Thus, data that appeared in the vast number of Chinese language publications have been missed due to the language barrier. Secondly, Greenfield R&D investment in the mainland China was only considered. However, despite these inherent limitations, the study nonetheless is a possible indicator of foreign R&D activity in India and China.

Findings

This study is an empirical survey of the total universe of foreign MNEs and their R&D units in India and China. The exploration of this paper focuses on the possible number of R&D units, their location and the factors behind the motivations to enter into these two emerging economies.

Table 4: Sector-wise distribution of foreign firm and their R&D units in China.

Rank	Sector	No of firms	Percentage of total	Number of R&D units	Percentage of total
1.	Information technology	136	42.10	287	48.72
2.	Industrials	46	14.24	68	11.54
3.	Consumer discretionary	44	13.62	93	15.78
4.	Healthcare	35	10.85	51	8.65
5.	Materials	34	10.52	41	6.96
6.	Consumer staples	17	5.26	34	5.77
7.	Telecommunication Service	5	1.54	7	1.18
8.	Energy	3	0.92	3	0.50
9.	Financials	3	0.92	5	0.84
	Total	323		589	

Number of foreign R&D units in India

MNEs have been setting up their R&D operations in India since the mid-1980s. It has already been mentioned that Texas Instruments was the first foreign firm to open an independent R&D unit in India, back in 1985. After that, until about the mid-1990s, there was a very sporadic inflow of foreign R&D units. The real momentum came only after the mid-1990s, after India's economic liberalization. From this self-designed and developed in-house database it is observed that there are about 515 firms with 733 R&D units in India. The sector-wise breakdown of the firms is shown in the Table 1.

As seen from Table 1, out of 515 firms 277 (53.78%) are from the IT sector, followed by the healthcare sector with 72 firms (about 14%). The others are as follows: Industrials 64 (12.42%), Consumer discretionary 41 (7.96%), Materials 29 (5.63%), Consumer staples 20 (3.88%), Telecommunication services 7 (1.35%), Energy 3 (0.58%), and Financial 2 (0.38%). It is also observed that IT firms constitute the largest part of the sample followed by healthcare firms. It is evident from Table 1 that firms in high technology, particularly in the ICT and healthcare sectors, select India as their preferred destination for doing R&D. Various government policies in different time intervals along with suitable incentives and India's strength in ICT, pharmaceutical and biotechnology infrastructure are perhaps the possible reasons for the firms to select India as their favoured destination for doing R&D.

Country of origin of foreign firms in India

Table 2 shows the country-wise origin of foreign firms in India. Among the 515 firms in the sample, 312 firms (60.58%) are from the US, followed by Germany 29 (5.63%) Japan 25 (4.85%) and the UK 25 (4.85%). The number of firms originating from other countries are follows; France (21), Sweden (12), Switzerland (12), the Netherland (7), China (7), Denmark (7), Korea (6), Canada (5), Finland (5), Belgium (4), Italy (4), Taiwan (4) and Israel (4). It can be observed from the data that many developed countries' firms, particularly from the triad regions, have established their R&D units in India in recent years.

Locations of foreign R&D units in India

MNE's subsidiaries establish external and internal innovation networks to facilitate the exploitation of external

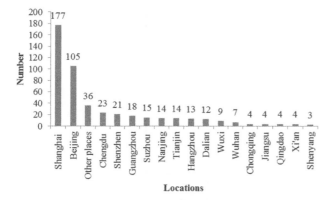

Figure 2: Locations of foreign R&D units in China.

knowledge from the globally dispersed technological centres. In today's globalized world, many technological centres of excellence are scattered across different geographical locations. To strengthen their existing competencies, firms generally extend those capabilities into new related fields of production and technology, and across a variety of locations (Cantwell and Iammarino 2003). Thus MNEs play an important role in the evolution of 'clusters', geographic concentrations of interconnected companies and institutions in a particular field (Birkinshaw 2000). Since knowledge has a tacit component, it becomes increasingly difficult to absorb it with increasing geographical distance. Thus, geographical proximity is required to absorb knowledge spillovers generated from globally dispersed knowledge hubs (Jaffe, Trajtenberg, and Henderson 1993; Jaffe, Trajtenberg, and Fogarty 2000). So, the knowledge base of a region plays an important role in the decisions of MNEs to locate their R&D activity. For these reasons, the local R&D by MNEs has a strong tendency to agglomerate at a sub-national and regional level (Cantwell and Piscitello 2005). Because of this tendency of agglomeration of firms in certain foreign locations, the location develops certain natural advantages. Also, firms can exploit spillovers and externalities from the knowledge base of a location (Zanfei 2000; Cantwell and Piscitello 2005). With increased globalization, particular locations can now become strategically important with their more specialized and differentiated resources and capabilities. In *The Competitive Advantage of Nations* Michael Porter focuses on dynamic local clusters. Local clusters with specialized industries and excellent educational institutions have an important role in MNEs to innovate (Porter 1990). They constitute 'home bases' for continuous upgrading of competitive advantage (Rugman and Verbeke 2003; Sölvell 2003).

Both, India and China have promoted high technology clusters to fuel their national economic development (Walcott and Heitzman 2006). Foreign firms have established their units in these strategic knowledge clusters. So, the foreign R&D units in India are mainly concentrated in the cities where there are knowledge hubs. The concentration of foreign R&D units in these cities is due to the abundant supply of skilled manpower.

The location of R&D units is as follows: Bangalore 253 (34.52%), Hyderabad 106 (14.46%), Pune 72 (9.85%), Chennai 61 (8.32%), Mumbai 59 (8.05%), and Delhi (Delhi, the national capital region or the NCR, includes Gurgaon in Haryana, Noida and Ghaziabad in Uttar Pradesh) 81 (11.05%). Bangalore, Hyderabad, NCR and Chennai are considered as Tier I cities and have a number of excellent educational and government research institutes. Apart from these six major cities, Kolkata 11 (1.5%), Goa 9 (1.23%), Thane 5 (0.68), Ahmedabad 4 (0.55%), Baroda 4 (0.55) and Coimbatore (0.55) are also deemed favourable destinations for foreign firms.

Figure 1 shows the distribution of ICT R&D centres in different cities. From Figure 1, it is evident that Bangalore hosts the largest concentration of R&D units. Regarding the location of R&D units, MNEs tend to prefer certain locations compared with others. Bangalore, Chennai, Pune-Mumbai and the NCR are the major knowledge-

Table 5: Country of origin of foreign firms in China.

	Consumer discretionary	Consumer Staple	Energy	Financial	Health Care	Industrial	Information Technology	Materials	Telecommunication	Total	percentage
United States	11	5	2	1	17	15	83	12	0	146	45.20
Japan	10	4	0	2	1	5	9	2	1	34	10.53
Germany	9	1	0	0	3	4	4	6	1	28	8.67
United Kingdom	0	1	0	0	4	1	9	0	1	16	4.95
France	3	1	0	0	1	4	3	1	1	14	4.33
Switzerland	0	1	0	0	4	3	3	2	0	13	4.02
Korea	3	0	0	0	0	3	2	2	1	11	3.41
India	2					2	5			9	2.79
Netherlands	0	0	1	0	0	1	2	2	0	6	1.86
Sweden	1	0	0	0	0	4	0	1	0	6	1.86
Taiwan	0						6			6	1.86
Belgium	0	1				1		2		4	1.24
Denmark	0	0			2	1		1		4	1.24
Finland	0					1	2	1		4	1.24
Canada	0	1					1	1		3	0.93
Hong Kong	1	1					1			3	0.93
Italy	2					1				3	0.93
Australia	0				1		1			2	0.62
Austria	1						1			2	0.62
Israel	0				1		1			2	0.62
Malaysia	1						1			2	0.62
Bermuda	0						1			1	0.31
China	0						1			1	0.31
Ireland	0				1					1	0.31
Saudi Arabia	0							1		1	0.31
Singapore	0	1								1	0.31

Table 6: Motives of firms for establishing R&D units in China.

Motives	Key words	Frequency	Percentage
Market-driven	Market size	199	88.84
	Market proximity	7	3.13
	Local sales	40	17.86
	Customer	25	11.16
	Speed	8	3.57
Production-driven	Manufacturing	34	15.18
Technology-driven	Talent	35	15.63
	University	16	7.14
	Qualified human resources	8	3.57
	Technology	97	43.30
Innovation-driven (push factors)	Ideas	3	1.34
	Products	87	38.84
	Process	10	4.46
	Location	12	5.36
Cost-driven	Cost	24	10.71
Policy-driven	Regulations		0.00
	Standards	4	1.79
	Tax	3	1.34
	IP	4	1.79
	Intellectual property	3	1.34

intensive city clusters in India. Many universities and research institutions are located in these cities. The presence of a number of educational and governmental research institutes promotes various forms of university-industry linkages in these cities (Basant and Chandra 2007). The availability of certain technological capabilities in the institutions promotes linkages among the various actors in the city cluster to exploit available opportunities. These foreign R&D units basically take advantage of the location and supply of highly skilled science and engineering graduates (Krishna 2012).

Motives to establish R&D units in India

Motivation is a combination of various factors. The motives for establishing R&D units in India can be broadly categorized into three types. These are *market-seeking, resource-seeking and efficiency-seeking* (World Investment Report 2005; World Investment Prospect Survey 2007–2009, 2007; Lundvall 2008). However, other factors such as host country's business environment are also important. The geographical location of R&D by MNEs can be analyzed in terms of two competing and contradictory forces i.e. 'centrifugal forces' (decentralization) and 'centripetal forces' (centralization) (Chiesa 1996; von Zedtwitz and Gassmann 2002). Although there are a number of studies which have identified various motives for internationalizing R&D, Gammeltoft (2006) divided the motives into six categories: *market-driven, production-driven, technology-driven, innovation-driven, cost-driven* and *policy-driven* (Gammeltoft 2006). This section is based on this framework to broadly classify the motives of foreign firms to establish R&D units in India and China. The responses collected from various press statements and newspaper reports are grouped into six major motives. However, many responses related to motives overlap and encompass more than one motive. For example, a firm's motives may be 'technology-driven' (to recruit the best talents available in these two countries) and 'market-driven' (to take advantage of the

vast local markets). Among the 515 firms in the database, about 330 firms have some information about their motives for establishing R&D units in India. These motives are categorized, based on the framework by Gammeltoft (2006) (Table 3).

Market-driven motives: these are exploitation of firm's assets in the local market, market size and market closeness; support to local sales operations and familiarity with the customer; to develop responsiveness in terms of both speed and relevance, and so on. Cumulatively, market-driven motives are the top ranked motive to establish R&D units in India. However, the market may not be only the local Indian market, but also South Asian regional markets or even the global markets. The market-driven motive can be explained by the following examples and related excerpts from press statement.

Japanese firm Hitachi have established 'Hitachi Global Solutions Center' in India in 2011 with the aim to double its turnover to US$2 billion in India by 2013. The Hitachi Asia Chief Executive Yasunori Taga stated in a press report that:

> Our vision is to have one R&D centre in India to address the global needs as well as to develop products for the local market ... Besides we also need to set up a new production site to meet the demands of the country and to have direct access to the market here. (Agencies 2011a)

Merck & Co. Inc., US invested about US$150 million in its Indian subsidiary MSD India. It started a R&D unit along with an Indian university on drug development and vaccine research in 2012. On this occasion, K G Ananthakrishnan, the managing director MSD India said in a press statement:

> For us India is the big market for growth. The population size, along with the opportunity for vaccines, drugs for infectious diseases and diabetes have led us to ramp up our presence here. (Vijay 2012a)

Technology-driven motives: the following factors may be considered as technology-driven motives: tapping into foreign S&T resources; technology monitoring (especially competitor analysis); acquire/monitor local expertise, knowledge and technologies. Technology-driven motives are the second-most predominant motive for firms to establish their R&D unit in India.

India has a strong skill base in the high-technology industry, particularly in software engineering. The country has developed world-class standards in software development, reflected in its rank in the worldwide ranking of the number of assessment and maturity levels reported to the Software Engineering Institute (SEI) (Balakrishna and Balakrishna 2003). As India is playing a significant role in both research and global operations, many R&D scientists located in the US are being encouraged to move to Indian facilities.

India has a huge pool of high-quality, technically-qualified and English-speaking manpower. India's more than 750 universities and 35,000 affiliated colleges produce about 30,000 PhDs every year (Krishna and Patra 2016). This huge reservoir of technically skilled manpower, available at much lower cost in comparison with developed countries, is a major attraction for foreign firms. Many foreign firms scout fresh talent from engineering colleges in India. So, the availability of high-skilled, low-cost manpower is also cited as a possible reason for the establishment of R&D units in India (Reddy 1997; Krishna 2012). The following example and statement show the importance of high-skilled manpower in India.

Freescale Semiconductor is a global player in the design and manufacture of embedded semiconductors for the automotive, consumer, industrial and networking markets. At the inauguration of the R&D centre in Hyderabad in 2011, Dr Lisa Su, senior vice president and general manager of Freescale said:

> ... Freescale's world class talent and resources in India play an important role in our global networking strategy, and the Hyderabad centre further reinforces our commitment to Freescale India. (Our Bureau 2011)

Production-driven motives: these arise due to an insufficient supply of skilled manpower and when other destinations provide the possibility of low-cost production. These factors direct firms to go beyond their home bases. Also, evolving learning processes and 'follow-the-leader' effects (Gammeltoft 2006) have led many firms to establish their production base in India.

Volvo Bus Corporation, a Swedish company, can be cited as an example for the production-driven motive. The firm considers India as an emerging global hub for exports and product development. Its India centre supplies Volvo products to the South Asian and other developing countries' markets. Recently, Volvo made a huge investment to increase its capacity in its manufacturing unit in Bangalore. In 2011, Hakan Karlsson, president and CEO of Volvo Bus Corporation said:

> The company is also expanding its R&D base in the country by setting up a new product development establishment in India. (Times News Network 2011)

Innovation-driven motives: many firms want to build R&D capacity at the home or host location. They may also decide on general expansion and sometimes corporate headquarters takes the strategic decision to conduct R&D at host locations. Good infrastructure available in the host location is another possible reason. In many high-technology sectors, product life cycle is short. To remain competitive in these sectors, (for example ICT, biotechnology and nanotechnology industries) it is necessary for firms to innovate frequently. These are the 'push factors' for firms to offshore their R&D to foreign locations.

The major Japanese IT firm in the computers and peripherals industry, Toshiba Corporation, planned to set up a manufacturing facility in India in 2011, to enable cost reduction in the highly competitive Indian market. Its R&D centre in India planned product development for the Indian market. Of their strategic decision to conduct R&D in India, Toshiba India Director, Wu Tengguo, said:

> We are determined to provide a broad array of high-value products that reflect the changing lifestyles in India, and to address market requirements. We are strengthening our product line up by introducing range of products across its all the categories including LCD, laptop and washing machines. (Agencies 2011b)

Cost-driven motives: The low costs of R&D operation are the fourth most important factors. It is reported in various newspaper articles and press briefing that companies shifted their R&D to India are saving a substantial amount of cost from their budget. Indian R&D operation cost is significantly lower. It is estimated that the cost of conducting R&D in India is one-fifth of the cost of developed countries like the US or Europe. For example, Dr P M Akbarali, managing director, Apotex Pharmachem India Pvt. Ltd., said in a press statement;

> India can be known as the Pharma Powerhouse going by the industry's maturity to meet the challenges of regulated regimes of the developed countries and provide cost competitive drugs. (Vijay 2012b)

Policy-driven motives: these are ranked sixth in the selected sample of the study. Policy-driven factors are national regulatory requirements or incentives; tax differentials; monitoring and exploitation of regulations and technical standards, and so on. The Indian government is very liberal in terms of foreign firms starting their R&D operations in India. Approvals are easily given for wholly-owned foreign subsidiaries or joint ventures. Particularly in the software sector, most development units are situated in software technology parks (STPs) and are entitled to tax breaks and other facilities, including duty-free imports.

In recent years, the Indian government has adopted many new policies which support firms by some or other form of concession. For example, India has one of the most favourable tax systems in the world to encourage firms to conduct their manufacturing and R&D operations in India. A firm can get tax benefits in the form of direct, indirect and other government incentives. Till 2009–10, these benefits were limited to manufacturing companies in certain specified sectors. As a result, the flow of foreign R&D was mainly concentrated in high-technology areas like software, electronics, telecommunication, automotive, pharmaceuticals, hardware and so on. However, since 2009–10, benefits are given to all manufacturing companies. The super deduction of tax at the early stage of business establishment has been enhanced from the earlier 150% to 200%. This means that during the start-up stage, firms are entitled to get a super deduction of 200%. This deduction is given to firms at this stage because they have high capital expenditure in the form of investments in R&D facilities and equipment (Business Standards 2011).

In another example of a policy-driven motive, the Indian government has recently indicated that it is likely to introduce new taxation rules known as 'safe harbor rules' to reduce lawsuits with MNEs on the applicability of transfer pricing norms. The new regulation is based on the recommendations of the Rangachary Committee Reports. The step has been welcomed by many MNEs (*The Hindu* 2013). The current benefits will be continued and are likely to attract more investments in R&D in India, as can be seen from the statement by a major global pharmaceutical firm:

> Bristol-Myers Squibb India is a subsidiary of Bristol-Myers Squibb – a global biopharmaceutical company headquartered in New York City. As per their website "… the company's strategy to transform to a next-generation BioPharma leader, Bristol-Myers Squibb established an R&D presence in Asia. India was chosen partly due to factors such as access to a large pool of skilled and knowledgeable talent, high-quality standards of academic research, increased commitment to protecting intellectual property (IP) rights, conducive infrastructure, established capabilities, committed work ethics, and a productive labor market.[1]

So, it is evident from the motives for establishing R&D units in India that the market-driven motives are the most predominant followed by technology-driven motives. Among the technology-driven motives, it is the search for new technologies and new talent at comparatively lower costs that predominate.

Number of foreign R&D units in China
Scholarly works on foreign R&D in China have reported that the number of offshore R&D units by MNEs in China is increasing. However, the actual estimates of R&D centres and their locations differ in studies. This is because of the different research focuses or different survey methodologies used. In addition, the definitions of R&D units are not uniform. For example, Prater and Jiang (2008) showed that difference in the number of reported R&D units is perhaps because different definitions of R&D units are used while compiling statistics. Further, sometimes, government reports do not discriminate between actual R&D units and manufacturing units. Also, the estimated number of foreign R&D organizations differs substantially depending on the sources of information (von Zedtwitz 2004; Lundin et al. 2008).

Table 4 gives sector-wise breakdown of R&D units in China. It is evident from Table 4 that majority of the R&D units of foreign firms are in the IT sector. It constitutes about 42% of the total population. About 136 IT firms have 287 R&D units in China which constitute about 50% of the total sample of this study. Industrial sector firms are in second position with 46 firms and 68 R&D units (11.54%). There are 44 firms from the consumer discretionary sector with 93 R&D units (15.78%). Healthcare firms are in fourth position with 35 firms and 51 R&D units (8.65%).

Sector-wise distribution of firms shows a difference between India and China. Although firms in the IT sector constitute the largest number in the sample of both countries, the firms in the healthcare sector constitutes the second largest group from the Indian sample. However, in the Chinese case, the industrial and consumer discretionary sector firms ranked second and third positions, respectively. The number of healthcare sector firms ranked fourth in the Chinese sample.

Country of origin of foreign firms in China
Table 5 shows the country-wise origin of firms with R&D units in China. In terms of parent countries of firms, the presence of MNEs from the United States is in top position with 146 firms (45.2%). Japanese MNEs are in second position with 34 firms (10.5%), followed by Germany. This finding confirms earlier research, according to which the United States, Western Europe and Japan are the main source of R&D investment in China. The study by Lundin et al. (2008) found that MNEs from North America had the largest number of R&D units in China, followed by Japan, Europe and Korea. According to the Chinese Ministry of Commerce, since 1995, Japan has been a major investor in China, with investment of about US$4.24 billion. In 2010, Japan ranked as the fourth-largest overseas investor in China, Hong Kong Taiwan, and Singapore (*China Daily* 2011).

This empirical study found an increasing presence of Korean MNEs' R&D investment in China. This phenomenon is related to the rise of high-technology industries, such as the ICT industry in Korea, with key players, such as Samsung, and the globalization of Korean companies. A number of firms originating from India also have R&D units in China. All major Indian IT firms, for example TCS, Wipro, Mahindra, Videocon, Satyam and so on, have R&D units in China.

Location of R&D units in China
Mainland China has become a major FDI destination since the mid-1990s. Deng Xiaoping during his famous southern visit in 1992 recommended the speeding up of economic reform and encourages inflow of FDI. Because of this 'open door policy', the central government as well as various local and provincial governments have formulated various favourable policies to attract more FDI. Although FDI has helped the overall economic development of China, the economic development is mainly concentrated on the coastal areas.

Foreign R&D units in China are highly selective in their locations and are mostly concentrated in few major cities. Most of them are clustered around a few coastal cities and provinces such as Guangdong, Shanghai, Jiangsu, Zhejiang and Beijing (Sun 2011). As a result, the regional disparity between the coastal areas and China's inland has increased, leading to a core-periphery structure (Luo et al. 2008). However, Shanghai and Beijing are the two of the most attractive locations of foreign R&D units, because of high-quality manpower, industrial clusters, high-technology parks, infrastructure, renowned universities and research institutes. Beside these two primary locations, many firms have concentrated their R&D units in Tier II cities in the southern part of the country like Guangdong, Shenzhen and so on. Von Zedtwitz (2004) found that 89% of all foreign R&D sites are concentrated in the Beijing-Tianjin and Shanghai-Suzhou clusters. Beijing is the preferred location for firms in the IT, telecommunications and electronics industries. It is also observed that US MNEs prefer Beijing because of its proximity to various engineering colleges and universities as well as government bodies that can facilitate critical government decisions (von Zedtwitz 2004).

Based on available empirical data, however, this study found Shanghai is a more preferred destination than Beijing. Figure 2 shows the location of foreign R&D units in China. There are altogether 177 (about 37%) firms situated in Shanghai. Among these 177 firms, 69 are from the ICT sector and 26 from the healthcare sector. A total of 105 firms are located in Beijing, of which 63 are from the ICT sector, followed by 12 from the healthcare sector.

Motives to establish R&D units in China

Prater and Jiang (2008) found that foreign firms had four major motives for starting R&D units in China, namely: *development of products for local Chinese market*, *market access strategy* by establishing linkages with government, *local talent*, and product development for *global markets*. Sun, Du, and Huang (2006) conducted interviews with 18 foreign R&D centres in Shanghai and found that they are mostly doing adaptive R&D of a tactical nature to serve the local Chinese market. However, there were exceptions, with a few R&D centres engaged in long-term research for the global market (Sun, Du, and Huang 2006).

This study used Gammeltoft's framework (discussed earlier) to categorize the firms' motives. Empirical observation has found that the market is the major motive for establishment of R&D units in China. This study confirms earlier works where the market was found to be the major motivating factor. In the following section, motivations are discussed in details (Table 6).

Market-driven motives: China is the largest potential market in the world. It has 1.3 billion customers with a growing middle-class population. The recent increase in FDI in China shows China's global position as a major market and manufacturing site. This market-driven driver further includes many factors. These factors are product customization according to local customers' needs, new product development for the local market, support to parent company's sales, manufacturing and other activities in Chinese market (Motohashi 2005, 2010; Prater and Jiang 2008). Li and Yue (2005) found that the growth of international R&D investments in China is part of a recent global trend. By locating R&D centres in China, MNEs can respond to local market

needs more quickly (Li and Yue 2005). The global market is another major target for foreign firms. Many firms have established their bases in Chinese markets and are now venturing into the global market with their products.

The major French pharmaceutical firm, Sanofi-Aventis, started its first regional R&D centre in China to tap growing demand in emerging markets. In this context, Thomas Kelly, vice president of Sanofi's operations in China, said:

> We hope to become the first multinational drug company that completes truly home-grown research activities and will cooperate with local partners.

Marc Cluzel, vice president of R&D at Sanofi, said: 'It is expected to accelerate the development of therapies and health solutions for the mass population in the region' (Yining 2010).

Production-driven motives: these include R&D units near manufacturing facilities, cooperation with local partners or suppliers and so on. Most international R&D investments follow local markets and customers, i.e. the need to develop more localized products. A foreign firm's Chinese R&D centre improves its credit both with the Chinese government and the Chinese population. Product localization can also be more efficiently done if firms locate their R&D units near to their markets. These motives are reflected in the following announcements by the firms in recent years. The South Korean manufacturing giant LG Electronics launched its China R&D centres in Beijing in 2002. John Koo, vice-chairman and chief executive officer of LG Electronics said,

> … China has become the largest manufacturing base for LG, and it will now become the largest R&D center for us …. biggest opportunity China offers LG following the nation's entry to the World Trade Organization (WTO) was not its huge market or cheap labor, but its large pool of talented professionals who are capable of inventing and developing first-class products.[2]

Technology-driven motives: with the opening up of China's market to the outside world, many MNEs from developed countries have established manufacturing activities in China to take advantage of low labour costs for manufacturing operations. The latest trend being observed is that many foreign MNEs are locating their R&D operations in China to take advantage of local Chinese talent in order to target global markets from China. Over the last couple of decades, the Chinese government has taken various policy measures to expand and improve the country's research infrastructure. As a result, China has become a significant global R&D player. The Chinese government is also continuously encouraging FDI in R&D, assuming that R&D performed by MNEs in China may be helpful in upgrading China's technological capability (Lundin et al. 2008).

Technology-driven motives are the second most major motive for establishment of R&D units in China. About 34 firms' motives may be categorized as technology-driven. These motives are the availability of a low-cost but high-quality talent pool, strategic location in a particular city because of proximity to universities or research institutes, the infrastructure available at a particular location, and competition derived from local or global competitors.

China's skilled human resource is a key factor in MNEs increasing R&D activities in China. Low-cost local talent available in China in comparison with developed Western countries is a primary motive behind foreign R&D. China offers one of the largest pools of human resources for conducting R&D. The fairly low cost of China's skilled labour also motivates MNEs to expand R&D activities in this burgeoning economy. Because of this reason, many MNEs develop certain technologies in China. It is both practical and reasonable to develop these technologies, rather than to use technologies developed in the home base (Prater and Jiang 2008).

The above observations can be substantiated by the following statements by executives of various firms. US IT firm Qualcomm Incorporated opened its R&D Centre in Shanghai in 2010. According to the firm's statement,

> The new R&D center is part of Qualcomm's ongoing efforts to both utilize the growing pool of telecommunications engineering talent and enhance local R&D capabilities for the increasingly important wireless communications market in China (2010).[3]

Innovation-driven motives: these are R&D capability building at home or in the host country, a general expansion decision taken by firms as part of their global expansion strategy and shorter product life cycles. For example, General Electric Company used to develop products for developed countries and then sometimes adapted them for markets in developing countries. Now, however, the company is developing products in its Chinese R&D units and selling them in the local market even before they are released onto traditional global markets. The firm decided to develop products quickly for local market because the Chinese market is more competitive (Bullis, April 8, 2010). Similarly, Siemens also develops products for the local market. According to the Metals Technology business head of the company, Mr Lueder,

> We differentiate ourselves through our strong capabilities in metallurgic equipment as well as in electro technology and automation. This R&D center is also aimed at technology transfer and sharing expertise to develop simple, maintenance-friendly, affordable, and reliable products for the local market. Strengthening the local value chain enables us to be closer to our customers and develop products that are tailored to their specific requirements ("Siemens Establishes Competence Center for Metallurgical Plants in Shanghai", 2011).

Cost-driven motives: in the high-technology sector, market demand exists before a product is launched and technology obsolescence is very high. So, firms prefer quicker R&D cycle time to develop products for the market. So, low-cost production at a quicker manufacturing time is another major motivator to respond to market demand. Continental AG, a major global automotive firm, followed a similar strategy. According to the company,

> ... Continental AG had opened a new Automotive Tech Center in Shanghai. Continental Automotive Tech Center plays an important role in expanding the firm's local R&D talent pool, building up local R&D competence and optimizing R&D cost for the Asian OEMs. It's designated to develop products and solutions with better quality, cost-effective high-performance and more innovative technologies, with the aim to further improve driving safety, reduce traffic fatalities, increase fuel efficiency and reduce CO_2 emissions for the local market ... ("Continental Opens New Automotive Tech Center in Jiading, China", 2009)

Policy-driven factors: initiated by the Chinese government in recent years, these are recognized as significant factors to attract foreign R&D. Most of the foreign MNEs are interested in the huge Chinese market potential. So, the government put conditions in place for the establishment of R&D centres, such as requiring MNEs to engage in technology transfers with local firms if they wanted to do business in China. China now spends about 2% of its GDP on R&D. The government is also increasing its budget in many new and emerging technologies like biotechnology, nanotechnology and so on. Moreover, there are plans to increase basic science research funding and infrastructure. The benefits of these emerging technologies are spreading across different sectors. These factors are not only making China a lucrative ground for foreign MNEs' R&D collaborations, but also making alliances feasible (Li and Yue 2005). Government also encourages Chinese universities and research institutes to collaborate with foreign R&D units. So, the well-established science and technology infrastructure, large pull of skilled human resources and the government's investment policies have encouraged R&D firm investments in the country (Li and Yue 2005). Beside this, Prater and Jiang (2008) assumed that China's entry to the WTO in 2001 is a major driver because its intellectual property regulations are on par with global standard. Also, the Chinese government's tax incentives and other policies are encouraging foreign investments in the country. For example,; foreign R&D centres in China are exempted from certain equipment duty.

Advanced Micro Devices invested about $100 million to expand its testing and manufacturing facilities in China. According to that firm,

> China is increasingly becoming a location of choice for these types of finishing operations because of lower operational costs there. Labor, electricity, power and other such expenses can be a substantial percentage of the overall outlay for testing and packaging facilities. By contrast, operations that actually process silicon involve expenses related to building the physical plant and buying equipment. As a result, local tax breaks are a larger factor than labor costs in determining where to build, and many manufacturers erect "fabs," or fabrication plants, in relatively costly labor markets. AMD, for instance, will build its next fab in Dresden, Germany. (Dan 2004)

In sum, the main motive of firms to start R&D units in China is local market driven. As noted earlier, China is a huge market of over a billion customers, with an increasingly middle-class population. To capture this huge market, many firms have established their R&D units in China. The second most important motives are technology-driven, which include tapping skilled manpower, foreign R&D resources technologies and knowledge.

Discussion

This study explored foreign R&D centres in India and China, using an in-house developed database from different primary and secondary sources. The main findings from the exploration carried out in this paper can be summarized in terms of the number of firms, their locations and motivating factors underlying MNEs' R&D units in India and China.

FDI plays an important role in the operation of MNEs in the developing and emerging economies (Edwards 2002; Lall 2003; Mody 2007; Beule 2010). Along with the increasing inflow of FDI in India and China, foreign firms also established their R&D units. Till the end of 1990s, MNEs usually explored new markets and developed products to adapt to local market conditions. However, this paradigm has significantly changed over the last two decades. Today, MNEs are developing external networks of relationships with local counterparts, through which foreign affiliates acquire external knowledge. MNEs enter emerging economies because these countries have developed a number of knowledge hubs. Firms found that S&T and R&D ecosystems are favourable clusters in which to locate their R&D operations.

The growth of MNEs' R&D units in India and China started in the early 1990s. Based on the data and empirical understanding generated in this study, it may be said that *along with the increasing inflow of FDI in India and China, MNEs have set up more and more R&D units in these two emerging economies.*

The exploration in this paper showed that US MNEs are the dominant firms operating in both these countries, followed by Europe and Japan, constituting the highest number in both India and China. German and Japanese firms constitute the second largest number in India and China, respectively. Close proximity and cultural bonding between China and Japan may the reason for the large presence of Japanese firms in China.

Foreign R&D units are mainly concentrated in Tier I cities in both countries. The preferred locations in India are Bangalore, Hyderabad, Delhi, Pune, Mumbai and Chennai. In China, the favoured destinations are Beijing and Shanghai. Bangalore is the favoured location for ICT firms, whereas Hyderabad is preferred by healthcare firms. Besides the Tier I cities, Tier II cities have also been attracting investments in recent years. Perhaps the major reasons of concentration are, *firstly*, the number of technology parks located in these Tire II city clusters; *secondly*, these city clusters along with government's favourable policies have attracted lots of foreign investment in R&D; and *thirdly*, the high concentration of educational institutes with an abundant supply of skilled manpower and excellent government research institutes have made these cities attractive locations for foreign firms to establish their R&D units (Basant and Chandra 2007; Bruche 2009; Basant and Mani 2012; Krishna, Patra, and Bhattacharya 2012). In city clusters, the availability of certain

capabilities in the universities and research institutions has led to the establishment of various types of R&D linkages and to the exploration of collaboration opportunities. Due to all these factors, the foreign firms selected these city clusters as their preferred R&D locations. Consequently, this paper asserts that MNEs set up R&D units in Indian and Chinese cities where knowledge hubs are evolving or emerging.

MNE's foreign-based subsidiaries used to play an increasing role in the generation, use and transmission of knowledge. Now, however, there appears to be a two-way transfer of knowledge between MNEs and their R&D units in emerging economies. It is evident from this study that both 'market-driven' and 'technology-driven' motives are the major reasons for MNEs' investment in R&D. MNEs locate their R&D units in India and China due to access to highly qualified R&D personnel, and the proximity of higher educational institutes and government research institutes. It is also important to note that the patent period for a number drugs is expiring within the next few years. So, both India and China will become low-cost manufacturing destinations for many pharmaceutical firms. Given these insights, it may be said that the empirical research in this study substantiates that MNEs located their R&D units in India not only for future market-related factors, but also for knowledge sources. Besides the cost advantages and availability of skilled manpower, there are also other reasons, for example, the time difference between the North American region and the Indian sub-continent. Perhaps the most important point to mention here is that both India and China are the world's most populous countries, and that their recent economic growth and well-developed infrastructure have made more resources available for MNEs' R&D units. In addition, other reasons such as host government incentives like tax breaks and public policy support to create knowledge hubs (ICT parks, R&D or science parks, etc.) have also played a part in luring MNE's R&D units to India and China.

Concluding Remarks

This study is an assessment of the number of R&D units in India and China, their country of origin and their motivation for establishing these R&D units in India and China. There is much scope for further research on foreign firms and their potential from the perspectives of both home and host countries. Further research would be helpful to map the benefits these two economies are getting from these R&D centres in terms of linkages with local universities, firms or research institutes. This perhaps can be mapped using joint patents, technology licensing, technology transfer and so on. This kind of exercise will help to map intended and unintended spillover from the foreign subsidiaries to the local firms or institutes. In addition, the amount of technology or knowledge being sourced by these firms from these countries through 'reverse technology transfer' requires further investigation. Reverse technology transfer may also have significant effects on the home country. If the knowledge and resources that are transferred back to the parent firm, spillover to the rest of the economy is going to happen.

Foreign R&D by MNEs in India and China is a very recent phenomenon. So, there are further avenues of research in this area.

Funding

This work was supported by the Tshwane University of Technology.

Notes

1. "Bristol-Myers Squibb India." Retrieved from http://www. bmsi.co.in/our%20company/Pages/AboutBMSIndia.aspx. Accessed on 31st March 2016
2. "LG Founds Overseas R&D Center." *China Daily*, December 11, 2002. Retrieved from http://www.china.org.cn/ english/scitech/50870.htm
3. "Qualcomm Opens Research and Development Center in Shanghai: Enhancing Local R&D Capabilities and Addressing China's Market Potential." *Qualcomm Press Release* (May 24, 2010). Retrieved from https://www.qualcomm. com/news/releases/2010/05/23/qualcomm-opens-research-and-development-center-shanghai.

ORCID

Swapan Kumar Patra ⓘ http://orcid.org/0000-0002-0825-7973

References

Agencies. 2011a. "Hitachi India Eyes $2 Billion Turnover." *The Indian Express,* February 9, http://indianexpress.com/article/news-archive/web/hitachi-india-eyes-2-billion-turnover/.

Agencies. 2011b. "Toshiba to Set up Production Unit." *The Hindu,* July 22, http://www.thehindu.com/todays-paper/tp-business/toshiba-to-set-up-production-unit/article2283670.ece.

Asakawa, Kazuhiro, and Ashok Som. 2008. "Internationalization of R&D in China and India: Conventional wisdom versus reality." *Asia Pacific Journal of Management* 25 (3): 375–394.

Balakrishna, P., and S. Balakrishna. 2003. "India: The emerging R&D hub." *Hindu Business Line,* August 6.

Basant, Rakesh, and Pankaj Chandra. 2007. "Role of Educational and R&D Institutions in City Clusters: An Exploratory Study of Bangalore and Pune Regions in India." *World Development* 35 (6): 1037–1055.

Basant, Rakesh, and Sunil Mani. January 2012. Foreign R&D Centres in India: An Analysis of their Size, Structure and Implications In *W.P. No. 2012-01-06*. Ahmedabad: Indian Institute of Management.

Beule, Filip de. 2010. "The Role of FDI in Technology Innovation in India and China." In *Indian and Chinese Enterprises : Global Trade, Technology and Investment Regimes*, edited by N.S. Siddharthan and K Narayanan, 102–116. London: Routledge.

Birkinshaw, Julian. 2000. "Upgrading of Industry Clusters and Foreign Investment." *International Studies of Management & Organization* 30 (2): 93–113.

Boutellier, Roman, Oliver Gassmann, and Maximilian von Zedtwitz. 2008. "Foreign R&D in China." In *Managing Global Innovation: Uncovering the Secrets of Future Competitiveness*, 61–75. Berlin Heidelberg: Springer.

Bruche, Gert. 2009. "The Emergence of China and India as New Competitors in MNCs' Innovation Networks." *Competition & Change* 13 (3): 267–288.

Business Standards. 2011. India emerging as a major centre for R&D projects for global multinationals. *Business Standard*, September 21, http://www.business-standard.com/article/press-releases/india-emerging-as-a-major-centre-for-r-d-projects-for-global-multinationals-111092100122_1.html.

Cantwell, John, and Simona Iammarino. 2003. *Multinational Corporations and European Regional Systems of Innovation*. London: Routledge.

Cantwell, J. A., and L. Piscitello. 2000. "Accumulating Technological Competence: Its Changing Impact on Corporate Diversification and Internationalization." *Industrial and Corporate Change* 9 (1): 21–51.

Cantwell, John, and Antwell Piscitello. 2005. "Recent Location of Foreign-owned Research and Development Activities by Large Multinational Corporations in the European Regions: The Role of Spillovers and Externalities." *Regional Studies* 39 (1): 1–16.

Chiesa, Vittorio. 1996. "Managing the Internationalization of R&D Activities." *IEEE Transactions on Engineering Management* 43 (1): 7–23.

"Continental Opens New Automotive Tech Center in Jiading, China.". Apr 2, 2009. Retrieved from http://www.continental-corporation.com/www/pressportal_com_en/themes/press_releases/3_automotive_group/chassis_safety/press_releases/jiading_tech_center_opening_02_04_09_en.html.

Dan, Wang. April 15, 2004. "AMD Expanding Facilities in China." http://www.zdnet.com/article/amd-expanding-facilities-in-china/.

Designed in Japan, but made in China. 2011. *China Daily*, June 9. http://english.peopledaily.com.cn/90001/90778/90861/7405095.html.

Drisko, James, and Tina Maschi. 2016. *Interpretive Content Analysis*. Oxford: Oxford Scholarship Online.

Dunning, John H., and Sarianna M. Lundan. 2009. "The Internationalization of Corporate R&D: A Review of the Evidence and Some Policy Implications for Home Countries." *Review of Policy Research* 26 (1-2): 13–33.

Edwards, Ron. 2002. "FDI : Strategic Issues." In *Foreign Direct Investment: Research issues*, edited by Bijit Bora, 28–45. London & New York: Routledge.

Gammeltoft, Peter. 2006. "Internationalisation of R&D: Trends, Drivers and Managerial Challenges." *International Journal of Technology and Globalisation* 2 (1-2): 177–199.

Gassmann, Oliver, and Zheng Han. 2004. "International R&D Activities in Emerging Markets - The Case of China." Engineering Management Conference, 2004. Proceedings. 2004 IEEE International 2: 601–605.

Gassmann, Oliver, and Marcus Matthias Keupp. 2008. "The Internationalisation of Western firms' R&D in China." *International Journal of Entrepreneurship and Small Business* 6 (4): 536–561.

Greatwall. 2002. *R&D Yongbao Zhongguo (R&D Embracing China)*. Nannin: Guangxi Renmin.

Jaffe, Adam B., Manuel Trajtenberg, and Michael S. Fogarty. 2000. "Knowledge Spillovers and Patent Citations: Evidence from a Survey of Inventors." *American Economic Review* 90 (2): 215–218.

Jaffe, Adam B., Manuel Trajtenberg, and Rebecca Henderson. 1993. "Geographical localisation of knowledge spillovers, as evidenced by patent citations." *Quarterly Journal of Economics* 108 (3): 577–598.

Kevin Bullis. April 8, 2010. "GE to Boost Research in China." *MIT Technology Review.* Retrieved from https://www.technologyreview.com/s/418367/ge-to-boost-research-in-china/

Krippendorff, Klaus. 2013. *Content Analysis: An Introduction to Its Methodology*. New Delhi: Sage Publishing.

Krishna, V. V. 2012. "Universities in India's National System of Innovation: An Overview." *Asian Journal of Innovation and Policy* 1 (1): 1–30.

Krishna, V. V., and Swapan Kumar Patra. 2016. "Research and Innovation in Universities." In *India Higher Education Report 2015*, edited by N. V. Varghese and Garima Malik, 163–196. New Delhi: Routledge; Taylor & Francis Group.

Krishna, V. V., Swapan Kumar Patra, and Sujit Bhattacharya. 2012. "Internationalisation of R&D and Global Nature of Innovation: Emerging Trends in India." *Science Technology & Society* 17 (2): 165–199.

Kuemmerle, Walter. 1999. "Foreign Direct Investment in Industrial Research in the Pharmaceutical and Electronics Industries—Results from a Survey of Multinational Firms." *Research Policy* 28 (2-3): 179–193.

Lall, Sanjaya. 2003. "Foreign direct investment, technology development and competitiveness : issues and evidence." In *Competitiveness, FDI and Technological Activity in East Asia*, edited by Lall Sanjaya and Urata Shujiro, 12–56. Cheltenham: Edward Elgar.

Li, Jianlin, and Jizhen Li. 2010. "The Internationalization of R&D and its Offshoring Process." *International Journal of Human and Social Science* 5 (2): 109–116.

Li, Jiatao, and Deborah R. Yue. 2005. "Managing Global Research and Development in China: Patterns of R&D Configuration and Evolution." *Technology Analysis & Strategic Management* 17 (3): 317–338.

Lundin, Nannan, Sylvia Schwaag Serger, Martin Berger, Lan Xue, and Zheng Liang. 2008. "China and the Globalization of Research and Development." In *OECD Reviews of Innovation Policy, China*, 263–304. Paris: Organization for Economic Cooperation and Development.

Lundvall, Bengt-Ake. 2008. "Preface." *International Journal of Technology and Globalization* 4 (1): 1–4.

Luo, Laijun, Louis Brennan, Chang Liu, and Yuze Luo. 2008. "Factors Influencing FDI Location Choice in China's Inland Areas." *China & World Economy* 16 (2): 93–108.

Mansfield, Edwin. 1975. "International Technology Transfer: Forms, Resource Requirements, and Policies." *The American Economic Review* 65 (2): 372–376.

Mansfield, Edwin, David Teece, and Anthony Romeo. 1979. "Overseas Research and Development by US-Based Firms." *Economica, New Series* 46 (182): 187–196.

Mody, Ashoka. 2007. "Is FDI integrating the world economy?" In *Foreign Direct Investment and the World Economy*, edited by Ashoka Mody, 1–31. New York: Routlegde.

Motohashi, Kazuyuki. 2005. R&D of Multinationals in China: Structure, Motivations and Regional Difference. RIETI Discussion Paper Series 06-E-005.

Motohashi, Kazuyuki. 2010. "R&D Activities of Manufacturing Multinationals in China: Structure, Motivations and Regional Differences." *China & World Economy* 18 (6): 56–72.

Mrinalini, N., Pradosh Nath, and G. D. Sandhya. 25 September 2013. "Foreign direct investment in R&D in India." *Current Science* 105 (6): 767–773.

Mrinalini, N., and S. Wakdikar. 2008. "Foreign R&D Centres in India: Is there any positive impact?" *Current Science* 94 (4): 452–458.

Nagpal, Aditi. 2010. "Caps are good, but then caps limit R&D." *Financial Express*.

Narula, R., and A. Zanfei. 2005. "Globalisation of Innovation: The Role of Multinational Enterprises." In *Handbook of Innovation*, edited by David Mowery, Jan Fagerberg and Richard R. Nelson, 318–347. Oxford: Oxford University Press.

New transfer pricing rules by early next month. 2013. *The Hindu*, August 15, http://www.thehindu.com/business/Economy/new-transfer-pricing-rules-by-early-next-month/article5022991.ece.

Our Bureau. 2011. "Freescale Opens R&D Centre in Hyderabad Sez." *The Hindu Business Line*, July 27. http://www.thehindubusinessline.com/info-tech/freescale-opens-rd-centre-in-hyderabad-sez/article2299140.ece.

Patra, Swapan Kumar. 2014. "Innovation Network in IT Sector: A Study of Collaboration Patterns Among Selected Foreign IT Firms in India and China." In *Collaboration in International & Comparative Librarianship*, edited by Susmita Chakraborty and Anup Kumar Das, 321–343. IGI Global: Hershey, Pennsylvania.

Patra, Swapan Kumar, and V. V. Krishna. 2015. "Globalization of R&D and open innovation: linkages of foreign R&D centers in India." *Journal of Open Innovation: Technology, Market, and Complexity* 1 (1): 1–24. doi:DOI 10.1186/s40852-015-0008-6.

Porter, M. 1990. *The Competitive Advantage of Nations*. New York: Free Press.

Prater, Edmund, and Bin Jiang. 2008. "The Drivers of Foreign R&D Investment in China." *Journal of Marketing Channels* 15 (2&3): 211–233.

Reddy, P. 1997. "New Trends in Globalization of Corporate R&D and Implications for Innovation Capability in Host Countries: A Survey from India." *World Development* 25 (11): 1821–1837.

Rugman, Alan, and Alain Verbeke. 2003. "Regional Multinationals: The Location-bound Drivers of Global Strategy." In *The Future of the Multinational Company*, edited by Sumantra Ghoshal, Julian Birkinshaw, Constantinos Markides, John Stopford and George Yip, 45–57. West Sussex, England: John Wiley & Sons Ltd.

Satyanand, Premila Nazareth. 2007. "Regions: Asia - Chemical attraction." *FT Business*, October 1.

"Siemens Establishes Competence Center for Metallurgical Plants in Shanghai.". December 11, 2011. http://www.siemens.com/press/en/pressrelease/?press=/en/pressrelease/2011/metals-technologies/imt201112072.htm&content[]=IMT&content[]=PDMT.

SiliconIndia. MNC R&D centers in India augment reverse brain drain 31 March 2010. Available from http://www.siliconindia.com/shownews/MNC_RD_centers_in_India_augment_reverse_brain_drain-nid-66703-cid-1.html.

Sölvell, Örjan. 2003. "The Multi-home-based Multinational: Combining Global Competitiveness and Local Innovativeness." In *The Future of the Multinational Company*, edited by Sumantra Ghoshal, Julian Birkinshaw, Constantinos Markides, John Stopford and George Yip, 34–44. West Sussex, England: John Wiley & Sons Ltd.

Sun, Yifei. 2011. "Location of foreign research and development in China." *GeoJournal* 76 (6): 589–604. DOI: 10.1007/s10708-009-9318-1:1-16.

Sun, Y. F., D. B. Du, and L. Huang. 2006. "Foreign R & D in Developing Countries: Empirical Evidence from Shanghai, China." *China Review - An Interdisciplinary Journal on Greater China* 6 (1): 67–91.

Times News Network. 2011. "India to Become Global Hub for Volvo Buses." *The Times of India*, January 14, http://timesofindia.indiatimes.com/business/india-business/India-to-become-global-hub-for-Volvo-Buses/articleshow/7284238.cms.

UNCTAD. 2005. Globalization of R&D and Developing Countries :Proceedings of the Expert Meeting, 24-26 January 2005, at Geneva.

Vijay, Nandita. 2012a. "MSD India Invests $150 Mn for R&D Centre at Jamia Millia Islamic Univ, to Focus on Vaccines, Novel Molecules." *Pharmabiz.com*, April 23, http://pharmabiz.com/PrintArticle.aspx?aid=68611&sid=1.

Vijay, Nandita. 2012b. "India on the Road to Becoming a Pharma Powerhouse: Apotex India Chief." *Pharmabiz*, January 17, http://pharmabiz.com/printarticle.aspx?aid=67038&sid=1.

von Zedtwitz, Maximilian. 2004. "Managing foreign R&D laboratories in China." *R&D Management* 34 (4): 439–452.

von Zedtwitz, Maximilian, and Oliver Gassmann. 2002. "Market Versus Technology Drive in R&D Internationalization: Four Different Patterns of Managing Research and Development." *Research Policy* 31 (4): 569–588.

Walcott, Susan M, and James Heitzman. 2006. "High Technology Clusters in India and China: Divergent Paths." In *Indian Journal of Economics and Business*. Special Issue on India and China, 113–146.

Walsh, Kathleen. 2003. *Foreign High-Tech R&D in China : Risks, Rewards, and Implications For U.S. China Relations*. Washington: The Henry L. Stimson Center.

Wen, Ke, and Zefu Lin. 2005. "The strategic evolution of foreign R&D investment in China." Engineering Management Conference, 2005. Proceedings. 2005 IEEE International.

World Investment Prospect Survey 2007–2009. 2007. New York and Geneva: United Nations Conference on Trade and Development.

World Investment Report. 2005. Transnational Corporations and the Internationalization of R&D. New York and Geneva, 2005: United Nations Conference on Trade and Development United Nations.

Yining, Ding. 2010. "Regional R&D Center Opens." *Shanghai Daily*, April 7. http://www.shanghaidaily.com/business/manufacturing/Regional-RD-center-opens/shdaily.shtml.

Zanfei, Antonello. 2000. "Transnational Firms and the Changing Organisation of Innovative Activities." *Cambridge Journal of Economics* 24: 515–542.

Exploring the Jawaharlal Nehru National Solar Mission (JNNSM): Impact on innovation ecosystem in India

Amitkumar Singh Akoijam ⊙ and V. V. Krishna

To make India one of the leaders in solar energy generation and to promote ecologically sustainable growth that addresses the nation's energy security challenge is one of the promising goals of the *Jawaharlal Nehru National Solar Mission (JNNSM)* or *National Solar Mission*. This paper presents the country's current solar energy scenario and explores ways in which various actors, agencies and policies shape the mission from the different perspectives on innovation literature. Innovation ecosystem is one of the perspectives where the sense of environment or ecology of various institutions, actors and other factors surrounds the activity of research and innovation. In this ecosystem, there is no single actor that can perform independently. The research outcomes, especially the patents, research publications and R&D investment, have become an increasingly essential area after the announcement of the JNNSM. The study also highlights that the number of research papers published in relation to solar energy has increased and there is a significant presence of productive R&D institutions, universities and supportive policy initiatives in the country.

Introduction

Energy is an important sector for the economic growth of a nation like India. The country's economy is one of the fastest growing economies in the world. Due to the rapidly growing population and growing economy, consumption in the energy sector has increased rapidly. There is a wide gap between the country's energy production and its energy demand (Krishna, Sagar, and Spratt 2015). About 300 million people in the country lack access to basic energy services, according to the World Bank Report (2014). On the other hand, the country has huge potential for the generation of solar energy. The government of India launched a mission, the *Jawaharlal Nehru National Solar Mission* (hereafter JNNSM or *the National Solar Mission*) in November 2009 (it officially took off in January 2010). It is a major initiative to promote ecologically sustainable growth while addressing India's energy security challenge. The goal is to make the nation one of the leaders in solar energy production in the world by 2022. The mission has three phases, phase-I (2010–2013), phase-II (2013–17) and phase-III (2017–2022), each with different target achievements. The mission also has other additional goals such as promoting R&D, providing public domain information, developing trained human resources for the solar industry and expanding the scope and coverage of earlier incentives for industries to set up solar photo voltaic (PV) manufacturing in India. This mission, for various reasons, has garnered a lot of attention and inspired fully-fledged research into itself, while, simultaneously, impacting the innovation ecosystem in the country.

Solar energy technologies and innovation promote ecologically sustainable growth while addressing the country's energy security challenge (MNRE 2013). JNNSM constitutes a major contribution by India to the global effort to meet the challenges of climate change. According to the Ministry of New and Renewable Energy (MNRE), solar energy has been an important component of the country's energy planning process and it is no longer an *'alternate energy'* but will increasingly become a key part of the solution to the nation's energy needs.[1] In June 2008, India released its National Action Plan on Climate Change (NAPCC) to promote development goals while addressing climate change mitigation and adaptation and to enhance ecological sustainability of the country's development path.[2]

The mission is the leading one among eight national missions[3] under the NAPCC because it fights the issue of climate change and it tries to answer the question of meeting India's energy demand while expanding development opportunities of different solar technologies throughout the country. There are different actors who constitute and shape the solar energy sector, like business enterprises or private firms, R&D institutions, universities, financial institutions, government ministries, non-governmental organization, etc. In our study, innovation ecosystem refers to the perspective where a sense of the environment or ecology of various institutions, actors and various other factors surrounds the activity of research and innovation.

Objectives and research methodology

The main objective of the study is to map the energy scenario in the country in order to understand the significance of renewable energy with a focus on solar energy. Others are:
- to study the various policies enunciated by the government with special reference to the first phase of *the National Solar Mission* and its impact on institution building; and

- to explore the dynamics of innovation by identifying various actors and institutions which determine the process of innovation in the solar energy sector.

Our study is based on both quantitative and qualitative data. The quantitative data primary means the number of scientific research publications related to solar energy and patents granted in various solar technologies. It was gathered first by reviewing the available literature related to renewable energy, solar energy and a few articles based on the mission. The concept of innovation that is used in the study was drawn from the innovation system perspectives of National Innovation Systems (Freeman 1987 and 1995; Nelson 1993; Lundvall 1997; Edquist 1997) and Sectoral Innovation Systems (Carlsson 1995; Breschi and Malerba 1997; Mowery and Nelson 1999). The information and data related to solar policies and programmes, various solar PV and solar thermal technologies, research and development activities and other useful information about various institutions, such as the Ministry of New and Renewable Energy (MNRE), the Solar Energy Centre (SEC), the Solar Energy Cooperation of India (SECI), the Indian Renewable Energy Development Agencies (IREDA), etc., were retrieved from their annual reports and websites. Some informal interviews were also undertaken to discuss various policy issues with professionals at the energy ministry.

To understand the research publications and patents' analysis, we used bibliometric analysis and databases available on USPTO (United States Patent and Trademark Office) and IPO (Indian Patent Office). Bibliometric is a set of online database tools for analyzing publication data. According to Norton, it defines the measure of texts and information associated with a publication, and includes author, affiliation, citations from other publications, co-citations with other publications, reader usage, and associated keywords (Norton 2001). With the help of analyzing the data available in the Scopus[4] database, the number of research publications related to the solar energy sector in different universities and R&D institutions in the country were calculated and analyzed accordingly. The information gathered from the libraries of the Indian Institute of Technology IIT, Delhi and the Energy Resources Institutes (TERI) and others have also been incorporated.

Theoretical framework and literature review

This study was undertaken to explore the ways in which various actors, agencies and policies influenced the JNNSM or solar energy sector under the theoretical framework of innovation literature. Since the development and deployment of different solar technologies are driven by different actors and agencies, a study on the JNNSM cannot be complete without bringing them into perspective. The concept of a system of innovation was developed in parallel at different places in Europe and the USA in 1980s. As Schumpeter (1939) defines innovation, it is the key driver to economic change and regional development of a nation because it is responsible for the setting up of new production functions that create new commodities as well new forms of organization. Schumpeter (1939) further mentions that invention, innovation and successful diffusion of new technologies are the major drivers of modern economies. The Schumpeterian concept of innovation also draws attention to the introduction of new products, process innovation that is new to an industry, the opening of new markets, the development of new sources of supply for raw materials or other inputs, and changes in industrial organizations (Schumpeter 1939).

Further, as Edquist (2001) argues, innovation is the new creation of economic significance which is normally carried out by firms or sometimes by individuals. The product or idea can be brand new, but is more often the new combination of existing elements. He further describes the category of innovation as extremely complex and heterogeneous. Similarly, Fagerberg, Mowery, and Nelson (2005) also stress that innovation is crucial for long-term economic growth and it tends to cluster in certain industries or sectors, which consequently grow more rapidly than others, implying structural changes in production and demand and, eventually, organizational and institutional change. Innovation is nowadays therefore perceived as the fundamental driving force behind both advanced and advancing economies.

In this study, we also drew on the system of innovation from Metcalf (1995) who perceives that institutions which jointly and individually contribute to the development and diffusion of new technology also provide the framework that governments use in forming and implementing policies to influence the innovation process. A system of innovation has also been defined as 'all important economic, social, political, organizational, and other factors that influence the development, diffusion, and use of innovations' (Edquist and Johnson 1997, 14). During the 1990s, it became increasingly common to regard the emergence of innovations as a complex process characterized by complicated feedback mechanisms and interactive relations involving science, technology, learning, institutions, production, public policy and market demand (Edquist and Johnson 1997). The development of innovations is seen as being characterized by a processes of interactive learning, i.e. there is often an exchange of knowledge between organizations involved in the innovation processes (Lundvall 1997). Various kinds of knowledge and information are exchanged between organizations and such exchanges often take the form of collaboration that is not mediated by a market. From the different perspectives of innovation, it could thus be studied in a national, regional or sectoral context such as the National Innovation System (NIS), the Regional Innovation System (RIS) or the Sectoral System of Innovation (SSI). The SSI perspective has been developed and increased in importance over time by Carlsson (1995), Breschi and Malerba (1997), Cooke et al. (1997), Mowery and Nelson (1999) and so on.

The concept of sectoral systems of innovation (SSI) is very popular today. Based on this framework, there are many studies that look at the dynamism of sectoral systems in many sectors. For instance, Turpin and Krishna (2007) view the dynamics underlying three promising sectors in India, namely ICT software, biotechnology and pharmaceuticals, through the lens of the SSI perspective. Krishna (2007) regards the three main building blocks of the innovation framework, namely sectoral boundaries, key elements of the SSI perspectives and transformation of the sectoral system, through the co-evolution of its constituent elements. He argues that the pharmaceutical sector is one of the most innovative of all sectors.

The SSI helps to draw our attention to the dynamics of innovation by identifying various actors and institutions which interact and determine the process of innovation in the solar energy sector. The SSI is composed of various agents, institutions, types and structures of interactions among firms and non-firm organizations in a sector. Malerba (2000; 2002b) defines the SSI as

> ... composed by the set of heterogeneous agents carrying out market and non-market interactions for the generation, adoption and use of (new and established) technologies and for the creation, production and use of (new and established) products that pertain to a sector (sectoral products).

Malerba (2004) mentions the basic elements of a sectoral system as products, agents, knowledge and learning processes, basic technologies, inputs and demand with links, interactions and institutions. In his approach, it has a knowledge base, technologies, input and (potential or existing) demand where the agents composing the sectoral system are organisations and individuals (Malerba 2002a).

The idea of innovation ecosystem that we employ here refers to the way in which various actors, agencies and policies are shaping the JNSSM or India's solar energy sector. In this ecosystem, there is a combination of public and private research institutions, universities and other technical and financial institutions in India. The country is trying to create its own innovation ecosystem by consolidation of various private or public institutions, universities, technical and funding bodies. The country can become a global innovation leader, if there is an effective innovation ecosystem with 'coherent synergy'[5] in the

field of science and technology and 'technology foresight' to make the right technology choices in the national perspective (Chidambaram 2007). Research, however, involves generation of new knowledge and innovation requires adding economic value (or societal benefit or strategic value or both) to that knowledge.

Key actors in the innovation ecosystem and installation capacity

System of innovation literature leads us to the concept of innovation ecosystem (Chidambaram 2007) which, in the context of this study, signifies different actors, agencies and polices among other features of the environment surrounding the solar energy sector in India. The JNSSM has initiated numerous steps to catalyze innovation ecosystem for the development and implementation of different technologies of solar energy in the country (MNRE, 2010a). Key actors in the ecosystem are depicted in Table 1.

India is a key emerging country in terms of solar power. During 2013–2014, the overall production was over 240 MW for solar cells and 661 MW for PV modules (MNRE 2015). The country's PV sector endeavours to focus on three key approaches which include licencing of patents, acquisitions and joint ventures, and in-house research and development, in a bid to match the sector's ongoing development levels of technology, production and knowledge capacity (Mallett et al. 2009).

Gujarat is the leading state in solar energy installation in India with 857.90 MW, followed by Rajasthan (552.90 MW), Maharashtra (100 MW), Madhya Pradesh (37.32 MW), Andhra Pradesh (23.35 MW) and so on. Gujarat, Rajasthan, Maharashtra, Madhya Pradesh and Andhra Pradesh respectively cover 50.87%, 32.74%,

Table 1: Actors in the innovation ecosystem in the Indian solar energy sector.

Key actors	Particulars
1. Business enterprises (solar firms)	Domestic manufactures (cells, modules, balance of systems): Tata BP Solar, Moser Baer, Solar Semiconductor, Photon Energy Systems, Central Electronics Laboratory (CEL), Reliance Industries Limited, Bharat Heavy Electricals Limited (BHEL), Lanco Solar, IndoSolar Ltd., Websol Energy System Ltd., Titan Energy Ltd. etc.
	Foreign-owned manufactures: SunEdison (US base), Trina Solar (China), etc.
	Project developers: Azure Power, Green Infra, Mahinder, Welspun, etc. Engineering, procurement & construction (EPC): Mahinder, Tatasolar, etc.
2. Policy support	Apex body and regulatory institutions: MNRE, Central Electricity Regulatory Commission (CERC), State Electricity Regulatory Commission (SERC), Ministry of Power, Ministry of Finance, Ministry of Environment, Forest and Climate Change, SECI, IREDA, National Thermal Power Corporation Vidyut Vyapar Nigam (NVVN)
	Policy instruments: Domestic Content Requirement (DCR), Generation Base Incentives (GBI), Accelerated Depreciation, Solar REC, state policies, Solar Viability Gap Funding (VGF), Direct Subsidies, tax incentives, etc.
3. Government Research Institutes (GRIs)	IITs, central universities, state university, 61 educational institutions (PG level) offering courses on renewable energy, National Physical Laboratory (NPL), Council of Scientific and Industrial Research (CSIR), National Institute of Solar Energy (NISE), Solar Energy Corporation of India (SECI), etc.
4. Financial institutions	Domestic sources: Scheduled commercial banks like SBI, Bank of Baroda, IDBI Bank, Axis Bank. Nonbanking financial services like L&T Infra Financial, IDFC, PFC Green Ventures, DFC, IREDA, etc.
	External sources: International Finance Corporation (IFC), Asian Development Bank (ADB), Overseas Promotion & Investment Corporation (OPIC), U.S EXIM Bank, EXIM Bank of China, etc.
5. Industry associations	Solar Energy Society of India (SESI), National Solar Energy Federation of India (NSEFI), Solar Thermal Federation of India (STFI), Indian Solar Manufactures Association, Solar Power Developers Association, Solar Energy Trade Association in India, etc.
6. NGOs and other organizations	Barefoot Engineers, Greenpeace, Centre for Science and Environment (CSE).

Source: Author's compilation; MNRE (2016) and TERI (2013)

5.9%, 2.1% and 1.3% of the total grid's connected solar energy capacity (MNRE 2015). Among the various renewable resources, wind has the maximum cumulative installed capacity which is around 59%, followed by solar (19.5%), bio-power (12%), small hydro (9%) and waste-to-power (0.5%). Table 2 shows the cumulative deployment of renewable energy sources (both grid and off-grid connected) at the end of November 2016. The cumulative installed solar energy capacity was about 9256.88 MW at the end of November 2016. Out of this, 8874.87 MW and 382.01 MW are from grid-connected and off-grid solar respectively (CEA 2016; MNRE 2016). The country stands in fourth position globally in terms of renewable energy market potential, just behind China, the US and Germany.

Renewable energy as a whole has the potential to meet 15% of the total contribution under the National Action Plan on Climate Change (NAPCC) by 2020[6] which will eventually reduce the emission of greenhouse gases (MNRE 2013). India has huge potential for solar power generation since about 58% of the total land area (1.89 million km^2) receives and annual average global insolation above 5 $kWh/m^2/day$ (Ramachandra, Jain, and Krishnadas 2011). The country's solar radiation is higher than countries like Germany where annual solar radiation ranges from 800 to 900 kWh/m^2. Moreover, there is around 1,000,000 km^2 of open and non-agricultural land which receives adequate radiation in the country and this land is available in Rajasthan, Gujarat, Madhya Pradesh and some parts of the Deccan plateau (Sukhatme 2011). About 1% of this land area (around 10,000 km^2) is sufficient to meet electricity needs of the country till 2031 as estimated by some sources. The challenge of providing energy access to the remote villages can be met only by solar PV, small hydro, wind or hybrid systems (Pillai and Banerjee 2009).

While assessing the potential of solar energy, the real issue is not the availability of solar radiation as much as the availability of open land.[7] Mitavachan and Srinivasan (2012) argue that the solar power plants require less land in comparison to hydro power plants, nuclear and coal including when life-cycle land transformations are considered. Currently, solar PV installations consist almost entirely of off-grid connectivity and small capacity appliances which are mostly used in public lighting such as street lighting, traffic lighting, and domestic power back-up in municipal areas and small electrification systems and solar lanterns in the rural areas. In recent years, it has also been used for powering water pumps for farming and small industrial areas (MNRE 2015).

Solar technology can broadly divided into two categories, such as solar PV technology and solar thermal technology (Bhargava 2001 and Kharul 2011). Several types of solar PV cells[8] are globally available, such as amorphous silicon, crystalline silicon, dye-sensitized cells as well as other newer technologies, such as silicon-nano particle ink, carbon nanotube and quantum dots (Willey and Hester, 2009). Kharul (2011) broadly categorized solar technology into four generations of technology. The first and second generation solar cells are commercially available globally, while the third and fourth generation solar cell technologies are in their initial stage.

Solar thermal technology is quite diverse in terms of its operational characteristics and applications in that it includes fairly simple technologies such as solar space heating and solar cooking as well as complex and sophisticated ones like solar air conditioning and solar thermal power generation (Purohit and Purohit 2010). As Asif and Muneer (2008) argue, solar thermal technologies are the most diverse and effective renewable energy technologies in the world.[9] The most successful application of solar thermal technologies is in the form of solar water heating (SWH).

Status of the Indian solar manufacturing sector

The Indian solar manufacturing sector consists of both crystalline silicon and thin films in PV technologies and thermal technologies. This sector is little over three decades old. Solar thermal technologies are still in the early stage of development compared with the PV technologies. Since the 1970s, BHEL and CEL have been engaged in making solar panels and other equipment, but were later joined by other companies that began small-scale manufacturing of modules which are limited to off-grid applications. With a 40 MW manufacturing capacity, Moser Baer Solar set up the first commercial-scale manufacturing plant in 2006 (Bhushan and Hamberg 2012). Since then, the domestic manufacturing industries have experienced nurturing growth owing to global demands and the ambitious national solar mission implications.

According to the World Bank (2013), most of the raw materials and consumables for solar cell and module manufacturing in India are imported. There is no poly-silicon or wafer manufacturing capability in the country. But more than 19 solar cell makers and 50 module makers have registered with the MNRE in the country[10] (Bhushan, and Hamberg, 2012). The Chinese are the

Table 2: Indian cumulative deployment in solar and other renewable resources (in MW), as on 30.11.2016.

Renewable resources	Grid-connected	Off-grid connected	Cumulative achievements
Wind	28,419.40	–	28,419.40
Solar	8,874.87	382.01	9,256.88
Bio-power*	4,932.33	838.79	5,771.12
Small hydro	4,324.85	18.81	4,343.66
Waste-to-power	114.08	161.12	275.20
Total renewable energy	46,665.53	1,400.73	48,066.26

Source: http://mnre.gov.in/mission-and-vision-2/achievements/ (accessed on 11.01.2017)
*Bio-power includes biomass-gasification and bagasse cogeneration which achieve maximum energy from the off-grid system as 651.91 MW and 186.88 MW, respectively.

leaders in both cell and module production globally (TERI 2013; Spratt et al. 2014). By the beginning of the national solar mission's second phase, about 1500 MW of cell manufacturing capability and around 2000 MW of domestic module manufacturing capability existed in India as compared with only 15 MW of ingots and wafers manufacturing (MNRE 2015). So far, the biggest solar manufactures in the country are Moser Baer Solar, Tata Power and HHV Solar. The main driver of solar energy investment is a perceived business opportunity and such an opportunity often arises because government provides incentives.

Policies on renewable energy with respect to solar energy in India

The flow of information and network linkages of solar energy technology among people, enterprises and institutions are the key that leads to an innovative process. It contains the interaction between the actors who are needed to turn an idea into a process, product or service on the market. There is no single actor that can perform independently. The diverse actors function as the linkages and networks in the innovation ecosystem for the development and growth of the solar energy sector in the country. The Government of India has established the Ministry of New and Renewable Energy (MNRE), the Indian Renewable Energy Development Agency (IREDA), State Nodal Agencies, the Central Electricity Regulatory Commission (CERC) and the State Electricity Regulatory Commission (SERC) for the respectable functioning of the ecosystem. These bodies are engaged with each one another coherently in order to maintain the development of the renewable energy sector in general and the solar energy sector in particular.

In the early 1980s, the government began to introduce policies to support the expansion of renewable energy in the country. In 1982, the Department of Non-Conventional Energy Sources (DNES) was established. The Commission for Additional Sources of Energy (CASE) was created in the then Ministry of Energy. In 1992, the Indian Government established the Ministry of Non-Conventional Energy Sources (MNES). Then, in 2006, the MNES was renamed the Ministry of New and Renewable

Energy (MNRE). Renewable energy promotion received a boost with the National Electricity Policy 2005, which provides measures for licensed utilities and producers of captive electricity to purchase certain amounts of renewable energy. In recent years, a number of specific federal and state-level incentive schemes have been created for specific purposes, ranging from rooftop PV installations to large-scale power plants. There are some states that have specific solar state policies[11] and some are in the pipeline in order to boost solar generation capacities and drive down costs through local manufacturing and R&D activities to accelerate the transition to clean and secure energy in the state itself. Figure 1 depicts the evolution of India's renewable energy policies with respect to solar energy.

There are a number of government institutions whose competence extends into the renewable energy sector. The Electricity Regulatory Commissions Act instituted independent regulatory bodies both at the central and state-level which are known as the Central Electricity Regulatory Commission (CERC) and the State Electricity Regulatory Commissions (SERCs) respectively. The CERC is responsible for regulating tariffs of generating companies owned or controlled by the central government and for promoting competition in the electricity industry, while the SERCs deal with matters concerning generation, transmission, distribution and trading of electricity in their respective state.

The MNRE is the nodal ministry responsible for all matters relating to new and renewable energy such as solar, wind, biomass, small hydro, hydrogen, biofuels, geothermal, etc. The broad aim of the ministry is to develop and deploy new and renewable energy to supplement the energy requirements of the country. The MNRE's role is to facilitate research, design and development of new and renewable energy that can be deployed in the rural, urban, industrial and commercial sectors (MNRE 2011). The MNRE undertakes policymaking, planning, and promotion of renewable energy including financial incentives, creation of industrial capacity, technology research and development, intellectual property rights, human resource development and international relations.

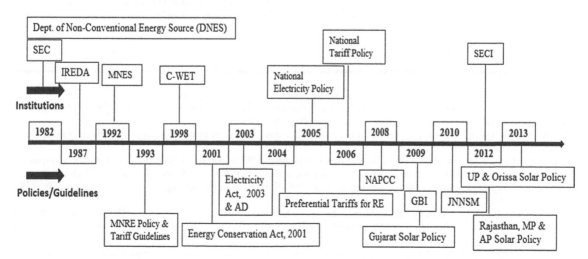

Figure 1: Timeline of various Indian energy policies and institutions set up to focus on the solar sector.
Source: Compiled by author; MNRE 2015

The vision of the MNRE is to develop new and renewable energy technologies, processes, materials, components, sub-systems, products and services on a par with international specifications, standards and performance parameters in order to make the country a net foreign exchange earner in the sector and deploy such indigenously developed and/or manufactured products and services in furtherance of the national goal of energy security.[12] Moreover, the MNRE supervises national institutions such as the Solar Energy Centre (SEC, recently renamed as National Institute of Solar Energy), the Centre for Wind Energy Technology (C-WET), and the Sardar Swaran Singh National Institute of Renewable Energy (SSS-NIRE). Apart from administering the institutions, the MNRE also affords financial assistance like supporting the Solar Energy Corporation of India (SECI). As a part of the Jawaharlal Nehru National Solar Mission (JNNSM), the Indian Institute of Technology (IIT) Delhi, IIT Mumbai, IIT Rajasthan, the Indian Institute of Science (IISc) Bangalore and the Indian Institute of Management (IIM) Ahmedabad are conducting several research and development activities in the area solar energy under the rubric of centre of excellence.

In the solar energy innovation ecosystem, the government is the main actor and it is responsible for the formulation of several renewable incentive policies that have increased the viability of increased deployment of solar energy technologies in the country, ranging from electricity sector reform to rural electrification incentives. *Domestic Content Requirement (DCR)* is a new set of guidelines under the JNNSM. To achieve solar capacity and cost targets, the JNNSM auctions Power Purchase Agreements (PPAs) to solar developers, and to ensure domestic solar manufacturing in the country the programme includes a DCR, developers must use solar cells and modules[13] manufactured in India. However, the guideline makes an exception for solar PV developers using thin film technologies, which may be imported. The majority of solar developers in the country currently use imported thin film modules.

Another policy, *Feed in Tariff*, is a mechanism designed to accelerate investment in renewable energy technologies in the country. They are minimum prices at which renewable energy projects can be purchased from the generating companies or private producers through contracts (PPAs) with transmission or distribution utilities or with trading licensees (Schmid 2012). Solar *Renewable Purchase Obligation* (RPO) is the minimum amount of solar energy that obligated entities, distribution licensees, open access and captive consumers have to deliver or consume as a percentage of their total available electricity. They can meet this obligation by purchasing the required quantity of solar power directly from producers. Alternatively, they can buy solar a *Renewable Energy Certificate* (REC) to fulfil their RPO. Many states are now establishing RPOs, which have stimulated development of a tradable REC programme (Altenburg and Engelmeier 2012).

Generation base incentives are provided to support small grid solar power projects connected to the distribution network under the solar *Generation Based Incentives* (GBI) scheme. The solar *Rooftop PV and Small Solar Power Generation Programme* (RPSSGP) is also an interesting scheme which was designed essentially to encourage states to implement grid connected projects focusing on the distribution network and to strengthen the tail end of the grid system. These *Feed in Tariff* and solar RPO, REC, GBI and RPSSPGP schemes have become notable policies or schemes during the implementation of the JNNSM.

Jawaharlal Nehru National Solar Mission and the status of Phase-I

The Jawaharlal Nehru National Solar Mission (JNNSM) is a major initiative of the government of India to promote ecologically sustainable growth while addressing the country's energy security challenge. The JNNSM seeks to kickstart solar generation capacities, drive down costs through local manufacturing, research and development to accelerate the transition to clean and secure energy (Deshmukh et al. 2011). Table 3 highlights the chronology of events in the JNNSM.

It aims to dramatically increase installed PV through attractive feed-in tariffs and a clear application and administration process. Under this mission, there are three phases: Phase-I (2010–2013), Phase-II (2013–17) and Phase-III (2017–2022). The mission aims installations of 20,000 MW of grid-connected solar power generation, 2000 MW of off-grid solar applications, 20 million

Table 3: Chronology of events in the *National Solar Mission*.

Date	Event
2007	The MNRE initiates the framing of an action plan or mission for solar energy internally.
2008–09	National Action Plan on Climate Change declared and the mission brought under the aegis of the Prime Minister's Office. The mission launched with an initial budget allocation of Rs.3850-million.
2009	NTPC Vidyut Vyapar Nigam Ltd. (NVVN) become nodal stakeholders in power purchase agreements through the National Thermal Power Corporation (NTPC).
2010	Phase-I initiates with two different batches. Asian Development Bank declares US$400m commitment. Around 418 project bids submitted for a cumulative target of 1–2 GW for the first batch of phase-I. Project sizes are small (5 MW cap) with DCR guidelines. Project developers prefer sourcing alternative equipment from foreign suppliers. Authorization of Rs.172.3 million to 37 solar cities.
2011	Solar Energy Industry Advisory Council is constituted to help attract investment, encourage R&D and make the Indian solar industry competitive. Allocated 350 MW in utility scale solar projects under the second batch of phase-I. 90% of the projects are in Rajasthan.
2012	NVVN replaced by the SECI under the supervision of the MNRE. Power purchase agreements directly signed with the SECI.
2013	Phase-I ends and Phase-II begins with a target on special focus on grid connected solar.

Source: Compiled by author; Krishna, Sagar, and Spratt (2015)

Table 4: Phase-wise and total targets in the JNNSM.

Segment	Phase-I 2010–13	Phase-II 2013–17	Phase-III 2017–22	Total
Grid-connected solar (MW)	1,100	3,000	16,000	20,000
Off-grid solar (MW)	200	800	1,000	2,000
Solar thermal collectors (million sq. meters)	7	8	5	20

Source: Ministry of New and Renewable Energy (MNRE) 2013

metres^2 of solar thermal collector area for industrial applications and 20 million solar lighting systems for rural areas by the year 2022.[14] The first phase, second phase and third phase have respective targets of 1100 MW, 3000 MW and 16,000 MW of grid-connected solar. In case of the off grid solar applications, 200 MW, 800 MW and 1000 MW are the targets in the first, second and third phases, respectively. And lastly, the three phases target to achieve 7, 8 and 5 million metres^2 of solar collectors (see Table 4).

The mission has also other additional goals, such as promoting R&D, disseminating public domain information, developing trained human resource for the solar industry, and expanding the scope and coverage of earlier incentives for industries to set up solar PV manufacturing in India. Under this mission, NTPC Vidyut Vyapar Nigam (NVVN) Ltd.[15] has been designated as nodal agency for procuring the solar power by entering into a Power Purchase Agreement (PPA) with solar power generation project developers who have been setting up solar projects during Phase-I. About 615 MW of different solar power projects, including both solar PV (145 MW) and thermal (470 MW) projects, were listed during according to NVVN. Out of them, the CSP projects are in Rajasthan (400 MW), Andhra Pradesh (50 MW) and Gujarat (20 MW), while solar PV projects are in Rajasthan (100 MW), Andhra Pradesh (15 MW), Karnataka (10 MW), Maharashtra (5 MW), Uttar Pradesh (5 MW), Orissa (5 MW) and Tamil Nadu (5 MW).[16]

The mission's phase-I is divided into batch-I and batch-II. In both batches, several schemes like NVVN scheme, migration scheme and RPSSGP scheme are introduced. Table 5 gives the status for both batch-I and batch-II under the JNNSM phase-I, including projects allotted under different schemes (see Table 5). About 1152.5 MW of grid connected solar projects have been allotted in batch-I and batch-II together. Out of the total 802.5 MW, 500 MW is allotted for solar thermal projects and 302.5 MW for solar PV projects in batch-I. On the other hand, 350 MW of solar PV projects are allotted

with respect to batch-II. Among the schemes, the purpose behind launching the migration scheme in 2010 was to provide transition to affordable solar projects from the existing arrangement to the one envisioned under the mission. The scheme is further subjected to the consent of state governments, the disposition of the project developer and the distribution licensee[17] (MNRE 2012). The bundling of solar power was introduced in this phase. As per this bundling, the cost of solar power is about Rs.5/kWh for which 500 MW capacities of both solar PV and thermal projects have been selected (MNRE 2012).

In the case of the off-grid connected/decentralized solar power in the JNNSM phase-I, around 27,841 solar lanterns, 53,588 home lights, 21,957 solar street lights, 1055 water pumping system and stand-alone solar PV power plants of 9365.39 KW capacity were installed during 2012–13 (MNRE 2013).

R&D activities and knowledge production in solar energy technologies

Promotion of R&D and increasing the knowledge production in solar energy technologies is one of the main objectives of national policies on solar energy and its innovation ecosystem. The various research and development programmes are designed to improve the efficiency, reliability and cost competitive performance of different solar energy technologies in the country. The National Institute of Solar Energy (NISE) and the Solar Energy Corporation of India (SECI) are the two most important R&D institutions that are established under the Solar Energy Research Advisory Council to address the existing research infrastructure in the domain of the solar energy sector and help to set up a framework which would incubate an environment for accelerating research and development activities in the country related to the goals of the National Solar Mission.

The NISE, which is the technical focal point of the MNRE, assists other research organizations and industry in implementing innovative ideas and development of

Table 5: Status of Batch-I and Batch-II of the JNNSM Phase-I.

I. For Batch-I (Schemes)	Projects allotted		Project commissioned	
	No.	MW	No.	MW
NVVN Scheme (Solar PV)	30	150	25	125
NVVN Scheme (CSP)	7	470	–	–
Migration Scheme (Solar PV)	13	54	11	48
Migration Scheme (CSP)	3	30	1	2.5
RPSSGP (Solar PV)	78	98.5	62	76.55
I. For Batch-II (Schemes)				
NVVN Scheme (Solar PV)	28	350	–	–
Total	159	1152.5	99	252.05

Source: Ministry of New and Renewable Energy (MNRE) 2012

new products by offering its facility and expertise for developmental testing on various solar technologies. Under this R&D programme, the evaluation of various emerging technologies and standardizing the technologies for applications suitable for various field conditions are important tasks. The institution is also responsible for conducting various training programmes, seminars, workshops and other solar energy technology courses with the objective of disseminating relevant knowledge. The MNRE has also introduced a fellowship, the national solar science fellowship programme. As part of the JNNSM, the SECI has taken up various projects or activities related to the solar sector across the country. One of the on-going schemes under the SECI is a pilot scheme for large scale grid-connected rooftop solar power generation which is 30% subsidy on the project cost made available from the ministry through the corporation.

Besides these, the government of India also allocates funds or provides other subsidies for various R&D activities in the country. The solar mission was launched in 2009 with a preliminary budget allocation of Rs.3.85 billion (IREDA 2012). India invests less money in its renewable energy sector than other countries globally. India has invested Rs 442.4 billion in the sector, which is only 2% of the global investment in the sector in recent year (MNRE 2015; BNF 2015). China has become the global leader in investment in the sector with around Rs.4532.4 billion invested, followed by the USA with about Rs.2450. These equal about 27% and 15% of global investment.

Knowledge production and comparisons with other countries

Solar energy has been harnessed mainly in two technologies. One is solar photovoltaic (PV) technology and the other solar thermal technology. The former technology is more advanced and developed than the latter (MNRE 2015). Among solar technologies, solar PV has emerged as the fastest growing renewable power technology worldwide (REN21 2015). It is one of the most promising ways to generate electricity in a decentralized manner at the point of use for providing electricity, especially for lighting and meeting small electricity needs, particularly in un-electrified households and unmanned locations.

The number of publications in various solar PV technologies such as amorphous cells, concentrating PV, dye-sensitized cells, mono crystalline, multi-junction

cells, poly crystalline and thin film cells in the country for 10 years (2006–2015) are shown in Table 6. The country has published the highest number of research publications globally in thin film, poly crystalline, amorphous and dye sensitized solar cells. Seven research organizations, namely the Indian Association of Cultivation of Science (Kolkata), the Indian Institute of Technology (Delhi), the University of Poona (Pune), the National Physical Laboratory (Delhi), the Indian Institute of Technology (Madras), the Indian Institute of Science (Bangalore) and the Indian Institute of Technology (Kharagpur) were involved in research on various aspects of amorphous material and process development (MNRE 2015). Central Electronics Ltd. and Rajasthan Electronics & Instruments Ltd. were the industrial organizations involved in the design and development of PV systems based on this technology (MNRE 2015).

In poly crystalline cells, in 2015 India ranked fourth globally with 981 publications. China had 2261 research publications in this particular technology leads the field globally the number of research papers published. In the case of dye sensitized, thin film cells and amorphous cells the country is ranked sixth and seventh, respectively. The rank-wise country in the publications of research in various solar technologies to 2016 is shown in Table 7.

In the case of number of patents granted in India related to solar energy, around 75 patents were granted by the Indian Patent Office (according to their website)[18] as at 2 July 2015. The online database on the patents was obtained by using the search engine 'Indian Patent Office' supported by the Office of the Controller General of Patents, Designs & Trade (CGPDT), the Ministry of Commerce and Industry and the government of India. But in the case of USPTO (United States Patent and Trademark Office) 53 patents[19] were granted in India in the different field of solar technologies. The highest number of patents were granted in the area of organic solar PV cells (18), followed by grid-connected applications (7), dye sensitized solar cells (7), concentrated PV (5), PV energy (3), solar thermal (3) and so on.

In the innovation ecosystem, universities are the one of the main actors because the interaction between industries/government institutions and universities is very important. The linkage between academia, industries and government institutions boost the quality of research output and development activities that explore solar energy worldwide. The productivity of various

Table 6: India's number of publication in various solar technologies for 10 years (2006–2015).

Technology	2006	2007	2008	2009	2010	2011	2012	2013	2014	2015
Amorphous	63	57	81	117	120	139	92	118	108	152
Concentrating PV	0	0	0	0	0	0	0	0	0	1
Dye-sensitized cells	1	5	13	18	29	65	71	95	153	170
Mono crystalline	3	4	7	9	6	6	10	8	8	13
Multi-junction cells	0	0	0	0	0	1	1	1	2	1
Poly crystalline	3	1	8	80	148	154	168	170	185	198
Thin film cells	63	45	61	66	82	112	120	156	220	203

Source: Researcher's data based on the Scopus Database**, 2016
**Based on the database, the number of publications related to a specific solar technology was analyzed. For instance, the keyword for dye sensitized cells are (TITLE-ABS-KEY (dye sensitized solar cells) AND PUBYEAR > 2005 AND PUBYEAR < 2016 AND ((LIMIT-TO (EXACTKEYWORD, 'Dye-sensitized Solar Cells') OR (LIMIT-TO (EXACTKEYWORD, 'Dye-Sensitized Solar Cell') OR (LIMIT-TO (EXACTKEYWORD, 'Dye-sensitized Solar Cell') OR (LIMIT-TO (EXACTKEYWORD, 'Dye Sensitized Solar Cell')) AND ((LIMIT-TO (DOCTYPE, 'ar')) AND ((LIMIT-TO (SRCTYPE 'j'))).

Table 7: Rank-wise country on the number of publications in various solar technologies.

Technology	Rank									
	1st	2nd	3rd	4th	5th	6th	7th	8th	9th	10th
Amorphous	China (4371)	USA (3207)	Japan (2169)	South Korea (1660)	Germany (1584)	France (1160)	India (**1072**)	Taiwan (919)	UK (774)	Italy (650)
Concentrating PV	USA (20)	China (18)	UK (12)	Germany (5)	Spain (5)	Israel (4)	Italy (4)	Australia (3)	Austria (2)	Netherlands (2)
Dye-sensitized cells	China (3016)	South Korea (1354)	USA (891)	Taiwan (752)	Japan (709)	India (**620**)	Switzerland (331)	Australia (254)	Germany (253)	Italy (251)
Mono crystalline	China (737)	USA (590)	Germany (402)	France (286)	Japan (226)	Russia (213)	Poland (131)	UK (126)	Italy (123)	South Korea (102)
Multi-junction cells	USA (103)	Spain (40)	China (37)	Japan (36)	Germany (33)	UK (15)	South Korea (13)	Taiwan (13)	Canada (11)	Australia (10)
Poly crystalline	China (2261)	USA (2150)	Japan (1152)	India (**981**)	Germany (892)	France (696)	South Korea (603)	UK (477)	Russia (359)	Spain (275)
Thin film cells	USA (3828)	China (3402)	South Korea (1950)	Germany (1851)	Japan (1660)	India (**1128**)	Taiwan (921)	France (892)	UK (827)	Italy (557)

Note: The figures within brackets indicate the number of publications by the respective country.
Source: Researcher's data based on the Scopus Database, 2016

universities and R&D institutions in India, compared with other countries globally, in terms of research publications related to the solar energy sector produced, was analyzed with the help of the data available in the Scopus database. To analyze the publications, we used the seven keywords[20] in the database. The total number publications in the solar energy sector as at 2016 was 287,853. Of this total, at 68,938 publications, the USA published the highest number of papers (24% of global publications), followed by China 37,562 (13%), Germany 20,601 (7%), Japan 18,844 (6.5%) and India 13,886 (5%). The country stands in fifth position. Figure 2 shows the number of publications by the top 10 countries in the world.

The country has also a significant presence of productive R&D institutions and universities. Figure 3 shows the year-wise growth of publications related to solar energy in India for a decade from 2006 to 2015. Out of the top 15 institutions and universities that published papers in India, the Indian Institute of Technology, Delhi, is topmost with 1186 publications (8.5% of the total publications), followed by the Indian Institute of Technology, Bombay, with 427 publications (3%), the Indian Institute of Science with 328 publications (2.4%) and others (see Figure 3). It is observed that among the top 15 affiliations there are four Indian Institutes of Technology and that all are the public research and development institutions and universities, except for the Tata Institute of Fundamental Research, Mumbai, which covers 1.5% of the total publications in the country (see Figure 4).

Conclusion

The innovation ecosystem in the solar energy sector is constituted of policies which regulate and provide stimulus to the industry as whole, which is endowed with R&D and technical support institutions, financial institutions, non-governmental organizations (NGOs) and business enterprises that are involved in manufacturing solar equipment. In an innovation system, there is no single actor that can perform independently, as all these above-mentioned actors are linked and connected in an ideal situation. In this ecosystem, the policy support is the main actor which is responsible for the formulation of several renewable incentive policies that have increased the viability of increased deployment and development of solar energy technologies in the country. In alignment with the goals of the *National Solar Mission*, the National Institute of Solar Energy (NISE) and the Solar Energy Corporation of India (SECI) are the most important R&D institutions. They are part of the Solar Energy Research Advisory Council which addresses the existing research infrastructure in the solar sector and helps to set up a framework that nurtures an environment for accelerating research and development activities in the country. Before the beginning of the *National Solar Mission*, the activities of the NISE were confined to solar thermal energy areas. After 2010, the main focus was on the development and promotion of solar PV technologies (mainly thin films and crystalline modules). Both the institutions provided an effective interface among the government, R&D institutions, industries and users of the technology

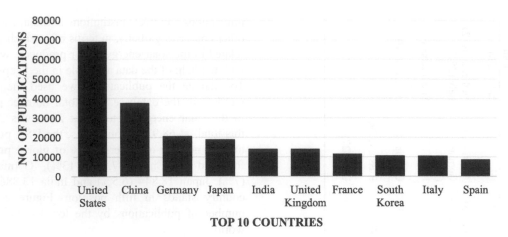

Figure 2: Publications by top 10 countries in the world.
Source: Researcher's data based on the Scopus Database, 2016

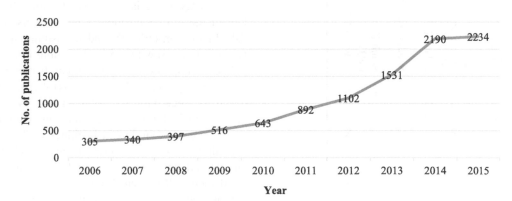

Figure 3. Number of publications in India for a decade (2006–2015).
Source: Researcher's data based on the Scopus Database, 2016

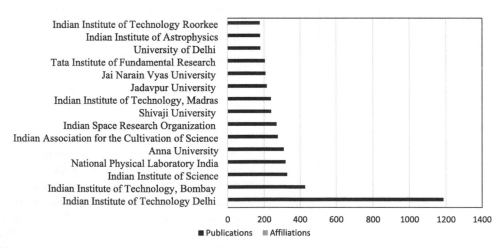

Figure 4: Publications by top 15 affiliations in India, 2016.
Source: Researcher's data based on the Scopus Database, 2016

for the development, promotion and widespread utilization of solar energy.

By far, the outcomes of phase-I have, to some extent, been ambiguous. The report given in the MNRE does not clearly show the mission's achievements in its first phase, as it is more of an overall cumulative report. Therefore, it is difficult to assess the individual outputs. In JNNSM phase-II, the mission identifies the need for international support in the form of technology transfer and financial assistance so as to meet its higher goals. Large-scale expansion of grid connected solar power is its main target. It is imperative for the central government to create a favourable environment for developing both solar PV and thermal technology to enhance its power sector. For phase-II, it would be mandatory to use cells and modules manufactured in India. So, some changes could be essential in domestic content requirements. Following this, new developments pertaining to trade relations of the country with the rest of the world are inevitable.

India ranks at fifth position in terms of knowledge production pertaining to the field of solar energy. The country has published extensively, particularly in the area of polycrystalline, thin-film, dye-sensitized and amorphous solar technologies. However, the country still lags behind countries such as the USA, China, Germany and Japan. Regarding the status of India's solar manufacturing sector, the country's expertise is in crystalline silicon and thin films under PV technologies. The country is more advanced in PV technologies than thermal technologies. Notwithstanding this, in recent years India has only contributed 2% of the global investment in renewable energy, whereas China has invested 27% of the global total, making China the global leader in terms of investment in renewable energy. Although India has created institutional R&D set ups and other actors in the innovation ecosystem, the country is still far behind China in terms of investments in solar R&D.

Apart from low level of investments, there are some infrastructure related problems. For instance, installation of thin films has already created some controversy. The acquisition of land and its investment is yet to be resolved. Subsequently, land acquisition has been an issue in terms of both its prices and investment for relevant authorities. Solar energy production in general must take into account the value of land. Resolving these minor issues are linked to the production of more efficient technology, whether thin films or crystalline modules or any other items.

Hence, the process of indigenizing solar films, including resolving infrastructural issues, seems imperative for the country. The import of films versus local manufacture deserves the utmost attention by decisionmakers. India has created all the necessary elements and actors relevant to a solar energy ecosystem, with a very high policy focus. However, further investment in research and strengthening innovation ecosystem will drastically improve India's comparative advantage.

Acknowledgements

This paper is derived from my M.Phil. dissertation work titled *Jawaharlal Nehru National Solar Mission (JNNSM): Impact on Innovation Ecosystem in India* which was submitted to the Centre for Studies in Science Policy (CSSP), Jawaharlal Nehru University. I would like to thank the editors of this journal and the anonymous reviewers of my paper for the valuable comments and suggestions. I would also like to acknowledge Michiko Iizuka and Rasmus Lema for constructive comments during the 13th GLOBELICS Conference held in Cuba, 2015, where the original form of this paper was first presented.

Disclosure statement

No potential conflict of interest was reported by the authors.

Notes

1. Ministry of New and Renewable Energy (MNRE), GoI. See http://www.mnre.gov.in/
2. Press Information Bureau, Government of India. See http://pib.nic.in/release/rel_print_page1.asp?relid=44098
3. The other seven national missions are National Mission for Enhanced Energy Efficiency, National Mission on Sustainable Habitat, National Water Mission, National Mission for Sustaining the Himalayan Ecosystem, National Mission for a Green India, National Mission for Sustainable Agriculture, and National Mission on Strategic Knowledge for Climate Change (MNRE, 2010b).
4. Scopus is one of the largest online abstract and citation database of peer reviewed literature with smart tools that extract, analyze and visualize research which that covers more than 205,000 titles from nearly 5000 publishers. See http://www.info.sciverse.com/scopus/.
5. According to Chidambaram, coherent synergy means coherence among the varieties of synergetic efforts; it may be Human Resource Development, R&D with prioritization, academia-industry interaction, international collaboration, made by the concerned parties in the field of science and technology for economic development in the country.
6. Annual report 2012–13, Ministry of New and Renewable Energy (MNRE), Government of India 2013.
7. Sukhatme (2011).
8. Solar PV cells are semiconductor devices that convert part of the incident solar radiation directly into electrical energy. The most common PV cells are made from single crystal silicon but there are many variations in cell material, design and methods of manufacture (Sharma 2011).
9. Asif and Muneer (2008). Solar thermal technologies. Encyclopaedia of energy engineering and technology. *Taylor and Francis,* New York, pp.1321–1330.
10. Bhushan and Hamberg (2012).
11. For instance, Gujarat State Policy, 2009 is the first solar specific state policy introduced in the country predating the JNNSM. Some others, like Rajasthan Solar Policy 2012, Andhra Pradesh Solar Policy 2012, Madhya Pradesh Solar Policy 2012, Tamil Nadu Solar Policy 2012, Uttar Pradesh Solar Policy 2013, Orissa Solar Policy 2013, etc., have come out with their own solar-specific policies.
12. Ministry of New and Renewable Energy, (MNRE); see the link: http://www.mnre.gov.in/mission-and-vision2/mission-and-vision/vision/ accessed 20 May 2013.
13. Solar cells and modules are the building blocks of solar PV (crystalline silicon) which are used to generate electricity.
14. Ministry of New and Renewable Energy (MNRE), Strategic Report; see http://www.mnre.gov.in/ accessed 22 February 2013.
15. NTPC Vidyut Vyapar Nigam (NVVN) Ltd. was formed by the National Thermal Power Cooperation (NTPC), as its wholly owned subsidiary to realize the potential of power trading, capacity utilization of power generation and transmission assets, and boost the development of the power market in the country.
16. Selected projects for phase-I JNNSM, NTPC Vidyut Vyapar Nigam (NVVN) Limited; see http://nvvn.co.in/Selected%20Projects%20List.pdf accessed 1 May 2013.
17. Jawaharlal Nehru National Solar Mission Phase II-Policy Document, 2012, Ministry of New & Renewable Energy. See the link: http://mnre.gov.in/file-manager/UserFiles/draft-jnnsmpd-2.pdf
18. See http://www.ipindia.nic.in/. The patent search was made in a double field search. The first field used was as search for titles with keyword "solar", and "solar energy" in the second field.
19. For example, the number of patents granted in solar thermal, tower concentrators, dish collectors, Fresnel lenses, trough concentrators, stirling solar thermal engines, thermal updraft, mounting or tracking, photovoltaic energy, PV systems concentrators, material technologies, CulSe2 material PV cells, dye sensitized solar cells, solar cells from group II-VI materials, solar cells from group III-V materials, micro-crystalline silicon PV

cells, poly-crystalline PV cells, mono-crystalline PV cells, amorphous silicon PV cells, organic PV cells, power conversion electric or electronic aspects, for grid-connected applications were obtained by using the CPC codes such as ICN/IN AND YO2E 10/40, ICN/IN AND YO2E 10/41, ICN/IN AND YO2E 10/42, … ICN/IN AND YO2E 10/63, etc.

20. Your query: ((TITLE-ABS-KEY(solar energy*) OR TITLE-ABS-KEY(solar photovoltaic*) OR TITLE-ABS-KEY(solar cell*) OR TITLE-ABS-KEY(solar thermal*) OR TITLE-ABS-KEY(solar power*) OR TITLE-ABS-KEY(solar panel*) OR TITLE-ABS-KEY(photovoltaic*)) AND PUBYEAR > 1979 AND PUBYEAR < 2017)

ORCID

Amitkumar Singh Akoijam ⑩ http://orcid.org/0000-0001-7938-3953

References

Altenburg, T., and T. Engelmeier. 2012. "Rent Management and Policy Learning in Green Technology Development: The Case of Solar Energy in India." Discussion Paper 12/2012, Bonn: German Development Institute, Accessed May 15, 2015 www.die-gdi.de/uploads/media/DP_12.2012.pdf.

Asif, M., and T. Muneer. 2008. "Solar Thermal Technologies." *Encyclopedia of Energy Engineering and Technology* Taylor and Francis, New York. 7: 1321–1330.

Bhargava, B. 2001. "Overview of Photovoltaic Technologies in India." *Solar Energy Materials and Solar Cells* 67: 639–646.

Bhushan, C., and J. Hamberg. 2012. "Facing the Sun: Policy for Sustainable Grid-Connected Solar Energy." New Delhi: Centre for Science and Environment.

BNEF (Bloomberg New Energy Finance). 2015. "Global Trends in Renewable Energy Investment 2015." Available at http://fs-unep-centre.org/publications/global-trends-renewable-energy-investment-2015

Breschi, S., and F. Malerba.1997. "Sectoral systems of innovation: Technological regimes, Schumpeterian dynamics and spatial boundaries." In *Systems of Innovation: Technologies, Institutions and Organizations*, edited by C. Edquist, 130–156. London: Pinter Publishers.

Carlsson, B. 1995. *Technological Systems and Economic Performance: The Case of Factory Automation*. Dordrecht: Kluwer.

Chidambaram, R. 2007. "Directed Basic Research." *Current Science* 92 (9): 1229–1233.

Cooke, P., M. Uranga, and G. Etxebarria. 1997. "Regional Innovation Systems: Institutional and Organisational Dimensions." *Research Policy* 26 (4): 475–491.

Deshmukh, R., G. Gambhir, and G. Sant. 2011. "India's Solar Mission: Procurements and Auctions." *Economic and Political Weekly* 46 (28): 22–25.

Edquist, C., ed. 1997. *Systems of Innovation: Technologies, Institutions and Organizations*. London: Pinter Publishers.

Edquist, C. 2001. "The systems of innovation approach and innovation policy: An account of the states of the art." DRUID Conference, theme on 'National systems of innovation, institutions and public policies' Aalborg, June 12–15, 2001, 1–24.

Edquist, C., and B. Johnson. 1997. "Institutions and Organizations in Systems of Innovation." In *Systems of Innovation: Technologies, Institutions and Organizations*, edited by C. Edquist, 41–60. London: Pinter Publishers.

Fagerberg, J., D. Mowery, and R. Nelson. 2005. *The Oxford Handbook of Innovation*. Oxford: Oxford University Press.

Freeman, C. 1987. *Technology Policy and Economic Performance: Lessons from Japan*. London: Pinter Publishers.

Freeman, C. 1995. "The 'National System of Innovation' in Historical Perspective." *Cambridge Journal of Economics* 19: 5–24.

IREDA (Indian Renewable Energy Development Agency). 2012. "Annual Report 2011-12, Delhi".

Kharul, R. 2011. "Grid-connected Photovoltaic." In *A Solar Future for India*, edited by G. M. Pillai, 118–140. Pune: World Institute of Sustainable Energy (WISE). ISBN 81-902925-2-8.

Krishna, V. V. 2007. "Dynamics in the Sectoral System of Innovation: Indian Experience in Software, Biotechnology and Pharmaceuticals." In *Science, Technology Policy and the Diffusion of Knowledge: Understanding the Dynamics of Innovation Systems in the Asia-Pacific*, edited by T. Turpin and V. V. Krishna, 193–233. Cheltenham: Edward Elgar.

Krishna, C., A. D. Sagar, and S. Spratt. 2015. "The Political Economy of Low Carbon Investments: Insights from the Wind and Solar Power Sectors in India." IDS University of Sussex. Working paper no. 104.

Lundvall, B. A. 1997. *"National Systems and National Styles of Innovation."* Paper presented at the fourth international ASEAT conference, "differences in "styles" of technological innovation", Manchester, UK, Sept. 1997.

Malerba, F. 2000. "Sectoral System of Innovation and Production." ESSY Working Paper No. 1.

Malerba, F. 2002a. "Sectoral System of Innovation and Technology Policy." Workshop on Frontiers of Innovation Research and Policy, Organised by Instituto de Economia/UFRJ and the Institute of Innovation at the University of Manchester, Rio de Janeiro, September 25–27, 2002.

Malerba, F. 2002b. "Sectoral Systems of Innovation and Production." *Research Policy* 31 (2): 247–264.

Malerba, F., ed. 2004. *Sectoral Systems of Innovation: Concepts, Issues and Analyses of six major Sectors in Europe*. Cambridge: Cambridge University Press.

Mallett, A., D. G. Ockwell, P. Pal, A. Kumar, Y. P. Abbi, R. Haum, G. MacKerron, J. Watson, and G. Sethi. 2009. "UK-India collaborative study on the transfer of low carbon technology:" Phase II final report. SPRU. Sussex University and Institute of Development Studies. Available at http:///www.sussex.ac.uk%2Fsussexenergygroup%2Fdocuments%2Fdecc-uk-india-carbon-technology-web.pdf

Metcalfe, S. 1995. "The economic foundations of technology policy: Equilibrium and evolutionary perspective." In *Handbook of the Economics of Innovation and Technological Change*, edited by P. Stoneman, 409–512. London: Blackwell Publishers.

Mitavachan, H., and J. Srinivasan, 2012. "Is land really a constraint for the utilization of solar energy in India?" *Current Science* 103 (2): 163–168.

MNRE (Ministry of New and Renewable Energy). 2010a. "Annual Report 2009-10, New Delhi".

MNRE (Ministry of New and Renewable Energy). 2010b. "Jawaharlal Nehru National Solar Mission, Towards Building Solar India, New Delhi".

MNRE (Ministry of New and Renewable Energy). 2011. "Strategic Plan for New and Renewable Energy Sector for the Period 2011-2017." p-9.

MNRE (Ministry of New and Renewable Energy). 2012. "Jawaharlal Nehru National Solar Mission Phase II-policy Document." See the link: http://mnre.gov.in/file-manager/UserFiles/draft-jnnsmpd-2.pdf.

MNRE (Ministry of New and Renewable Energy). 2013. "Annual Report 2012-13, New Delhi".

MNRE (Ministry of New and Renewable Energy). 2015. "Annual Report 2015-14, New Delhi".

MNRE (Ministry of New and Renewable Energy). 2016. "Annual report 2016-15, New Delhi".

MoP (Ministry of Power) CEA (Central Electricity Authority). 2016. "Load Generation Balance Report, 2016-2017." New Delhi.

Mowery, D. C., and R. R. Nelson, eds. 1999. *Sources of Industrial Leadership: Studies of Seven Industries*. Cambridge University Press: Cambridge.

Nelson, R. R., ed. 1993. *National Innovation Systems: A Comparative Study*. Oxford: Oxford University Press.

Norton, M. J., ed. 2001. *Introductory Concepts in Information Science*. New Jersey: Information Today, Inc.

Pillai, I., and R. Banerjee. 2009. "Renewable Energy in India: Status and Potential." *Energy* 34: 970–980.

Purohit, I., and P. Purohit. 2010. "Techno-Economic Evaluation of Concentrating Solar Power Generation in India." *Energy Policy* 38: 3015–3029.

Ramachandra, T.V., Rishabh Jain, and Gautham Krishnadas. 2011. "Hotspots of Solar Potential in India." *Renewable and Sustainable Energy Reviews* 15: 3178–3186.

REN21. 2015. "Renewable 2015 Global Status Report." Annual Reporting on Renewable; 10 years of Excellence, Paris. ISBN, 978-3-9815934-6-4.

Schmid, G. 2012. "The Development of Renewable Energy Power in India: Which Policies Have Been Effective?" *Energy Policy* 45: 317–326.

Schumpeter, J. A. 1939. *Business Cycles: A Theoretical, Historical, Statistical Analysis of the Capitalist Process*. New York: McGraw-Hill.

Sharma, A. 2011. "A Comprehensive Study of Solar Power in India and World." *Renewable and Sustainable Energy Reviews* 15: 1767–1776.

Spratt, S., W. Dong, C. Krishna, A. Sagar, and Q, Ye. 2014. "What drives wind and solar energy investment in India and china?" IDS University of Sussex. Working paper No. 87.

Sukhatme, P. 2011. "Meeting India's Needs of Electricity Through Renewable Energy Sources." *Current Science* 0.101 (5): 624–630.

TERI (The Energy and Research Institute). 2013. "TERI Energy Data Dictionary and Yearbook (TEDDY) 2012-2013 Report, New Delhi".

Turpin, T. and V. V. Krishna, eds. 2007. *Science, Technology Policy and the Diffusion of Knowledge: Understanding the Dynamics of Innovation Systems in the Asia-Pacific*. Cheltenham: Edward Elgar Publishers.

Willey, T., and S. Hester. 2009. "Solar electric technologies and applications." *Cogeneration and Completitive Power Journal* 18 (2): 37–47.

World Bank. 2013. "Paving the way for a transformational future: Lessons from JNNSM phase-1." Energy sector management assistance programme (ESMAP). Washington DC: World Bank.

Technological appropriability and export performance of Brazilian firms

Graziela Ferrero Zucoloto, Julio Raffo and Sergio Leão

This paper aims to evaluate the strategies of Brazilian manufacturing firms in their use of intellectual property (IP) and its impact on their export performance. Although the correlation between exports and innovative activities is already consolidated in the existing literature, this study contributes by analyzing the extent to which export performance of innovative firms is related to their different IP-related appropriation strategies. In order to determine this, we analyzed the export behaviour of innovative industrial firms, aiming to identify the relevance of each IP appropriation instrument, including invention patents, utility models, industrial designs and trademarks.

The paper presents an overview of previous findings about innovation, technological appropriability and export performance. It also discusses the relationship between innovation and exports in Brazil, showing that innovative Brazilian firms tend to export more than non-innovative ones, which corroborates the main literature findings. Using cross-section and panel data, the impact of technological appropriability on export performance of Brazilian innovative firms is evaluated.

Introduction

This paper aims to evaluate the relationship between technological appropriability and export performance of Brazilian industrial firms.

The correlation between exports and innovative activities is already consolidated in the international literature. Innovative firms tend to export more intensively than firms that do not innovate. In addition, both exporting and innovative firms are, in general, larger than firms that do not innovate, as well as more productive and more intensive in skilled labour.

Although the relationship between innovation and exports has been thoroughly explored in many studies, the role of appropriability is a controversial issue, especially in developing countries. On the one hand, if a firm appropriates the results of these innovations, its competitiveness may become even more significant. Since technological appropriations can boost market leadership and monopolistic advantage, they may improve their ability to compete abroad. On the other hand, Hall et al. (2012) show that innovative firms in many countries use no appropriability method at all. In Brazil, the low-tech sectors are a relevant part of industry, and the largest part of innovative expenditure is directed at the acquisition of machinery and equipment. These factors may contribute to the minimization of the relevance of appropriability methods.

Since Brazilian data also suggest a correlation between innovation and exports, this study raises the following question:

Is the better export performance of innovative firms related to technological appropriability?

In order to answer to this question, we analyze the behaviour of innovative industrial firms, aiming to identify the relevance of formal appropriability methods to the export performance of these firms.

However, technological appropriability is a broad concept, which includes different types of methods. In this paper, we focus our analysis on the formal methods, which include invention patents, utility models, industrial design and trademarks.[1]

According to the definition presented by the Brazilian Patent and Trademark Institute (INPI 2017):[2]

> Patent is a temporary title to an invention or utility model, granted by the State to inventors or other persons who have rights over the creation. Based on this right, the inventor or the patent holder has the right to prevent third parties, without their consent, from producing, using, selling or importing a product subject to their patent and/or a process or a product obtained directly by a process patented by him. In return, the inventor undertakes to disclose in detail all the technical content of the matter protected by the patent.

Patents and utility models have distinct requirements and are protected for different periods. According to the definition presented by INPI, patents of invention (IP) are products or processes that meet the requirements of inventive activity, novelty and industrial application, and utility model patents (UM) are objects of practical use, or part of it, susceptible to industrial application, presenting a new form or arrangement involving an inventive act that results in a functional improvement in its use or in its manufacture. Utility models are considered especially relevant for developing countries, whose innovative capabilities are less advanced. As argued by Kim et al. (2012), in developing countries, innovations are adaptive and imitative, and can therefore be protected especially through utility models. Such innovations may not have the 'inventive step' to generate an invention patent, but they may be sufficient to generate a utility model, allowing their firms to gain knowledge and protect it. The authors also point out that the deposit of utility models is generally less expensive, as it does not require substantive examination.

As the required inventive step of utility models is smaller than that of invention patents, they tend to be suitable for meeting the local needs of these countries, acting as a stage for further complex technological developments.

The industrial design constitutes the ornamental or aesthetic aspect of an article. It may consist of three-dimensional features, such as the shape of an article, or two-dimensional features, such as patterns, lines or colour. Industrial designs are applied to a wide variety of industrial and handicraft products and items, from packages and containers to furnishing and household goods, from lighting equipment to jewellery, and from electronic devices to textiles. Industrial designs may also be relevant to graphic symbols, graphical user interfaces and logos. In Brazil, industrial design is protected by registration and not by patent as in other countries. Brazilian law provides protection for up to 20 objects per application since they are variations of the same object or others that compose a set with the same preponderant distinctive features, that is, part of the same 'family', maintaining a visual identity. The application is only an expectation of rights (but not acquired rights). Once granted by the State, the industrial design registration is valid in the country and gives the holder the right, during the term, to prevent third parties from manufacturing, marketing, importing, using or selling the protected material without permission. The duration of this protection is for ten years from the date of filing, renewable for three successive periods of five years (WIPO 2016; INPI 2016).

A trademark is a sign capable of distinguishing the goods or services of one enterprise from those of other enterprises. The possibilities are almost limitless and may consist of drawings, symbols, three-dimensional features such as the shape and packaging of goods, non-visible signs such as sounds or fragrances, or shades of colour used as distinguishing features. The registration of a trademark confers an exclusive right to the use of the registered trademark, which implies that the trademark can be exclusively used by its owner, or licensed to another party for use in return for payment. In Brazil, trademark registration is for ten years but can be renewed indefinitely (WIPO 2016; INPI 2016).

These definitions show that the appropriability methods protect different types of inventions. So, their impact on economic indicators is not the same. Their relevance also changes according to the socioeconomic profile of the countries. In developed countries, invention patents are more relevant, while in developing nations, utility models and industrial design are used more often. In Brazil, invention patents are especially applied for foreign companies, while utility models and industrial design are proportionally more relevant to national companies (Zucoloto 2010). Thus, it is more important to evaluate the different impacts of each appropriability method than to analyze the general impact of using IP protection.

To understand the relationship between appropriability methods and export performance, we used different databases. The main firm characteristics, innovative expenditures and appropriability information are provided by the Technological Innovation Survey, applied by the Brazilian Statistic Institute (PINTEC/IBGE). Export data are consolidated by the Foreign Trade Secretariat (SECEX)/ Ministry of Development, Industry and Foreign Trade. The study also includes firm age, taken from the Ministry of Labour and Employment (RAIS database).

Regarding methodology, we used fixed effect models to deal with endogeneity problems. As innovation and exports traditionally present an endogenous relationship, it is hard to determine whether export performance boosts patenting or if patent use stimulates exports. Besides, omitted variables, which were not included as variables of control, can impact both on exports and on patenting. Also, lag control variables were applied in the models. The analysis is concentrated in large firms, as this is the only group of firms that is present in all Pintec editions.

The rest of this article is organized as follows. In the next section, an overview of previous findings about innovation, technological appropriability and firm competitiveness, measured by exports, is provided. In the section following that, we present the data and some descriptive analysis of innovation, appropriability and exports of Brazilian firms. In the penultimate section, the impact of technological appropriability on export performance of Brazilian innovative firms is discussed, using pooled and fixed effect models. The final section presents the main conclusions.

Existing evidence on innovation and exports

In his seminal study, Posner (1961) found that when firms develop a new product they create a monopoly in its country of origin, until the entry of imitators into the market. The author suggests that the technical change created in one country induces its trade until the rest of the world imitates its innovation. Posner's work allowed the development of a number of concepts which became the basis for the theory of technology gaps. He assumed that technology is not a free good which can be freely acquired and reproduced without any cost to firms. Therefore, there are substantial advantages to being the first one to innovate. Following this line of thought, Vernon (1966) argued that the competitive advantages of American firms were linked to their innovative capacity in terms of products and processes. Similarly, Freeman et al. (1963), when studying the plastic industry, concluded that technical progress leads to productive leadership. When the innovative product starts being imitated, it is more likely that the traditional production factors – which are more cost related – will determine the trade flows. He emphasized that the technology gap between innovating and imitating countries may last long, but he also stressed the importance of patents and trade secrets for postponing the process of technological diffusion and guaranteeing monopoly profits.

After them, several empirical studies have attempted to explain sector productivity according to the model of technology gaps. For instance, Soete (1987) observed whether sector exports were determined by technological performance – measured by patents – in a sample of 22 OECD countries. The results indicated the crucial role of the technology variable in explaining variations in export performance in 28 of 40 industries. Dosi, Pavitt, and Soete (1990) extended this analysis in a dynamic version of technology gaps model at aggregate level. Among other results, these authors found that technological asymmetries are a main determinant of trade flows. Interestingly, the authors also measured innovation using patenting activity, which was emphasized not to be an entirely appropriate indicator to represent the process of technological innovation, as many innovations may not be patentable. Other examples of these empirical studies are Amable and Verspagen (1995), Amendola, Guerrieri, and Padoan (1998), Breschi and Helg (1996), Laursen (1999), Laursen and Drejer (1999), Laursen and Meliciani (2002), Montobbio (2003), and Andersson and Ejermo (2008), among many others. Most of these studies highlighted the relevance of technological progress to explain trade patterns.

Calvo (2003) estimated the influence of firms' innovation activities on export performance using a sample of Spanish manufacturing firms, in 2000. He found that size, age and innovation activities affect the decision to export, but export propensity was independent of both firm size and innovative behaviour. At the same time, the presence of foreign capital positively influenced both decisions. Focusing on small firms, Nassimbeni (2001) presented the results of an empirical study conducted on a sample consisting of 165 small manufacturing companies in the furniture, mechanics and electro-electronics sectors. The aim of the study was to point out which technological and innovative capacity-related factors mostly differentiate export from non-export small enterprises. He was motivated by the fact that, in the case of small businesses, many studies had failed to produce consistent results when examining the relationship between technology, innovation capacity and export performance. The author concluded that technology, and, more generally, process innovations play a secondary role compared with product innovation.

From a dynamic point of view, the technological and commercial performances interact, since, to remain competitive, firms are encouraged to adopt efficient processes and to invest in innovation. In this sense, participation in foreign trade would not only result from innovation, but also boost technological improvements, in a virtuous circle. As innovation and exports may be strongly correlated, some studies go further to identify whether there is some causal relation between them, or if both activities are boosted by external variables.[3] Bernard and Bradford Jensen (1999) asked whether good firms become exporters or whether exporting improves firm performance. For the authors, the evidence is quite clear on one point: good firms become exporters, since both growth rates and levels of success measures are higher ex-ante for exporters. However, the benefits of exporting for the firm are less clear. Being aware of this possible reverse causality, Lachenmaier and Woßmann (2006) empirically tested whether innovation fosters exports in German manufacturing firms. Their empirical strategy identified variation in innovative activity that occurs because of specific impulses and obstacles for innovative activity, which were treated as exogenous to firms' export performance. Using the innovation impulses and obstacles as instrumental variables, they found that innovation emanating from these variations

leads to a share of exports in a firm's total revenue that is roughly seven percentage points higher on average. Therefore, their results support the hypothesis that innovation is a driving force for exports in industrialized countries. The effect is heterogeneous across sectors, being hardly detectable in relatively traditional sectors. Also, Damijan, Kostevc, and Polanec (2010) investigated the bidirectional causal relationship between firm innovation and export activity in Slovenian firms between 1996 and 2002. They found no evidence for the hypothesis that either product or process innovations increase the probability of a firm becoming a first-time exporter, although they found evidence of a causal link in the case of process innovation of medium and large firms. However, no such link was found among small firms.

The existing literature has also focused on the association between firm productivity and export performance. Empirical studies show that one of the most important sources of productivity heterogeneity at firm level is related to R&D and innovative activities. Cassiman, Golovko, and Martínez-Ros (2010) argue that the positive association found between productivity and exports in the literature is related to the firm's innovative decisions. Using a panel of Spanish manufacturing firms, they found strong evidence that product innovation affects productivity and induces small non-export firms to enter the export market. R&D and innovative activities seem to play an important role in explaining a firm's decision to export and export volumes. Damijan, Kostevc, and Polanec (2010), mentioned above, also explored the links between productivity and export, finding both related to firm innovation activities. Using plant-level data for the Taiwanese electronics industry, Aw, Roberts, and Xu (2011) estimated a dynamic structural model that captures both the behavioural and technological linkages among R&D, export and productivity. Among its conclusions, the report shows that the marginal benefits of both export and R&D increase with the plant's productivity, and that high-productivity plants derive particularly large benefits from exports. Also, Clerides, Lach, and Tybout (1998) analyzed the causal links between export and productivity using plant-level data, and identified that export firms are more efficient, although they did not find a positive impact of export on productivity in Colombia or Morocco. The authors also observed a positive association between export and efficiency, which is explained by the self-selection of the more efficient firms into the export market.

Relevance of technological appropriability

More recently, the literature has sought to deepen the understanding about technological innovation, through the analysis of the importance of knowledge technological appropriation in economic and trade performance. The appropriability of innovation is a concern for inventors since one of the outputs of inventive activity is often knowledge, which is difficult to exclude others from using it due to its intangibility (Arrow 1962).

According to Hanel (2008), the successful completion of an innovation process alone is not a sufficient condition for obtaining the expected benefits from innovation. A firm also must be able to appropriate these benefits, i.e. to prevent its competitors from imitating them, which can be achieved through IP rights or other strategies, such as secrecy or lead time. Other authors (Teece 1986; Levin et al. 1987; Cohen, Nelson, and Walsh 2000) have also argued that the benefits of product innovations depend on the ability of firms to use appropriation methods.

Several studies have focused on the impact of appropriability on firm performance. According to Hall et al. (2012), the main performance variables used in these studies are profits (Hanel 2008), percentage of sales of new products, productivity and market value. Also, Amendola et al. (1993) and Laursen and Meliciani (2000, 2002) showed that technological factors, measured by patent indicators, appear to be the main determinant of a country's export performance in the long run, while non-technological factors (labour costs and lagged export performance) are only significant in the short run. Dosso (2011) investigated, from an empirical perspective, the relative importance of technological vis-à-vis non-technological determinants of the dynamics of international productivity in manufacturing industries over the period of 1980–2005. He found that patent shares have a positive and significant impact on relative export performance in the long run. The adoption of technology also presented in some cases a positive and significant effect.

Appropriation methods differs among sectors and technological specificities, and also depends on the strategic behaviour of firms. In general, large, R&D-performing firms as well as multinational ones prefer patents. As mentioned in Hall et al. (2012), most studies have found that the use of patents is primarily associated with product rather than process innovations. Arundel and Kabla (1998) pointed out that the importance of patents increases with the relevance of global markets. They argue that patents are more important for firms exporting to the US or Japan. They also observed that patents play an important role in the ability of firms to enter foreign markets.

Levin et al. (1987) carried out a seminal study in this area, which was later updated by Cohen, Nelson, and Walsh (2000). Both analyzed the extent to which firms in different industries choose IP and other methods to secure returns from their innovations. These studies showed that, on average, patents are not the most frequent mechanism of appropriation. Instead, secrecy and lead time advantages are the most frequently used strategies. However, this does not apply equally across all industries or innovation types, among other characteristics. In general, product innovators use more often patents than process innovators. Similarly, some specific industries – such as the pharmaceutical and the chemical ones – do use patents more often to secure their returns on technological investments.

Furthermore, as discussed by Graham and Somaya (2006), IP rights and other protection methods are often complementary rather than substitutes. In most empirical studies, it is difficult to determine which appropriation strategy – or which IP instrument – is protecting each innovation outcome. Different protection methods can be used at different stages of the innovative process. For example, secrecy may be applied in the early stages of

the innovative process, whereas patents are likely to be used to protect the innovation when it is close to commercialization (Basberg 1987). After the invention has entered into the market, however, patents and secrecy are mutually exclusive because of the patent disclosure requirement. In this sense, Hall et al. (2012) argue about what determines a firm's decision to choose between patents and trade secrets. A fundamental question raised by these authors is why an innovative firm able to use patent protection would choose not to. On the one hand, applying for patent protection requires direct and indirect financial expenses even before any certainty of grant. And, when granted, there is a considerable financial burden relating to maintenance fees – which is to be multiplied by every protected jurisdiction – in order to keep the patent in force. Moreover, patents are only valuable if enforceable, which can also be substantially costly not only because of the expenses related to legal action, but also because it requires active monitoring of potential infringements. Besides, a patent also requires full disclosure of information in its application, which may be useful to competitors.

Indeed, patent costs have been suggested as one of the main reasons why firms avoid patent applications. Similarly, financially constrained firms tend to prefer other appropriation methods than IP rights. Therefore, the benefits that arise from excluding competitors and licensing patents must offset these costs. Moreover, these benefits must be compared with other available alternatives. For example, in contrast to patents, which last 20 years, trade secrecy can potentially protect the invention indefinitely. Secrecy is also applicable to a much broader range of inventions than patents, as there is no restriction like that of patentable subject matter. However, secrecy also does have costs, including confidentiality agreements.

In a similar direction, Llerena and Millot (2013) assessed the interrelated effects of patents and trademarks. Based on a data set encompassing the IP activities of French firms, they found that patents and trademarks are complementary in life science sectors (pharmaceutical products and health services), but substitutes in high-tech business sectors (computer, electronic and optical products and electrical equipment). The temporal substitutability effect occurs while the patent is in force, reducing market competition, and the firm's need to use trademarks to protect its reputation; while the complementarity effect is present as the trademark enables the firm to extend the reputational benefits of the monopoly period beyond the expiration of the patent. Following the conclusion of Teece (1986) that the profit gained from innovation depends on the possibility of the firm to use complementary assets, their model states that the relationship between the various assets is itself dependent on the context in which the firms operate.

Academic and policy debates have largely focused on the effects of IPRs, but few studies consider the different impacts according to the countries' degree of development. Lee et al. (2012) and Kim et al. (2012) are central studies to understand the relevance of different types of IP to economic development. An important lesson from their results is that what matters to innovation and growth is not so much the strength of IP rights, but the type of protection used.

An accurate comparison of invention patent and the utility model was made by Kim et al. (2012). Based on econometric analysis of Korean firms' microdata, they found that when companies were technologically advanced, patentable innovations had positive and significant impacts on business growth, while the use of utility models was not significant. The opposite occurred in the case of technologically less advanced firms: innovations protected by utility models were relevant for firm growth, while patentable innovations had no impact. According to the authors, the main lesson of their article was that what is important for innovation and growth is not only the strength of intellectual property rights, but also the type of protection used:

> For example, the availability of legal protection for minor, adaptive inventions should be most useful to firms with low technological capacities and limited resources. In developing markets, patents raise the cost of doing business and innovation. This cost tends to be more onerous for lower income economies. In contrast, a utility model system provides an alternative way for such economies to create incentives for innovation, albeit incremental, without affecting the cost of doing business adversely, and while providing the technological inputs appropriate for local needs. (Kim et al. 2012, 374)

Besides utility models, the authors also stressed the importance in developing countries of promoting the utilization of other forms of IPRs such as trademarks, copyrights and industrial designs (Kim et al. 2012; Lee et al. 2012).

Innovation and exports in the Brazilian context

One limitation of the above-mentioned strands of literature is that most of these studies have focused on developed countries (Avellar and Carvalho 2013). There is however evidence about the Brazilian case for some of these which are worth noting.

De Negri (2005) examined the relationship between technological standards and foreign trade of Brazilian firms, concluding that technology is an important factor in their export performance, considering both their inclusion in the international market and the expansion of their export volumes. Raffo et al. (2011) empirically tested the link between product and process innovation and export performance using a sample of industrial firms from Argentina, Brazil, Mexico, France, Spain and Switzerland. Similarly, Avellar and Carvalho (2013) tested the relationship between innovative efforts – measured as new products, R&D expenditures or a cooperation index – and export performance using a sample of industrial firms from Brazil, India and China. In both cases, the innovation – either effort or output – increased the export behaviour of firms. Conversely, Gonçalves, Lemos, and de Negri (2007) assessed the impact of exports on innovation in Brazil and Argentina, observing a positive impact of trade integration on both countries' propensity to innovate. Mais and Amal (2011) evaluated the relevance of institutional factors and innovation on the export performance of firms, based on a multicase study of five firms in the metalworking industry in the state of Santa Catarina (southern Brazil). They concluded that in countries with intermediate levels of technological development, such as Brazil, the effects of the institutional framework on export performance involve mainly tax benefits and financial incentives. Regarding intellectual property, although firms used to register them, they see no commercial advantage in doing so.

Regarding IP specifically, Luna and Baessa (2007) analyzed the impact of patents and trademarks on the economic performance of firms. Their results are not unequivocal, as they observe a positive relation of both trademarks and patents with labour productivity, but these are not robust across different estimation strategies.

Data

The data used in this report were consolidated from three different statistical sources. The innovation indicators were sourced from the Brazilian Innovation Survey (PINTEC), carried by the Brazilian Institute of Geography and Statistics. Export related information was sourced from the data compiled by the Foreign Trade Secretariat (SECEX), under the Industry and Trade Ministry. Some additional information was sourced from the RAIS database of the Ministry of Labour.

PINTEC includes firms' indicators, such as origin of capital and number of employees (proxy for firm size), which are available for all surveyed firms.[4] But only those firms which had engaged in innovative activities during the surveyed period answered the questions regarding innovative expenditure and the use of appropriability methods, among others. This group, which we call 'innovative firms', includes not only those which introduced product or process innovations but also those with incomplete or abandoned innovation projects during each surveyed three-year period.

The survey classifies the firms' origin of capital as national, foreign or mixed, according to the following definitions:
- National: the firm is under direct or indirect ownership of individuals or legal entities resident and domiciled in the country;
- Foreign: the firm is under direct or indirect ownership of individuals or legal entities resident and domiciled abroad;
- National and foreign (mixed): both domestic and foreign ownership have similar shareholdings.

Regarding innovative expenditures, PINTEC include:
- R&D expenditures;
- External acquisition of R&D;
- Acquisition of other external knowledge;
- Machinery and equipment acquisitions to innovate;
- Training;
- Introduction of technological innovations in the market;
- Production and distribution of innovations.

PINTEC also contains basic information about appropriation methods used in order to protect innovation outcomes. Following the standard innovation survey structure, it distinguishes between IP-related methods of appropriation – invention patents, utility models, industrial design and trademarks – from non-IP related ones.[5] In concrete terms, the questions can be summarized as: 'Did the firm use

Table 1: Export performance of innovative and non-innovative firms.*

Exchange rate = 2,33, em 31.12.2008			tx cambio media = 1.835 (2008)		
	Number of firms		Exporting	Exports: Average	Exports:
2008	Non-exporting	Exporting	firms/Total (%)	values per firm (US$)	Firm/Sector (%)
Non-innovative	56,422	5,020	8.2	3,51,047	0.12
Innovative	32,744	5,617	14.6	33,10,078	0.43

*Average exchange rate (2008): 1.835
Source: IBGE/PINTEC 2008 and MDIC/Sexec 2008 (elaborated by the authors)

any of the methods ... [invention patents/utility models/ industrial designs/trademarks] ... to protect product and/ or process innovations? 'Yes/No'.

We make use of three editions of the PINTEC, covering the periods: 2001–2003, 2003–2005 and 2006–2008, for which the appropriability variables are available. Some of the variables refer to the whole period – e.g. instance product or process innovation, while others correspond to the period's last year – e.g. variables in nominal value.

We also limit the coverage exclusively to manufacturing industries, as the services' sectors have been only included since 2005 and their coverage is limited to a few sectors

One singularity of the Brazilian sampling method is that only industrial firms with 500 employees or more, which we call large firms, were included in the survey (census), while firms with less than 500 employees were sampling. In practice, this implies that we applied sample weights when using the whole survey.

In addition, from the export database, tree variables were designed and built:

• Dummy for export firms;
• Exported value per firm;
• Firm share on sectorial exports (value of exports_firm i/ value of exports_sector of firm i)

RAIS database complements the firms' characteristics with the following information: sector and firm age.

An overview of innovation, IP and export performance in Brazil

This section presents the general relationship between innovation, IP and exports in Brazil. The first part presents the correlation between innovation and export performance. It then discusses the relevance of IP methods to

innovative firms. Finally, based on econometric analysis, the section details the relationship between IP and the export performance of large innovative firms.

As shown in the literature, there is a positive correlation between export performance and innovative activities. Innovative firms tend to be more intensive in exports compared with firms that do not innovate. Added to this fact, both exporting and innovative firms are, in general, larger, more productive and more intensive in skilled labour.

In Brazil, the results confirm the literature findings. Innovative Brazilian firms are more likely to export and they do it in a greater extent (Table 1). Indeed, 14.6% of innovative firms are exporters, while only 8.2% of non-innovative firms are. On average, the export sales of innovative firms (more than US$3.3 million) represent almost ten times those observed among non-innovative ones (US$0.35 million). Innovative firms also have higher participation in sectorial exports: on average, 0.43% compared to 0.12% in the case of non-innovative firms.

Jointly, these numbers suggest a correlation between export and innovation variables, as innovative firms present a better performance among Brazilian exporters.

Even using some control variables, this correlation is observed in a simple econometric exercise (Table 2):

$$Y = f(\text{innovative firm; number of employees; foreign; mixed, sector}) + u_t$$

Dependent variables (Y):
• dummy_export: dummy = 1, if firm exported; otherwise, dummy = 0
• Log (exp): log of value of exports
• Firm share: share of firms' exports per sector (3 digits)

Table 2: Innovation and export performance.

Dependent variable	Dummy export	Log (export)	Export share
	Logit	OLS	OLS
	(1)	(2)	(3)
Innovative firm	0.0224***	0.293***	0.00109***
	(0.00664)	(0.0846)	(0.000230)
Log (number of employees)	0.0715***	1.532***	0.00582***
	(0.00180)	(0.0475)	(0.000309)
Foreign	0.164***	7.153***	0.0368***
	(0.0154)	(0.446)	(0.00337)
Mixed	0.201***	6.668***	0.0155***
	(0.0380)	(1.212)	(0.00399)
Observations	13,945	13,945	13,841
R-squared		0.310	0.101

Explanatory variable:
• innovative firm: dummy = 1, if firm has innovated among 2006 and 2008

Control variables:
• log of number of employees, a proxy for firm size;
• foreign: dummy = 1 if the firm is foreign; otherwise, dummy = 0
• mixed: dummy = 1 if the firm has foreign and national capital; otherwise, dummy = 0
• sector: dummies for sectorial controls

As showed in the descriptive analysis, the results indicate a positive and significant correlation between innovation and (a) propensity to export, (b) exported value and (c) firm share of sectoral exports. Additionally, we observe a positive correlation between firm size and foreign origin of capital, which suggests that foreign or mixed firms present a higher propensity to export than national ones.

According to column (1), an innovative firm has 2.2 percentage points more chance to export than a non-innovative firm. Column (2) also indicates that the average value exported is significantly higher (29% higher) for an innovative firm. Finally, column (3) points out that an innovative firm has a larger share in the value exported by its sector than a non-innovative one.

Appropriability and export performance of innovative firms

Since Brazilian data suggest a correlation between innovation and exports, this study raises the following question:

Is the better export performance of innovative firms related to technological appropriability?

In order to seek an answer to this question, we analyzed the behaviour of innovative industrial firms, aiming to identify the relevance of appropriability methods to the export performance of firms. In other words, the relevance of technological appropriability to export performance of innovative industrial firms was investigated.

The figures presented in Tables 3–6 refer exclusively to innovative Brazilian industrial firms between 2006 and 2008. Tables 3 and 4 present the export profile by firm size and origin of capital. Most foreign controlled firms – either fully or mixed – are exporters (Table 3). Similarly, export firms are larger than non-exporters, considering their number of employees or net sales. On average, they have 10 times more employees and 40 times more sales than non-exporters Moreover, the average export firm is also more knowledge intensive. They have 30 times more skilled labour exclusively associated with R&D activities (Table 4) and they expend more in innovation-related activities (Table 5). This is particularly true in the case of R&D expenditures – either internal or external – for which export firms not only expend in excess of 100 times more than non-export ones but they also do so three times more intensively. This is not the case in the acquisition of external knowledge, for which export firms expend quantitatively more than non-export ones, but almost the same in relative terms. Interestingly, we observed that non-export firms have higher intensities in the other innovation-related

Table 3: Export firms by origin of capital.

2008	Innovative industrial firms		
	National	Mixed	Foreign
Non-export firms	32,447	55	242
Exporting firms	4,552	254	811
% Exporting firms	12.3%	82.2%	77.0%

Source: IBGE/PINTEC 2008 and MDIC/SECEX 2008 (elaborated by the authors)

Table 4: Firm size and exports.

Innovative industrial firms	Average values		
	Number of employees	Highly-skilled employees	Net sales
Non-export firms	42.9	0.02	2,710.2
Exporting firms	476.2	0.63	1,11,180.5

Source: IBGE/PINTEC 2008 and MDIC/SECEX 2008 (elaborated by the authors)

Table 5: Innovative expenditures and export propensity.

Innovative industrial firms	Average values (1000 US$)		Innovative expenditures/Net sales	
	Non-export firms	Exporting firms	Non-export firms	Exporting firms
R&D expenditures	8.27	986.94	0.31%	0.89%
External acquisition of R&D	1.22	163.02	0.04%	0.15%
Acquisition of other external knowledge	2.26	102.82	0.08%	0.09%
Acquisition of machinery and equipment	108.19	1439.19	3.99%	1.29%
Training	6.17	53.88	0.23%	0.05%
Introduction of technological innovations in the market	6.18	208.11	0.23%	0.19%
Other preparations for production and distribution	17.93	278.60	0.66%	0.25%

Source: IBGE/PINTEC 2008 and MDIC/SECEX 2008 (elaborated by the authors)

Table 6: Percentage of firms that uses technological appropriability.

Percentage of innovative firms that used methods of technological appropriability			
	Number of firms	Non-exporting firms	Exporting firms
Invention patent	1,936	2.9%	17.7%
Utility model	1,167	2.1%	8.3%
Industrial design	1,637	3.6%	8.1%
Trademarks	9,205	21.2%	40.5%

Source: IBGE/PINTEC 2008 and MDIC/SECEX 2008 (elaborated by the authors)

activities. This is particularly true for acquisition of machinery and equipment, which represents on average the largest innovation expense for both exporters (1.3%) and non-exporters (4%).

Not surprisingly, there are higher shares of export firms making use of each appropriation method as an effective means to protect their innovation (Table 6). The interesting result here is that this is the case regardless of the type of appropriation strategy. In proportional terms, we observe the highest differences for invention patents.

In short, descriptive statistics show that export firms are generally larger (number of employees and net sales) than non-export firms, are more intensive in innovative activities and use more technological appropriability instruments. However, it is not possible to determine a causal relationship between the mentioned variables. Therefore, the next section presents an econometric analysis to provide a better understanding of the relationship between export performance and IP, controlled by the remaining indicators.

Regression analysis: IP and exports of Brazilian firms

The results from the previous descriptive analysis indicate a link between innovation and different measures of export activities. Nevertheless, we observed in the same analysis that export firms also relate to other firm characteristics such as origin of capital and size. We also observed that sector heterogeneity plays some role in these metrics and it needs to be considered more thoroughly.

In order to account for these issues, we analyze the relationship between IP-related appropriation and export variables.

Methodology

The models presented in this section search for a causality relationship between technological appropriability and export variables. As the literature has shown, there are significant differences in the use of appropriation methods among industries, which may impact differently on exports (among other economic variables). Besides technological appropriation, it is also important to control the relationship between IP and exports through innovative expenditures, such as in R&D, technology embedded in equipment or other innovative activities. In addition to innovative activities, there are other firm characteristics related to export performance, such as firm size, origin of capital and sectors.

As in the descriptive section, we analyze the export performance of Brazilian firms through three different dependent variables: (i) export firm (yes/no), (ii) value of exports (in logs) and (iii) the firm's share of the sector exports (calculated at the three-digit level of ISIC). The main explanatory variables of interest are four dummy variables that capture whether the firm has used (i) invention patents, (ii) utility models, (iii) industrial designs or (iv) trademarks to protect its innovation. In addition, we include control variables for sector (ISIC, two-digit level), origin of capital, size and innovative expenditures.

As briefly mentioned in the data section, we forced temporal lags in most independent variables as a way out of any simultaneity bias arising in our main variables of interest.[6] This lag was not applied to the appropriation variables, as they already refer to a previous three-year period. Also, the survey includes methods that are already in use (e.g. granted intellectual property rights), so their benefits should have been incorporated by the firms. The inclusion of temporal lags requires firms to be present in at least two editions of the PINTEC survey, a condition that was met only by those firms with 500 employees or more. Thus, the results relate exclusively to the performance of large Brazilian firms.

As part of the econometric analysis, we merged these two cross-sections into a two-period unbalanced panel, which includes only those firms with 500 or more employees. Therefore, the database contains large firms present in both PINTEC 2003 and 2005 as well as firms present in both editions of 2005 and 2008.

The model can be briefly described as:

$$Y_{it} = f(\text{variables of interest}_{it}; \text{variables of control}_{i,t-1;\alpha_i}) + u_{it},$$

where t: 2008/2005; t-1: 2005/2003; α_i: fixed effect. More detailed, the dependent variables (Y_{it}) are:
- Dummy export: propensity to export (dummy = 1, if firm exported; otherwise, dummy = 0);
- Log (exports): log of value of exports
- Firm share on sectorial exports: share of firms' exports per sector (ISIC three-digit)

The variables of interest are:
- Invention patent
- Utility model
- Industrial design
- Trademark

And the variables of control are:
- Origin of capital:
  - foreign: dummy = 1 if firm is foreign; otherwise, dummy = 0

- mixed: dummy = 1 if firm is 'national and foreign'; otherwise, dummy = 0
- Firm size: logarithm of number of employees
- Innovative expenditures: logarithm of the following innovative expenditures: R&D; acquisition of technology; machinery and equipment and others (training, introduction of innovation on the market and preparation for the production and distribution of innovation)
- Sectorial controls (ISIC two-digits)
- Firm age.[7]

As the literature on innovation has shown, there are significant differences in the use of appropriability methods among industries and firms. As technological appropriability can increase the competitiveness of firms, a positive relation between technological appropriability and export variables is expected to be found.

Innovation and exports traditionally present an endogenous relationship, which means it is hard to determine whether export performance boosts patenting or if patent use stimulates exports. In fact, omitted variables, such as firm productivity, which are not included as variables of control, can impact both on exports and on patenting. The fixed effect model assumes that the causal effect of patenting on exports is measured by the association between individual changes in exports and individual movements related to patenting. In this case, an individual's propensity to use a patent or the export sales may be endogenous, but the unobserved component of the effect of this propensity on exports is constant over time.[8]

Descriptive statistics
Table 7 shows the summary statistics of the export variables and controls for the database used in the analysis. The mean of the variable dummy export reports the proportion of export firms in the sample of firms with 500 or more employees. For this group of large firms, the proportion of export firms is much higher than in the whole sample: 87% against 15%.

Next, for each export variable, we compare the averages by groups of large firms according to the use of each appropriability method: invention patent, utility model, industrial design and trademark. According to Table 8, the group of firms using any type of appropriability method has a better export performance. In 2008, only for the measure of firm share on export, the group of firms that used trademarks is not statistically different from the groups that did not use the method. In 2005, non-significance was found for the same export variable in the case of industrial design and utility model.

Results
We first investigated whether the use of each described formal method of appropriability is associated with a higher propensity to export. Table 9 presents the results of a logit regression where the outcome variable indicates if a firm exports (y = 1) or not (y = 0) and the explanatory variables are dummies for each appropriability method and other controls, depending of the specification. In column (1), where no controls are included, the results indicate that a firm using an invention patent has 13% more chance to export when compared with a firm that does not use any of the appropriability methods. This specification also shows that a firm using a utility method or an industrial design increases its export chances by about 6%. As we add controls, we observe that the impact of invention patent decreases to 8–9% but remains statistically and economically significant, while the impact of the utility model and industrial design loses its significance.

In column (2) the controls added intend to capture characteristics of firms as size, origin of capital and age, in addition to sectorial controls, in order to distinguish large and mature firms, since those features are associated with higher export chances. The estimated coefficients for all the controls show that size and the foreign origin of capital are associated with a higher propensity to export. Adding firm controls significantly increase the explanatory power of the regression according to the pseudo R-squared statistic.

Table 7: Summary statistics – sample of large firms.

Variable	Mean	Std. dev.	p5	p95	p50	N
2008/2005						
Log (exported value)	14.637	6.193	0.000	20.681	16.755	827
Dummy export	0.869	0.337	0.000	1.000	1.000	827
Firm share on sectorial exports	0.239	0.000	0.000	1.000	0.078	827
Log (R&D expenditures)	4.905	3.651	0.000	9.938	5.858	827
Log (technology transfer expenditures)	5.653	3.641	0.000	10.352	6.771	827
Log (machinery and equipment expenditures)	2.767	3.201	0.000	8.455	0.000	827
Log (other innovative expenditures)	5.098	3.409	0.000	10.065	5.730	827
2005/2003						
Log (exported value)	14.549	6.168	0.000	20.442	16.718	812
Dummy export	0.868	0.338	0.000	1.000	1.000	812
Firm share on sectorial exports	0.232	0.310	0.000	1.000	0.076	812
Log (R&D expenditures)	5.197	3.237	0.000	9.599	5.893	812
Log (technology transfer expenditures)	6.218	3.341	0.000	10.597	7.090	812
Log (machinery and equipment expenditures)	1.889	2.920	0.000	7.824	0.000	812
Log (other innovative expenditures)	5.274	3.152	0.000	9.649	5.859	812

Table 8: Averages of appropriability method by group of large firms.

		2008			2005		
Panel A: Invention patent							
Variable name		IP = 0	IP = 1	p-value	IP = 0	IP = 1	p-value
Log (exports)	mean	13.689	16.519	0.000	13.729	16.660	0.000
	std error	0.286	0.270		0.273	0.277	
Dummy export	mean	0.827	0.953	0.000	0.832	0.960	0.000
	std error	0.016	0.013		0.015	0.013	
Firm share on sectorial exports	mean	0.211	0.293	0.000	0.201	0.314	0.000
	std error	0.014	0.020		0.012	0.021	
Panel B: Utility model							
Variable name		UM = 0	UM = 1	p-value	UM = 0	UM = 1	p-value
Log (exports)	mean	14.211	16.659	0.000	14.211	15.823	0.001
	std error	0.248	0.331		0.257	0.343	
Dummy export	mean	0.849	0.965	0.000	0.847	0.947	0.000
	std error	0.014	0.015		0.014	0.017	
Firm share on sectorial exports	mean	0.231	0.272	0.083	0.226	0.255	0.147
	std error	0.012	0.026		0.012	0.023	
Panel C: Industrial design							
Variable name		ID = 0	ID = 1	p-value	ID = 0	ID = 1	p-value
Log (exports)	mean	14.339	16.027	0.001	14.384	15.920	0.014
	std error	0.245	0.405		0.237	0.413	
Dummy export	mean	0.855	0.938	0.003	0.857	0.966	0.002
	std error	0.014	0.020		0.013	0.020	
Firm share on sectorial exports	mean	0.226	0.296	0.009	0.230	0.252	0.261
	std error	0.012	0.028		0.011	0.034	
Panel D: Trademark							
Variable name		TM = 0	TM = 1	p-value	TM = 0	TM = 1	p-value
Log (exports)	mean	14.112	15.184	0.006	13.894	15.230	0.001
	std error	0.316	0.289		0.331	0.272	
Dummy export	mean	0.846	0.894	0.021	0.829	0.910	0.000
	std error	0.018	0.015		0.019	0.014	
Firm share on sectorial exports	mean	0.232	0.246	0.272	0.205	0.261	0.006
	std error	0.016	0.016		0.015	0.016	

In column (3), we isolate the impact of the use of an appropriability method from the innovation expenditures. The main argument is that because we are interested in assessing whether the use of invention patents increases a firm's propensity to export, we need to control for previous investment that is both positively correlated with the chance to use an appropriability method and that may also make the firm more productive and therefore increase its propensity to export. Results indicate that reported expenditures in R&D and in other innovative actives have a positive and statistically significant impact on the propensity to export. Again, we observe an increase of pseudo R-squared by adding firms' expenses as controls.

In Table 10, we extended the analysis to consider the relationship between the technological appropriability and the export revenue of large firms. Columns (1) – (3) report the coefficients of a linear regression of the log of export sales on the same explanatory variable as used in the analysis of export decision presented in Table 9. The coefficients estimated in the pooled panel indicate a statistically significant correlation between export sales and the use of invention patent. Similar to the previous econometric test, as we add controls to the regression, the magnitude of the invention patent correlation decreases. Even though not statistically significant, the size of the other appropriability methods' coefficients also reduces with the addition of controls. The controls have the same signal presented before. Firm size (log of the number of employees) and the foreign origin of capital are associated with larger export sales. Also, expenditures in R&D and in other innovative activities have a positive and statistically significant impact on export values. We also observe an increase in the regression explanatory power as we add controls to this regression. Employing a Fisher-test to compare the specifications (2) and (1), we find the F statistics is 14.320, significant at 1%. Comparing model (3) to (2), the F statistics is 9.012, also significant at 1%.

Columns (4) – (6) control for firm fixed effect. It is well established in the literature that innovation, exports and productivity are closely related (Aw, Roberts, and Xu 2008; Aw, Roberts, and Xu 2011; Greenaway and Kneller 2007). So, without controlling for firm productivity, the correlation between our export variable and the dummies indicating the use of each appropriability method may be biased, as the literature documents that firms that are more productive export more and that they have a larger chance to innovate and therefore to use formal methods to appropriate their innovations. Using fixed effect estimation, we control for all time invariant firm characteristics, and we expect that most of the omitted variable biases are treated with this strategy.

The results in columns (4) – (6) indicate that an omitted variable is indeed a serious concern in the pooled regression. The magnitude of the impact of invention patent on the export sales reduces significantly compared with all pooled regressions. As we add controls, the coefficients of interest do not change. Moreover, none of the controls is statistically significant.

Table 9: Technological appropriability and probability to export of large firms.

Dependent variable:	Dummy export		
	(1)	(2)	(3)
Invention patent	**0.133***	**0.0941***	**0.0815***
	(0.0263)	(0.0263)	(0.0257)
Utility model	**0.0617***	0.0373	0.0321
	(0.0320)	(0.0337)	(0.0344)
Industrial design	**0.0588***	0.0536	0.0399
	(0.0342)	(0.0359)	(0.0357)
Trademark	0.0264	0.0250	0.0119
	(0.0174)	(0.0175)	(0.0177)
Foreign		0.131***	0.117***
		(0.0257)	(0.0261)
Mixed		0.219**	0.231***
		(0.0893)	(0.0832)
Log (number of employees)		0.0335**	0.0234*
		(0.0140)	(0.0126)
Log (R&D expenditures)			0.00631**
			(0.00306)
Log (technology transfer expenditures)			0.00489
			(0.00353)
Log (machinery and equipment expenditures)			−0.000846
			(0.00283)
Log (other innovative expenditures)			0.00620*
			(0.00339)
Firm age		0.000326	0.000177
		(0.000647)	(0.000611)
Observations	1,639	1,556	1,556
Pseudo R-squared	0.0571	0.1657	0.1825
Firm fixed effect	No	No	No
Dummy period	Yes	Yes	Yes
Dummy ISIC	No	Yes	Yes

Note: The sample comprises all firms with 500 or more employees surveyed by Pintec in 2005 and 2008. Export data are from Foreign Trade Secretariat (SECEX)/Ministry of Development, Industry and Foreign Trade. The number of employees are provided by the Minister of Labour and Employment (RAIS). An innovative firm is defined as a firm that implemented at least one product or process innovation between 2006 to 2008. Standard errors are in parentheses. Column (1) reports marginal effects of the logit regression. Columns (2) and (3) report the coefficients of OLS regression. Type of sector is controlled in all regressions.

As we discard the cross-section variation in order to explore just the within-firm variation between the years 2005 and 2008, our estimation is less precise: the impact of the use of invention patent and utility model is statistically significant at 10% and the trademark coefficient is statistically significant at 5%. The results indicate that the use of invention patent or utility model is associated with an increase in export sales (more than 50% for invention patent and more than 40% for utility model), while firms that start using trademarks end up exporting substantially less compare with firm that do not start or stop using any appropriability method. The industrial design coefficient is not significant in fixed effect analysis.

To sum up, even if only statistically significant at 10%, we observe that firms which were not using patents in the first period increased their exports on average by more than 51% if they start using them in the second period. In the case of utility models, this increase is more than 42%, but also barely statistically significant. Curiously, trademarks have the most statistically significant effect but it is negative. Firms which did not use trademarks in the first period exported on average approximately 40% less when they started using them.

We turn now to the third indicator of export activities: the share of the sector exports at the ISIC three-digit level. Using the same methodology discussed above, we test the impact of the four different appropriation methods on the firm export sales share in its sector (Table 11). The coefficients estimated in the pooled panel indicate a statistically significant correlation between firm share on exports and the use of invention patent. Similar to the previous econometric test, as we add controls to the regression, the measured magnitude of the invention patent correlation decreases. In column (3), the correlation between the use of utility model and the log of export sales turns to negative. Size measure (log of the number of employees) and the foreign origin of capital are once again associated with larger export sales, as well as expenditures in R&D and in other innovative actives, and have a positive and statistically significant impact on the export variable. Comparing model (2) with (1), the F statistics is 14.086, significant at 1%, while comparing model (3) with (2) the F statistics is 4.417, also significant at 1%. Both results indicate that the control variables included in each step improved the specification of the model.

The results for fixed effect, presented in columns (4) – (6), confirm that omitted variable is a serious concern in the pooled regression. The magnitude of the impact of invention patent on the export variable once again reduces the observed coefficient value in the pooled analysis. As we add controls, the coefficients of interest do not change. Moreover, none of the controls is statistically significant.

Table 10: Technological appropriability and exported value of large firms (pooled and fixed effect).

| | Log (exports) | | | | | |
| | Pooled | | | Fixed effect | | |
Dependent variable:	(1)	(2)	(3)	(4)	(5)	(6)
Invention Patent (dummy)	**2.517***	**1.278***	**1.103***	**0.569***	**0.543***	**0.516***
	(0.305)	(0.297)	(0.296)	(0.305)	(0.300)	(0.300)
Utility model (dummy)	0.413	0.148	−0.0116	**0.404***	**0.429***	**0.420***
	(0.343)	(0.319)	(0.323)	(0.243)	(0.243)	(0.246)
Industrial design (dummy)	0.502	0.398	0.281	−0.131	−0.138	−0.157
	(0.364)	(0.340)	(0.339)	(0.231)	(0.236)	(0.235)
Trademark (dummy)	0.447	0.369	0.0925	**−0.543****	**−0.534****	**−0.505****
	(0.319)	(0.298)	(0.299)	(0.235)	(0.229)	(0.223)
Foreign (dummy)		2.758***	2.468***		−1.071	−1.006
		(0.299)	(0.304)		(0.872)	(0.872)
Mixed (dummy)		2.826***	2.670***		−0.0789	0.00741
		(0.423)	(0.433)		(0.877)	(0.884)
Employees (logs)		1.560***	1.223***		−0.213	−0.265
		(0.219)	(0.218)		(0.560)	(0.550)
R&D (logs)			0.154***			0.0227
			(0.0575)			(0.0538)
Tech. transfer (logs)			0.0407			−0.0210
			(0.0467)			(0.0426)
Tech. equipment (logs)			0.0357			−0.0465
			(0.0502)			(0.0291)
Oth. innov. (logs)			0.130**			0.0451
			(0.0619)			(0.0517)
Firm age		0.00326	1.54e-05			
		(0.0116)	(0.0113)			
Observations	1,639	1,638	1,638	1,639	1,639	1,639
R-squared	0.050	0.223	0.240	0.022	0.027	0.031
Firm fixed effect	No	No	No	Yes	Yes	Yes
Dummy period	Yes	Yes	Yes	Yes	Yes	Yes
Dummy ISIC	No	Yes	Yes	No	No	No

Robust standard errors in parenthesis.

Table 11: Technological appropriability and firm share on sectorial exports.

| | Firm share on sectorial exports (3 digits) | | | | | |
| P5 | Pooled | | | Fixed effect | | |
Dependent variable:	(1)	(2)	(3)	(4)	(5)	(6)
Invention patent (dummy)	**0.0982***	**0.0684***	**0.0604***	**0.0406****	**0.0394****	**0.0386****
	(0.0193)	(0.0193)	(0.0198)	(0.0172)	(0.0173)	(0.0173)
Utility model (dummy)	−0.0314	−0.0335	**−0.0384***	−0.00854	−0.00797	−0.00785
	(0.0228)	(0.0221)	(0.0222)	(0.0148)	(0.0148)	(0.0147)
Industrial design (dummy)	0.0325	0.0293	0.0235	−0.00340	−0.00317	−0.00371
	(0.0260)	(0.0254)	(0.0256)	(0.0178)	(0.0173)	(0.0173)
Trademark (dummy)	0.0126	0.00625	−0.00418	**−0.0249***	**−0.0245***	**−0.0242***
	(0.0163)	(0.0161)	(0.0162)	(0.0138)	(0.0138)	(0.0137)
Foreign (dummy)		0.0820***	0.0707***		−0.0968	−0.0974
		(0.0184)	(0.0186)		(0.0677)	(0.0671)
Mixed (dummy)		0.150***	0.140***		−0.0644	−0.0650
		(0.0467)	(0.0480)		(0.0557)	(0.0558)
Employees (logs)		0.0477***	0.0366***		0.0133	0.0140
		(0.00981)	(0.0101)		(0.0344)	(0.0342)
R&D (logs)			0.00520*			−0.000705
			(0.00295)			(0.00371)
Tech. transfer (logs)			0.00377			−0.000728
			(0.00289)			(0.00226)
Tech. equipment (logs)			−0.00145			−0.000977
			(0.00261)			(0.00200)
Oth. innov. (logs)			0.00575*			0.000447
			(0.00296)			(0.00322)
Firm age		0.00133**	0.00119**			
		(0.000519)	(0.000515)			
Observations	1,639	1,638	1,638	1,639	1,639	1,639
R-squared	0.022	0.062	0.072	0.013	0.019	0.020
Firm fixed effect	No	No	No	Yes	Yes	Yes
Dummy period	Yes	Yes	Yes	Yes	Yes	Yes
Dummy ISIC	No	Yes	Yes	No	No	No

Table 12: Correlation of variables – sample of large firms.

	Export dummy	Log (exports)	Invention patent	Utility model	Industrial design	Trademark	Log (number of employees)	Log (R&D expenditures)	Log (technology transfer expenditures)	Log (machinery and equipment expenditures)	Log (other innovative expenditures)
Export dummy	1.0000										
Log (exports)	0.9180	1.0000									
Invention patent	0.1728	0.2146	**1.0000**								
Utility model	0.1249	0.1271	**0.4181**	1.0000							
Industrial design	0.0961	0.0919	**0.2361**	**0.4367**	1.0000						
Trademark	0.0952	0.0973	**0.2541**	**0.2641**	**0.2266**	1.0000					
Log (number of employees)	0.0974	0.2449	0.1750	0.0865	0.0644	0.0988	1.0000				
Log (R&D expenditures)	0.2046	0.2783	0.3185	0.2504	0.1861	0.2480	0.2786	1.0000			
Log (technology transfer expenditures)	0.1022	0.1590	0.1119	0.0915	0.1093	0.0970	0.2136	0.3679	1.0000		
Log (machinery and equipment expenditures)	0.0826	0.1660	0.0759	0.0694	0.0338	0.1030	0.1993	0.3260	0.2932	1.0000	
Log (other innovative expenditures)	0.1593	0.2318	0.2030	0.1492	0.1231	0.2091	0.2373	0.5578	0.3903	0.5201	1.0000

Table 13: Variables of interest.

VARIABLES	
IP	invention patent
UM	utility model
ID	industrial design
TM	Trademark
Interactions: if the firm used any of the following combinations	
IPUM	invention patent and utility model
IPID	invention patent and industrial design
IPTM	invention patent and trademark
UMID	utility model and industrial design
UMTM	utility model and trademark
IDTM	industrial design and trademark
IPUMID	invention patent, utility model and industrial design
IPUMTM	invention patent, utility model and trademark
IPIDTM	invention patent, industrial design and trademark
UMIDTM	utility model, industrial design and trademark
IPUMIDTM	invention patent, utility model, industrial design and trademark

Table 14: Technological appropriability and probability to export of large firms – including interactions (logit model).

Dummy export (logit)	Dummy export			
Dependent variable	(1)	(2)	(3)	(4)
IP	**0.133***	**0.115***	**0.0815***	**0.0705***
	(0.0263)	(0.0399)	(0.0257)	(0.0396)
UM	**0.0617***	0.166	0.0321	0.0763
	(0.0320)	(0.113)	(0.0344)	(0.0995)
ID	**0.0588***	0.177	0.0399	0.149
	(0.0342)	(0.112)	(0.0357)	(0.108)
TM	0.0264	0.0157	0.0119	0.00182
	(0.0174)	(0.0187)	(0.0177)	(0.0189)
IPUM		−0.163		−0.107
		(0.145)		(0.132)
IPID		**1.148***		**0.875***
		(0.148)		(0.142)
IPTM		0.129*		0.100
		(0.0766)		(0.0730)
UMID		−0.236		−0.175
		(0.196)		(0.176)
UMTM		0.0339		0.0859
		(0.159)		(0.153)
IDTM		−0.0418		−0.0599
		(0.138)		(0.136)
IPUMID		**−1.066***		**−0.876***
		(0.257)		(0.233)
IPUMTM		−0.0832		−0.100
		(0.203)		(0.194)
IPIDTM		**−1.413***		**−1.104***
		(0.201)		(0.190)
UMIDTM		−0.0288		−0.0213
		(0.248)		(0.237)
IPUMIDTM		**1.499***		**1.252***
		(0.335)		(0.313)
Foreign			0.117***	0.118***
			(0.0261)	(0.0261)
Mixed			0.231***	0.230***
			(0.0832)	(0.0822)
Log (number of employees)			0.0234*	0.0221*
			(0.0126)	(0.0125)
Log (R&D expenditures)			0.00631**	0.00594*
			(0.00306)	(0.00309)
Log (technology transfer expenditures)			0.00489	0.00506
			(0.00353)	(0.00353)
Log (machinery and equipment expenditures)			−0.000846	−0.000812
			(0.00283)	(0.00278)
Log (other innovative (Expenditures)			0.00620*	0.00642*
			(0.00339)	(0.00334)
Observations	1,639	1,639	1,556	1,556
Pseudo *R*-squared	0.0571	0.0675	0.1825	0.1908

Standard errors in parentheses.
***$p < 0.01$, **$p < 0.05$, *$p < 0.1$

Table 15: Technological appropriability and export value of large firms – including interactions.

Lexp – exported value (OLS pooled) – inclui firmas náo exportadoras LINGO

Model	OLS				Fixed effect
	Exported value				
Dependent variable	(1)	(2)	(3)	(4)	(5)
IP	2.517***	2.705***	1.103***	1.578***	1.142*
	(0.305)	(0.577)	(0.296)	(0.561)	(0.628)
UM	0.413	2.983***	−0.0116	0.787	0.332
	(0.343)	(0.986)	(0.323)	(0.903)	(0.563)
ID	0.502	1.609*	0.281	1.440	0.210
	(0.364)	(0.937)	(0.339)	(0.918)	(0.362)
TM	0.447	0.167	0.0925	−0.0840	−0.267
	(0.319)	(0.450)	(0.299)	(0.414)	(0.285)
IPUM		−3.246**		−1.933	0.0578
		(1.492)		(1.375)	(0.996)
IPID		−1.586		−3.527**	−1.457*
		(1.471)		(1.488)	(0.866)
IPTM		0.872		0.215	−0.829
		(0.743)		(0.711)	(0.573)
UMID		−3.428*		−1.792	0.414
		(1.960)		(1.814)	(0.680)
UMTM		−0.698		0.566	0.913
		(1.245)		(1.181)	(0.951)
IDTM		0.680		−0.389	−0.206
		(1.253)		(1.192)	(0.632)
IPUMID		3.064		3.299	0.786
		(2.728)		(2.518)	(1.587)
IPUMTM		0.531		−0.0335	−0.874
		(1.731)		(1.625)	(1.274)
IPIDTM		−0.919		1.632	1.553
		(1.980)		(1.862)	(0.953)
UMIDTM		0.296		0.649	−1.664
		(2.452)		(2.348)	(1.374)
IPUMIDTM		0.296		−0.954	0.0703
		(3.317)		(3.091)	(2.056)
foreign			2.468***	2.475***	−1.028
			(0.304)	(0.309)	(0.873)
mixed			2.670***	2.699***	0.118
			(0.433)	(0.435)	(0.909)
Log (number of employees)			1.223***	1.232***	−0.280
			(0.218)	(0.220)	(0.546)
Log (R&D expenditures)			0.154***	0.145**	0.0234
			(0.0575)	(0.0580)	(0.0550)
Log (technology transfer expenditures)			0.0407	0.0439	−0.0133
			(0.0467)	(0.0470)	(0.0430)
Log (machinery and equipment expenditures)			0.0357	0.0407	−0.0467
			(0.0502)	(0.0502)	(0.0291)
Log (other innovative expenditures)			0.130**	0.132**	0.0454
			(0.0619)	(0.0620)	(0.0529)
Observations	1,639	1,639	1,638	1,638	1,639
R-squared	0.05	0.058	0.240	0.244	0.040

Robust standard errors in parentheses.
***$p < 0.01$, **$p < 0.05$, *$p < 0.1$

As we discard the cross-section variation in order to explore only the within-firm variation between the years 2005 and 2008, our estimation for invention patent is less precise: the impact of the use of invention patent reduces the statistical significance to 10%. On the other hand, the negative coefficient related to the use of trademarks becomes statistically significant at 5%.

Summing up, in all cases we identified a positive impact of invention patent on all export variables, while in most cases a negative relationship between the use of trademark and exports was observed.

Robustness of results

Adding interactions

The results presented above show the relationship between export performance and the use of formal appropriability methods. However, as discussed in the international literature (Llerena and Millot 2013), these methods can be used simultaneously by firms, that is, firms tend to use more than one method of technological appropriation. Therefore, the risk of collinearity between IP-related appropriation methods is relevant. The positive correlation among all these variables seems to confirm this concern (Table 12). We observe the highest Pearson correlation

Table 16: Technological appropriability and export value of large firms – including interactions.

Model Dependent variable	Export/sector (total)				
	OLS				Fixed effect
	Firm share on sectorial exports				
	(1)	(2)	(3)	(4)	(5)
IP	**0.0982*****	**0.103*****	**0.0604*****	0.0693*	0.0287
	(0.0193)	(0.0349)	(0.0198)	(0.0361)	(0.0283)
UM	−0.0314	0.00836	**−0.0384***	−0.0574	−0.0402
	(0.0228)	(0.0571)	(0.0222)	(0.0546)	(0.0726)
ID	0.0325	0.122	0.0235	0.114	0.00255
	(0.0260)	(0.0853)	(0.0256)	(0.0885)	(0.0416)
TM	0.0126	0.00264	−0.00418	−0.0122	−0.0133
	(0.0163)	(0.0203)	(0.0162)	(0.0203)	(0.0162)
IPUM		−0.0693		−0.0248	0.0716
		(0.0860)		(0.0815)	(0.0823)
IPID		0.0947		0.0505	−0.0264
		(0.259)		(0.271)	(0.0544)
IPTM		0.0201		0.00773	−0.0103
		(0.0476)		(0.0471)	(0.0341)
UMID		−0.162		−0.0781	0.0688
		(0.130)		(0.132)	(0.143)
UMTM		−0.0259		0.0290	−0.0611
		(0.0770)		(0.0739)	(0.0863)
IDTM		−0.0455		−0.0669	−0.0693
		(0.102)		(0.104)	(0.0610)
IPUMID		−0.139		−0.164	−0.0987
		(0.287)		(0.297)	(0.156)
IPUMTM		0.0632		0.0280	0.0286
		(0.108)		(0.103)	(0.106)
IPIDTM		−0.212		−0.165	0.0964
		(0.271)		(0.282)	(0.0735)
UMIDTM		0.142		0.0778	0.0861
		(0.160)		(0.159)	(0.160)
IPUMIDTM		0.207		0.241	−0.0594
		(0.311)		(0.318)	(0.186)
Foreign			0.0707***	0.0710***	−0.0993
			(0.0186)	(0.0186)	(0.0674)
Mixed			0.140***	0.142***	−0.0612
			(0.0480)	(0.0485)	(0.0563)
Log (number of employees)			0.0366***	0.0369***	0.0160
			(0.0101)	(0.0101)	(0.0338)
Log (R&D expenditures)			0.00520*	0.00524*	−9.26e-05
			(0.00295)	(0.00299)	(0.00381)
Log (technology transfer expenditures)			0.00377	0.00372	−0.000993
			(0.00289)	(0.00290)	(0.00229)
Log (machinery and equipment expenditures)			−0.00145	−0.00143	−0.00117
			(0.00261)	(0.00264)	(0.00201)
Log (other innovative expenditures)			0.00575*	0.00591**	0.000402
			(0.00296)	(0.00296)	(0.00328)
Observations	1,639	1,639	1,638	1,638	1,639
R-squared	0.022	0.027	0.072	0.078	0.031

		Regressão pooled: reg LINGO			
		(1)	(2)	(3)	(4)
F(a,b)	a	5	16	14	#
	b	1633	1622	1623	#
	F(a,b)	6.87	2.7	9.73	6
	Prob>F	0.0000	0.0003	0.0000	#
R-squared		0.0215	0.0275	0.0722	0
Root MSE		0.31354	0.31363	0.30623	0
			(2) × (1)	(4) × (3)	
F comparado ((R2new - R2old)/(a_new-a_old))/(1-R2new)/b_new)			0.910	0.890	
F - 1% significancia			2.259	2.418	
F - 10% significancia			1.574	1.635	
				(n sign)	

Robust standard errors in parentheses
***p < 0.01, **p < 0.05, *p < 0.1

Table 17: Technological appropriability and selected economic variables (pooled regression).

OLS Pooled Dependent Variable	export firms	export/sales	mkshare	innov/sales	innov. market/sales
Invention patent	**0.293****	**0.0222***	**0.0349****	**0.0306*****	**0.00867*****
	(0.147)	(0.0121)	(0.0174)	(0.0115)	(0.00323)
Utility model	−0.232	−0.0221	**−0.0345***	0.0174	**0.0179****
	(0.159)	(0.0135)	(0.0197)	(0.0153)	(0.00707)
Industrial design	−0.0954	−0.0167	0.0178	−0.00761	−0.00505
	(0.167)	(0.0131)	(0.0218)	(0.0158)	(0.00624)
Trademark	−0.150	**−0.0298*****	−0.000581	**0.0185****	**0.00362***
	(0.125)	(0.00997)	(0.0141)	(0.00915)	(0.00217)
Foreign	1.115***	0.0373***	0.0602***	0.0208*	−0.000598
	(0.144)	(0.0112)	(0.0169)	(0.0112)	(0.00473)
Mixed	1.017***	0.0645**	0.100**	0.0277	0.00125
	(0.319)	(0.0284)	(0.0419)	(0.0296)	(0.00522)
Log (number of employees)	1.068***	0.0432***	0.0616***	0.00829	−0.00191
	(0.101)	(0.00730)	(0.00979)	(0.00604)	(0.00213)
Log (R&D expenditures)	0.0435*	−0.00296	0.00641**	0.00306*	9.86e-05
	(0.0255)	(0.00193)	(0.00259)	(0.00170)	(0.000558)
Log (technology transfer expenditures)	0.000197	0.00212	0.00122	−0.00192	−0.000962
	(0.0222)	(0.00171)	(0.00243)	(0.00161)	(0.000649)
Log (machinery and equipment expenditures)	0.0512**	0.00246	−0.00294	−0.000331	0.000214
	(0.0206)	(0.00169)	(0.00222)	(0.00144)	(0.000577)
Log (other innovative expenditures)	0.0312	−0.00287	0.00669***	0.00265	0.000238
	(0.0260)	(0.00201)	(0.00249)	(0.00175)	(0.000600)
Observations	1,423	1,634	1,636	1,638	1,638
Pseudo R-squared	0.339	0.266	0.188	0.093	0.045

coefficients between utility models and industrial designs (0.4367) and between invention patents and utility models (0.4181). These, however, are far from being a severe case of collinearity.

In order to control for that problem, we estimated a saturated model including all interaction among the use of appropriability methods.[9] The interactions of variables are described in Table 13.

We reproduce all the previous specifications including these interacted variables. The results are presented in Tables 14 – 16. Although the joint use of appropriability methods is traditionally identified in the literature, the inclusion of the interactions did not significantly improve the explanatory power of our models. This can be noted by comparing columns (1) and (2). When dummy export was used as the dependent variable, the

Table 18: Technological appropriability and selected economic variables (fixed effect).

Fixed Effect Dependent variable	Exporting firms	Export/sales	Market share	New product to the firm/sales	New product to the market/sales
Invention patent	−0.139	0.00459	**0.0224***	**0.0302***	0.00420
	(0.0861)	(0.00689)	(0.0124)	(0.0172)	(0.00503)
Utility model	0.0815	−0.00354	−0.0168	**0.0542****	0.0119
	(0.155)	(0.00900)	(0.0135)	(0.0243)	(0.0108)
Industrial design	−0.105	0.000505	0.00282	**−0.0506****	0.000901
	(0.117)	(0.00685)	(0.0144)	(0.0230)	(0.00797)
Trademark	**−0.232*****	−0.00856	−0.00848	0.0217	−0.000244
	(0.0899)	(0.00686)	(0.0105)	(0.0133)	(0.00334)
Foreign	0.196	0.0140	0.00988	0.0432	−0.00647
	(0.412)	(0.0282)	(0.0148)	(0.0329)	(0.00550)
Mixed	−0.112	0.00833	0.0145	−0.0452	−0.00208
	(0.408)	(0.0236)	(0.0181)	(0.0435)	(0.00776)
Log (number of employees)	0.00616	0.0186*	0.0145	−0.00581	0.00612
	(0.189)	(0.0109)	(0.0197)	(0.0257)	(0.00744)
Log (R&D expenditures)	0.0325	0.00229*	0.00226	−0.000920	−0.00184*
	(0.0239)	(0.00137)	(0.00202)	(0.00287)	(0.00111)
Log (technology transfer expenditures)	−0.0247*	−0.00309***	−0.00124	−0.000986	−0.000415
	(0.0129)	(0.00119)	(0.00137)	(0.00250)	(0.00110)
Log (machinery and equipment expenditures)	−0.0104	0.000174	−0.000511	0.00264	0.000690
	(0.0154)	(0.000967)	(0.00121)	(0.00222)	(0.000681)
Log (other innovative expenditures)	0.0228	−0.000759	−0.000189	−0.00587**	−0.000614
	(0.0225)	(0.00136)	(0.00158)	(0.00298)	(0.00145)
Observations	1,424	1,635	1,637	1,639	1,639
Pseudo R-squared	0.040	0.083	0.013	0.053	0.020

pseudo R^2 presented a slightly improvement when interactions were added.

Next, we present the inclusion of interactions when the dependent variable is exported value. Aiming to verify whether the interactive variables of interest improve the performance of the OLS models, we estimated the incremental contribution of these new explanatory variables. Employing a Fisher-test to compare the specifications (2) and (1) presented in Table 15, we find the F statistics is 1.220, non-significant at 10%. Comparing model (4) with (3) the F statistics is 0.765, also non-significant at 10%. Column 5 shows the fixed effect model, in which the IP variable remains statistically significant.

Lastly, we repeat the analysis using firm share on sectorial exports as a dependent variable. Once more, comparing the specifications (2) and (1), we find the Fisher test is 0.910, non-significant at 10%. Comparing model (4) with (3) the Fisher test is 0.890, also non-significant at 10%. In the fixed effect analysis, all variables of interest lost their significance.

Despite what the literature states about the joint use of these appropriation methods, the inclusion of the interactions did not significantly change the results or improve the explanatory power of our specifications. Particularly, the main effect remains that from patents, which stays statistically significant in most specifications. This adds robustness to our results, regarding the statistical significance between IP and the export performance variables analyzed in this paper.

Preliminary analysis of other dependent variables
The main goal of our study was to analyze the relationship between technological appropriability and export performance of manufacturing firms in Brazil. However, we can expect IP-related appropriation methods to affect innovation and economic performance more broadly defined. A careful analysis of such a link is beyond the scope of this study, but basic results are presented in the annex using a similar framework just for robustness purposes. Here we present a preliminary analysis of the following dependent variables:
• Only export firms: it includes only export firms in the sample;
• Export/sales: percentage of firm exports over sales
• Market share: firm share in sectorial market share
• New product to the firm/sales: percentage of sales that come from innovative products (products that are new to the firm)
• New product to the market/sales: percentage of sales that come from innovative products (products that are new to the market)

Table 17 presents the main results of an OLS pooled regression. As observed in the previous sections, we verified a positive and significant relationship between invention patent and all dependent variables. The coefficient is most expressive when only export firms are included in the sample, although it is significant in all cases. Unlike the results presented in the previous sections, here we verified a positive and significant relationship between trademarks and innovations/sales, which suggests that firms that use

trademarks present a higher participation of innovative products on their sales. Results are less apparent in statistical significance when controlling for firm fixed effects (Table 18). In some cases, the IP coefficient lost significance, although it was maintained when market share and innovation/sales are the dependent variables.

Conclusions
In this paper, we evaluated the relationship between technological appropriability and export performance of Brazilian industrial firms. Information about formal methods to protect innovation was obtained from the Technological Innovation Survey, applied by the Brazilian Statistic Institute (PINTEC/IBGE). Although Brazil is historically characterized by the export of commodities (de Negri 2005), the data presented in this paper show that innovative firms tend to export more, and that some types of technological appropriability are relevant to promote exports.

Descriptive statistics showed that, on average, innovative manufacturing firms are more likely to export and present a higher average export value than non-innovative firms. In a sub-sample including only innovative industrial firms, we found that export firms are larger (when the size is measured by the number of employed persons or by net sales) and invest proportionately more in R&D. Moreover, a proportionally higher percentage of firms with foreign capital or using appropriability methods are exporters.

Aiming to evaluate the impact of appropriability methods on export performance, the econometric analysis concentrated on large firms. It enabled the use of temporal lag in most independent variables. This lag was not incorporated in the appropriability variables, since, according to the PINTEC survey, they refer to methods that are already in use (e.g. granted intellectual property rights), so their benefits should have been incorporated by the firms. Also, the fixed effect panel data model was used, as it deals with endogeneity problems.

The results showed a positive and significant impact of invention patent on export performance. This was the only variable of interest that presented a consistent result in all tests. Besides, in some tests, a negative and significant relationship between exports and trademark secrets was found. This result differs from that observed in Kim et al. (2012), which emphasizes the importance of utility models to developing countries. However, our econometric analysis includes only large innovative firms that are more intensive in R&D activities and for which invention patents tend to be more accessible. Thus, our results are not representative of all Brazilian industries, but are exclusively relevant to the most sophisticated group of firms: the larger and more innovative ones.

Aiming to understand the joint use of appropriability methods, we included these interactions among the variables of interest. However, although the joint use of appropriability methods is traditionally identified in the literature, the inclusion of these interactions did not improve the explanatory power of our models.

These results suggest that the incentive to patent may protect technologies developed by firms and expand their competitiveness, boosting their ability to compete in foreign markets.

Notes

1. There are also strategic methods of appropriation, such as industrial secrets and lead time, which are not explored in this study.
2. Translated by the authors.
3. For a summary of the recent literature, refer to Greenaway and Kneller (2007).
4. Just like the Community Innovation Survey, PINTEC is based on the Oslo Manual.
5. This is often referred as *strategic* or *informal* means of appropriation.
6. And if a firm performs R&D today, it takes some time for it to innovate, launch a new product into the market and export it.
7. Firm age helps control for efficiencies due to entrepreneurs learning about their abilities over time (Kim 2012). Also, firm size helps control for economies of scale in generating patents due to the fixed costs of maintaining a legal department that handles intellectual property matters.
8. Some studies based on instrumental variables do deal with these problems. However, it is not trivial to find a correct instrument. Among the available data, it was not possible to find a good instrument, so we decided not to include this method.
9. The model is not, however, strictly saturated once we don't insert the interaction between the dummies of interest and the dummy of year.

Acknowledgement

The author wishes to thank the IBGE team for the data availability; Alessandro Pinheiro, manager of the PINTEC/IBGE for his helpful clarifications and Glaucia de Sousa Ferreira and Larissa Pereira for their assistance in the preparation of this report.

Disclosure statement

No potential conflict of interest was reported by the authors.

References

Amable, B., and B. Verspagen. 1995. "The Role of Technology in Market Shares Dynamics." *Applied Economics* 27: 197–204.

Amendola, G., P. Guerrieri, and E. Padoan. 1998. "International Patterns of Technological Accumulation and Trade." In *Trade, Growth and Technical Change*, edited by D. Archibugi and D. Michie, 83–97. Cambridge: Cambridge University Press.

Andersson, M., and O. Ejermo. 2008. "Technology Specialization and the Magnitude and Quality of Exports." *Economics of Innovation and New Technology* 17 (4): 355–375.

Arundel, A., and I. Kabla. 1998. "What Percentage of Innovations are Patented? Empirical Estimates for European Firms." *Research Policy* 27: 127–141.

Avellar, A. P., and L. Carvalho. 2013. Esforço Inovativo e Desempenho Exportador: Evidências para Brasil, Índia e China. Disponível em: http://anpec.org.br/encontro/2011/inscricao/arquivos/000-5ef324d543484edf62acaed0c200194c.pdf

Aw, B. Y., M. J. Roberts, and D. Y. Xu. 2008. "R&D Investments, Exporting, and the Evolution of Firm Productivity." *American Economic Review* 98 (2): 451–456.

Aw, B. Y., M. J. Roberts, and D. Y. Xu. 2011. "R&D Investment, Exporting, and Productivity Dynamics." *The American Economic Review* 101: 1312–1344.

Basberg, B. L. 1987. "Patents and the measurement of technological change: A survey of the literature." *Research policy* 16 (2–4): 131–141

Bernard, A. B., & Jensen, J. B. 1999. "Exceptional exporter performance: cause, effect, or both?." *Journal of international economics* 47(1): 1–25.

Breschi, S., and R. Helg. 1996. "Technological Change and International Competitiveness: The Case of Switzerland." *Liuc Papers*, 31. serie Economia e Impresa 7, junho.

Calvo, J. L. 2003, August. The Export Activity of Spanish Manufacturing Firms: Does Innovation Matter?. In 23th Congress of the European Regional Science Association.

Cassiman, B., E. Golovko, and E. Martínez-Ros. 2010. "Innovation, Exports and Productivity." *International Journal of Industrial Organization* 28 (4): 372–376.

Clerides, S. K., S. Lach, and J. R. Tybout. 1998. "Is Learning by Exporting Important? Micro-Dynamic Evidence From Colombia, Mexico, and Morocco." *The Quarterly Journal of Economics* 113 (3): 903–947.

Cohen, W. M., R. R. Nelson, and J. Walsh. 2000. Protecting Their Intellectual Assets: Appropriability Conditions and Why U.S. Manufacturing Firms Patent (or Not). Cambridge, MA: NBER Working Paper No. 7552.

Damijan, J. P., Č Kostevc, and S. Polanec. 2010. "From Innovation to Exporting or Vice Versa?" *World Economy* 33 (3): 374–398.

de Negri, F. 2005. Inovação Tecnológica e Exportações das Firmas Brasileiras. Anais do XXXIII Encontro Nacional de Economia – Anpec. Natal.

Dosi, G., K. Pavitt, e L. G. Soete. 1990. *The Economics of Technical Change and International Trade*. London: Harvester Wheatsheaf.

Dosso, M. 2011. Sectoral dynamics of international trade and technological change. http://www.lem.sssup.it/WPLem/documents/papers_EMAEE/dosso.pdf

Freeman, C., M. A. Young, and J. Fuller. 1963. "The plastics industry: a comparative study of research and innovation." *National Institute Economic Review* 22–62.

Gonçalves, E., M. B. Lemos, and J. de Negri. 2007. Condicionantes Da Inovação Tecnológica Na Argentina E No Brasil. In Anais do XXXV Encontro Nacional de Economia [Proceedings of the 35th Brazilian Economics Meeting] (No. 117). ANPEC-Associação Nacional dos Centros de Pósgraduação em Economia [Brazilian Association of Graduate Programs in Economics].

Graham, S. J. H., and D. Somaya. 2006. Vermeers and Rembrandts in the same attic: Complementarity between copyright and trademark leveraging strategies in software. http://papers.ssrn.com/sol3/papers.cfm?abstract_id=887484

Greenaway, D., and R. Kneller. 2007. "Firm Heterogeneity, Exporting and Foreign Direct Investment." *Economic Journal* 117 (517): F134–F161.

Hall, B. H., C. Helmers, M. Rogers, and V. Sena. 2012. "The Choice Between Formal and Informal Intellectual Property: a Literature Review (No. w17983)." Cambridge, MA: National Bureau of Economic Research.

Hanel, P. 2008. "The use of Intellectual Property Rights and Innovation by Manufacturing Firms in Canada." *Economics of Innovation and New Technology* 17 (4): 285–309.

INPI. (Instituto Nacional de la Propiedad Industrial). 2017. http://www.inpi.gov.br/servicos/perguntas-frequentes-paginas-internas/perguntas-frequentes-patente#patente

IBGE. (Instituto Brasileiro de Geografia e Estatística); PINTEC (Pesquisa de Inovação). 2008. Available at www.pintec.ibge.gov.br

Kim, Y. K., K. Lee, W. Park, and K. Choo. 2012. "Appropriate Intellectual Property Protection and Economic Growth in Countries at Different Levels of Development." *Research Policy* 41 (2): 358–375.

Lachenmaier, S., & Wößmann, L. 2006. "Does innovation cause exports? Evidence from exogenous innovation impulses and obstacles using German micro data." *Oxford Economic Papers* 58(2): 317–350.

Laursen, K. 1999. "The Impact of Technological Opportunity on the Dynamics of Trade Performance." *Structural Change and Economic Dynamics* 10 (3): 341–357.

Laursen, K., and I. Drejer. 1999. "Do Inter-Sectoral Linkages Matter for International Export Specialisation?" *Economics of Innovation and New Technology* 8 (4): 311–330.

Laursen, K., and V. Meliciani. 2002. "The Relative Importance of International vis à vis National Technological Spillovers for Market Share Dynamics." *Industrial and Corporate Change* 11 (4): 875–894.

Lee, K., J. Kim, J.-B. OH, and K. Park. 2012. Economics of Intellectual Property in the Context of a Shifting Innovation Paradigm: A Review from the Perspective of Developing Countries.

Levin, R. C., A. K. Klevorick, R. R. Nelson, S. G. Winter, R. Gilbert, and Z. Griliches. 1987. "Appropriating the Returns From Industrial Research and Development." *Brookings Papers on Economic Activity* 1987 (3): 783–831.

Llerena, P., and V. Millot. 2013. Are Trade Marks and Patents Complementary or Substitute Protections for Innovation (No. 2013-01). Bureau d'Economie Théorique et Appliquée, UDS, Strasbourg.

Luna, F., and A. Baessa. 2007. Impacto das marcas e das patentes no desempenho econômico das firmas. In Anais do XXXV Encontro Nacional de Economia [Proceedings of the 35th Brazilian Economics Meeting] (No. 155). ANPEC-Associação Nacional dos Centros de Pósgraduação em Economia [Brazilian Association of Graduate Programs in Economics].

Mais, I., and M. Amal. 2011. "Determinants of Export Performance: an Institutional Approach." *Latin American Business Review* 12 (4): 281–307.

Montobbio, F. 2003. "Sectoral Patterns of Technological Activity and Export Market Share Dynamics." *Cambridge Journal of Economics* 8 (3): 435–470.

Nassimbeni, G. 2001. "Technology, Innovation Capacity, and the Export Attitude of Small Manufacturing Firms: a Logit/Tobit Model." *Research Policy* 30 (2): 245–262.

Posner, M. V. 1961. "International Trade and Technical Change." *Oxford Economic Papers* 13 (3): 323–341.

Soete, L. 1987. "The Impact of Technological Innovation on International Trade Patterns: the Evidence Reconsidered." *Research Policy* 16 (3-5): 101–130. julho.

Teece, D. J. 1986. "Profiting From Technological Innovation: Implications for Integration, Collaboration, Licensing and Public Policy." *Research Policy* 15: 285–305. junho.

WIPO. (World Intellectual Property Organization). 2016. Available at www.wipo.int

The contribution of creative industries to sustainable urban development in South Africa

Oluwayemisi Adebola Oyekunle ⓘ

South African policymakers are increasingly advocating for creative industry entrepreneurship to play a vital role in the economic development of cities, for example, by creating pathways for job creation and encouraging urban regeneration. The question, 'How can we use creative industries for the economic regeneration of urban regions?' is dealt with in this study. This paper addresses the question in three ways: examining the international debates about the effectiveness of creative projects as tools for urban development; deliberating on the factors that cause city authorities to put more importance on policies guiding creative industries; and describing the evolution of creative industries in two South African cities: Johannesburg and Cape Town.

In this paper, I argue that the role of creative industries in South Africa, as a promoter of urban development as an essential component of broader development plans, has been dynamic, but also problematic at times. I interrogate the role of creative industries in a range of areas: urban regeneration; the integration of creativity in urban development projects; the current creative economy indicators; policies to address creative diversity and social inclusion issues within the communities; and the potential of creative projects for community empowerment and involvement.

Introduction

In the last two decades, much policy interest and research on the role of creative industries in economic development, job creation and urban regeneration have been a point of discourse. The United Nations Conference on Trade and Development (United Nation 2010) argues the creative industries can be a medium to assist government endeavours to incorporate a country into the world economy by building the competitive efforts of its creative goods and services, and together promoting creative diversity, social inclusion and job creation. Policy development and academic research emphasizing the role of creative industries in local, regional and national economic development have evolved in the midst of many pertinent debates (c.f. O'connor 2004, 2007, 2009; Oakley 2009; Oakley and O'connor 2015; Pratt 2015). Bontje et al. (2011) have argued that creative industries, particularly because they are 'knowledge-intensive industries', are relevant to those economies going through an industrial decline because creative industries sometimes represent an increasing share of the overall economy.

Most importantly, creative industries can perform a significant role in the future development of **urban** areas. The development of creative industries, e.g. tourism, arts, music, film and fashion stimulates the urban economy and exemplifies new models of economic process and development. Scott (1997) notably predicted, at the beginning of the twenty-first century, the intersection of creative economic development. In South Africa, the integration of creative industries in urban regeneration projects has been common (e.g. Rogerson 2005, 2006; Joffe and Newton 2008). Suggested impacts from the expansion of regional economies have been citizens' satisfaction with the cultural conditions under which they live and with the city's image. In addition to economic and personal outcomes, innovative national policies are being developed to integrate and emphasize creative diversity, social inclusion, community involvement and empowerment. The acknowledgment of creativity in economic development must be deliberated in the broader context of the recent development agenda of South African economies. This study outlines some lessons that have been learned from experiences using creative industries for economic regeneration.

This paper contributes to debates about the role of creative industries in the economic development of major South African cities. It addresses general views that creative urban regeneration will naturally produce both economic and social benefits, such as wide-ranging job creation and the enhancement of creative diversity. This paper also reflects on the limitations of this model of urban development.

Definition of creative industries

The term 'creative industries' comprises a wide-range of activities that link together innovation, professional services, information economy, research and development, and creative activities. The United Nations Conference on Trade and Development (United Nations 2010) categorized the following activities as encompassing the creative industries:

• Graphic design	• Advertising	• Film and video
• Music	• Performing arts	• Fashion and jewellery
• Product and surface design	• Industrial design	• New media
• Publishing	• Radio and television	• Visual art
• Architecture	• Crafts	

A related term, *creative economy*, is defined in different ways. For example, UNCTAD (United Nations 2010) defines it as broader than the creative industries and centres it on the use of creative assets potentially to produce economic development and growth. UNCTAD suggests that the creative economy can boost job creation and income generation, while also building creative diversity, social inclusion and human development. In contrast, the United Kingdom's Department of Culture, Media and Sport (DCMS) (2014) describes the creative economy with greater focus on the people who do such work, including all those employed in the creative industries, whether for creative work or non-creative work (e.g. support for creative workers such as accounting), as well as those in creative jobs outside the creative industries.

South Africa creative industries: An overview

In South Africa, the arts, culture, heritage and creative industries were influenced by apartheid, as were all other facets of the people's political, economic and social life (Oyekunle 2014). Under apartheid, the majority of South Africans were subjected to unfair and discriminatory skills development programmes, including those in the creative industries, with the consequence that the sustained participation of previously disadvantaged persons at entire levels within the creative industries (i.e. production, creation, transmission, dissemination, and consumption) has yet to be accomplished (South Africa 2013b).

Johannesburg and Cape Town are the most creative 'city hubs' in South Africa, as well as the biggest metropolitan cities in the country (Evans 2009). Both cities regard creativity and culture as the driver for urban regeneration and economic development (Rogerson 2006). Cape Town has been increasingly marketed and promoting itself as a creative city (City of Cape Town, 2013). Over the past decade, these cities have experienced central city regeneration. Simultaneously, there has been growth in the service sectors, even though the manufacturing sector has declined. Yet, the creative industries in South Africa have long been neglected in conventional industrial and trade policy, despite the fact that these industries are known to make an important contribution to the economy of developing economies (Joffe and Newton 2008).

Oyekunle (2015c 11) argues that

[i]n connection with the prospect of design intervention in social life, the design process used when manufacturing goods and products allow sustainable development; this comprises the processes and structure of fundamental interaction and communication, which is used to strategies and produce activities and services within social life.

Similarly, Buchanan (cited in in Oyekunle 2015c) emphasizes the importance of these interactions in the social and economic structure of South Africa. Buchanan additionally affirms that design plays an essential role in the creation of building complex frameworks of human arts. The South African Department of Cultural Affairs and Sport (**DCAS**) has implemented many initiatives to create design awareness and improve design. The department has dedicated itself to developing a design sector strategy, which focuses on the following:

- developmental initiatives through a partnership with South Africa Fashion Week to address job creation
- operation with the National Product Development Centre at the CSIR, to improve the contributions in the field of design technology
- the CSIR Computer-Assisted Design (CAD) initiative linked to the technology station
- an award through Create SA to support emerging designers
- the Product Development Skills Programme and Thuthuka Jewellery to address the jewellery sector
- the Young Designers Simulcast and Annual Emerging Creatives Programme (WESGRO 2013a).

Africa is South Africa's primary destination market for the export of creative goods (61%), followed by Europe (13%). Table 1 shows the United States, the United Republic of Tanzania, Zambia, Zimbabwe and Mozambique as South Africa's leading export destinations in 2012.

In 2012, the export of creative goods stood at US$454bn, which was down by 3.20% from 2011. A positive trade increase of 8.74% was recorded between 2003 and 2008. However, trade decreased in 2009 by 8.39%, but increased by 8.2% in 2010. Global imports totalled US$417.62bn, and experienced a total increase of 9.23% from 2010. This is shown in the Table 2.

The trade performance of all the creative industries in 2003 had a total value of US$410.06 m for the country's exports and US$595.70 m for its imports, indicating a

Table 1: Top 10 export partners for creative goods, 2003 and 2012.

	2003				2012			
	Values in US$ millions				Values in US$ millions			
Rank	Country	Exports	Imports	Balance	Country	Exports	Imports	Balance
1	United States	84.5	64.1	20.42	United States	58.2	125.5	67.28
2	United Kingdom	68.7	95.7	26.98	United Republic of Tanzania	52.2	0.2	51.97
3	Zambia	14.5	0.0	14.47	Zambia	32.8	0.4	32.35
4	Mozambique	9.9	11.2	9.83	Zimbabwe	30.5	2.4	28.35
5	Australia	9.7	0.0	1.59	Mozambique	21.7	0.1	21.64
6	Nigeria	9.6	0.0	9.57	United Kingdom	16.8	177.5	160.63
7	Angola	9.5	11.3	9.51	Angola	15.9	0.0	15.85
8	France	9.4	0.2	1.87	Nigeria	10.6	0.1	10.48
9	United Republic of Tanzania	8.6	0.2	8.44	Germany	10.3	54.7	44.40
10	Germany	7.8	25.4	17.61	Malawi	9.8	0.1	9.69

Source: UNCTAD Stat (2013)

Table 2: Growth rates of creative goods imports and exports, between, 2003–2012.

MEASURE	Annual average growth rates		ECONOMY	South Africa	PARTNER World
PERIOD	2003–2012			2008–2012	
FLOW	Exports	Imports		Exports	Imports
PRODUCT					
All creative goods	0.37	12.81		−1.03	8.46
Art crafts	−7.25	10.72		−11.50	10.89
Carpets	−4.52	10.21		−11.73	10.22
Celebration	9.54	13.74		−7.18	12.15
Other	6.97	5.29		2.69	1.58
Paper ware	−4.69	7.35		9.04	0.18
Wicker ware	1.65	4.15		−3.84	5.62
Yarn	−14.78	12.63		−18.41	13.37
Audio-visuals	30.98	23.25		33.37	−0.19
Film	−15.20	−2.89		−42.69	−4.68
CD, DVD, Tapes	32.46	23.92		34.22	-0.15
Design	−4.73	13.48		−4.67	11.93
Architecture	9.78	−15.52		−59.71	10.97
Fashion	1.71	13.45		−1.23	13.68
Glassware	6.79	4.25		6.87	−3.47
Interior	−3.96	12.97		−0.59	10.76
Jewellery	−9.12	15.68		−12.01	14.59
Toys	8.34	13.77		3.27	10.98
New media	20.66	19.44		−0.36	20.99
Recorded media	28.86	29.57		6.06	29.67
Video games	10.47	8.63		−10.99	10.59
Performing arts	8.16	7.62		0.62	7.24
Musical Instruments	8.34	7.74		0.83	7.51
Printed music	−5.23	3.60		−28.49	−1.26
Publishing	3.15	4.32		6.63	2.50
Books	5.37	2.87		5.53	0.73
Newspaper	2.52	20.31		6.94	19.18
Other printed matter	2.75	6.24		8.02	4.01
Visual arts	5.74	1.50		−12.34	−2.04
Antiques	7.47	4.66		−21.49	−1.83
Paintings	11.19	−3.28		2.68	−11.19
Photography	1.54	2.97		−2.48	10.32
Sculpture	−1.96	8.90		−7.35	8.57

Source: UNCTAD Stat (2013).

total balance of US$184.64 m for the year. In 2012, there was an increase of 14.4% in the trade performance of all creative industries in South Africa. Likewise, in 2003, the trade of all creative goods exported totalled US $362.9 m with imports at US$591.5 m, giving a trade balance of $241.6 m for the year. During 2012, all creative goods exported totalled US$362.9 m with imports at US $2077.9 m, showing a trade balance of US$1715 m. Since 2003, a positive trade performance has been recorded with an increase of 17.2% in 2012. This is illustrated in Figure 1.

The role of creative industries in urban regeneration

A considerable body of literature has been developed on the connections between creative industries, culture and the city (e.g. Landry 2000; Lazzeretti, Boix, and Capone 2008; O'connor and Kong 2009; Pratt 2010; Canada, Policy Research Group 2013; Rutten 2016). Around the globe, a growing number of cities use the idea of 'creative cities' in urban development strategies designed to strengthen growth through creative activities (Canada, Policy Research Group 2013). Accra, Dakar, Luanda, Cape Town, Johannesburg, Douala, Lagos, Nairobi and Maputo – these are all African cities where urban

regeneration using creative and culture enterprises has been or is still being planned (United Nations 2013). Amalgamation presents an opportunity to re-vision and re-structure the urban planning future and its priorities (Duxbury 2004). The creative city's first official plan was developed to fulfil some of the sustainable development goals/ millennium development goals (SDGs/ MDGs), in which environmental, social, economic and cultural issues would be held in balance. The heritage and arts sectors were placed as a pillar in the regeneration of the cities of Johannesburg and Cape Town, central to their future development. The concept of sustainable development is vital for cities.

Landry (2000), in his seminal work on the creative city, argues that the most vital resource a city has is its people, who express desires, cleverness, imagination and motivation. He argues that creativity is replacing location, access to markets and natural resources as a form of urban resource. Landry further stresses that the creativity of the people living in and running cities will determine their future success. But, the realization of a creative city is multimodal and can originate from any citizen through any profession or sector. Landry points out that a creative city consists of creative individuals, a creative

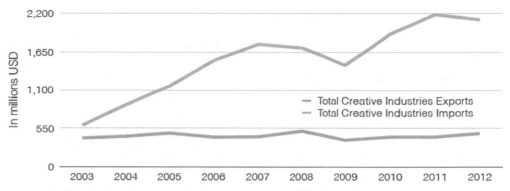

Figure 1: South Africa's creative industries trade performance by sector, 2008 and 2012.
Source: UNCTAD (2016)

bureaucracy, creative organizations, creative communities, creative schools and universities, and so forth. Therefore, broadly speaking, increased promotion of creativity is the key to creative city strategies. SDG goal 11 focused on making cities inclusive, safe, resilient and sustainable. This agenda views cities as hubs for ideas, culture, commerce, science, social development, productivity and much more. At their greatest capacities, cities allow people to advance economically and socially.

Since apartheid, most South African cities, faced with the issue of planning for urban regeneration, have gradually used creative policies within urban regeneration plans to enhance the change to a post-industrial city, to improve the quality of life for all urban users, and to re-establish the image of the city (Booyens 2012). One of the main problems in analyzing *urban creativity* is the term itself. There has been a profusion of debate about the concept of urban creativity and the concept of creativity more generally (Hall 2000; Landry, 2000; Scott 2000; Florida 2004; Florida and Tinagali 2004; Hutton 2004; Lloyd 2006; O'connor and Kong 2009; Scott 2007; Wood and Landry 2007; Pratt 2010). The problem with these authors' writings is that they use the term in different ways and with different assumptions. Urban creativity policies represent different social, political and practical objectives: that is, the urban creativity concepts are positioned for political and social intonation in social, urban or economic policy.

In Evans (2005) the term 'urban regeneration' is defined as the transformation of a commercial or residential area that has shown the symptoms of social, physical, and/or economic decline, but into which new life has been or is being breathed, causing sustainable, lasting improvements in the local quality of life, including socially, economically and environmentally. It happens that there is no precise plan for an urban policy or strategy using creative industries for economic regeneration (Rutten 2016). However, this does not indicate that all strategic objective potentials are realistic alternatives for every city.

David Hesmondhalgh (2007) claims that creativity is a crucial factor in urban regeneration, on the assumption that twenty-first century industries will depend progressively more on the generation of knowledge through innovation and creativity. Some authors argue that the higher the level of a creative class in a city, the greater the city is connected with stability, economic prospects and competitive

advantage (e.g. Florida 2002; Miszlivetz and Markus 2013). In connection with creating a creative class, Hesmondhalgh (2007) argues that the creative class is strongly interested in creative goods of several types, and cities should therefore invest in culture. Abisuga-Oyekunle and Fillis (2016) in their study identify that creative industries increase employment and assist in the generation of sustainable livelihoods. Florida (2002) argues that in addition to employment opportunities, cultural amenity and liberal social values are important attractors to cities, in deciding where to live. In general, these authors argue that cities are the key social and economic organizing component of the present age, marrying people with jobs.

Creative cities and South Africa

There is a persistent need for South Africa to grow its economy and improve social conditions (Oyekunle 2015c). According to Shand (2010), South Africa has since the late-nineties progressively been using 'culture' and creative urban policies for urban regeneration strategies to design industrial cities for the transition to post-industrial cities. As a result, cities now cultivate great interest in becoming 'creative' and in embarking on big creative projects and hosting big cultural institutions (Aber 2009). Two cities, in particular, in South Africa, represent the concept of creative cities, namely, Johannesburg and Cape Town.

South African creative cities: Johannesburg and Cape Town

Histories

The city of Johannesburg located in the Gauteng province was founded in 1886. The city is home to about 38% of Gauteng's population and to more than 7% of South Africa's population (City of Johannesburg Metropolitan Municipality 2004). It is the most well-developed creative economy in South Africa and home to the highest concentration of creative enterprises in the country. The City of Cape Town was established by Provincial Notice 479 of 2000 dated 22 September 2000 in terms of section 12 of the Local Government: Municipal Structures Act, 1998 (Act No. 117 of 1998), and includes any committee or sub-council established by the City or duly authorized agent or any employee of the City, acting in connection framework virtue of a power vested in the City and delegated to such committee, sub-council, employee or agent (City of Cape Town 2014). The city is located in the Western Cape Province.

Current creative policies

The Johannesburg Metropolitan Municipality states in its growth and development strategy that since December 2000 the City of Johannesburg has acknowledged the necessity for a forceful policy and a long-term and fresh innovative strategy to established Johannesburg on an equitable growth path for economic sustainability. In 2002, this strategy helped the design launching of the Joburg 2030 plan (City of Johannesburg Metropolitan Municipality 2004). The City of Johannesburg has been dealing with unemployment through the Johannesburg 2030 plan, and particularly by supporting labour-intensive sectors, for instance, creative industries and call centres (South Africa 2006). The City of Cape Town is aiming to be a global city in order to increase its competitiveness on the global platform (Lemanski 2007). The city's culture is regarded as a driver for urban regeneration and local economic development (Rogerson 2006; Evans 2009). In addition, Cape Town has recently begun advertising itself as a creative city. In addition, for the past decade, Cape Town has experienced inner city renewal.

According to Rogerson (2006), the creative industry strategies in Johannesburg were built on the foundations of the Creative Strategy Group, which argued for changing the name of the sector in Johannesburg to 'creative industries'. Newton (2003), in his research, critically highlights the fact that the creative industries sector in Johannesburg controls the national profile and that the sub-sectors that control the local landscape are the craft, performing arts, visual arts, music and film. The focus of the strategies, hence, was primarily on TV and film, with Johannesburg (alongside Cape Town) as the key national centre for performing arts, music, visual arts, design and crafts.

Likewise, the exciting creative sector in Cape Town is a blend of creativity, appealing to both local and foreign investors and integrating local creative culture with international trends. Cape Town was declared the 2014 Design Capital of the World in 2011, an award set to place the city on the global design map and poising the Western Cape as a world-class destination for creative industries. The various provincial and city authorities have combined their efforts in the direction of creative clustering in the city to showcase the role of design and the creative economy in job creation, community empowerment and social cohesion. In association with Cape Town, the Creative Cape Town theme was created which promotes creative industries as a portion of the central city's economic strategy. National events like the advertising sector's Loerie Awards and the Design Indaba, in conjunction with the country's leading fashion week, take place in Cape Town. The city also provides inexpensive production costs for video production and film (20% cheaper than Europe and 15% cheaper than Australia), along with the leading visual effects studio in South Africa.

Creative urban development

Past research shows Johannesburg and Cape Town as citadels for creative urban development. For instance, research conducted by Create SA in 2003 showed that over 40% of all creative enterprises are found in the Gauteng province, with one of its major cities, Johannesburg, boasting the highest group of companies in many sectors of the creative industries (South Africa 2006). In addition, research carried out by the City of Cape Town showed that in 2003 the Cape Town International Jazz Festival contributed an extra R58 million to the local economy. This event alone, which is in fact the biggest event in the Western Cape province, accounts for almost R120 million in income for the city. Also, the City of Johannesburg established its first Growth and Development Strategy (GDS) in 2006, as a long-term strategy that articulates Johannesburg's future development track (South Africa, City of Johannesburg 2011).

The creative industry is being used more and more as a vehicle for urban regeneration, a means of development adopted by Johannesburg with its huge investment in, for example, the Newtown Cultural Precinct. In order to develop a creative response to urban problems in cities such as Cape Town and Johannesburg, whether business development, rejuvenation of city areas or addressing the loss of identity in communities, a new system of knowledge must be developed, using a different approach to re-visit all existing systems, responses and procedures. City and urban planners, managers and politicians need to ascertain that different levels of innovation are appropriate for different areas (South Africa, Department of Sports, Arts, Culture and Recreation 2006).

It is necessary for cities to implement and develop a comprehensive development strategy to help build creative industries, open economic opportunities for creative industries and provide opportunities for showcasing their work to the community. However, this should not exclude the traditional arts and culture sectors. Most of the support, promotion and facilitation of creative industries' development in the direction of the government's socioeconomic objectives have come through the arts and creative industries. Also, museums and heritage concentrate on those interventions that preserve, develop, promote and identify heritage resources, which, in turn, contribute to urban economic growth, especially in Johannesburg and Cape Town, and in nation-building.

According to Simmie et al. (2006), in Harris (2007), sustainable urban development can continually help to upgrade the business environment, as well as the physical, workforce, social and creative infrastructure. This helps ensure high growth, and lucrative and innovative firms with an educated, creative and entrepreneurial workforce attracted and retained. This, in turn, allows cities to accomplish high productivity and growth rates, low levels of inequality, high employment, wages and gross domestic product (GDP) *per capita* and social exclusion (Simmie et al. 2006; in Harris 2007). Hlobo and Sanan (2014) provide the following policy aims for sustainable urban development:

• Celebrate the city's rich history and diversity, for inclusive, social cohesion and city positioning.
• Maximize the city's cultural and creative assets and resources.

- Maximize opportunities to support economic growth and sustainability through creative industry events.
- Honour and support arts and culture entities and relevant infrastructure.
- Support urban and community regeneration through public arts, monuments, and the development of cultural centres.
- Preserve and promote our tangible and intangible heritage.

Main policy directives:
- Facilitate citizen engagement with the public life within the city, including its creative diversity, public space, history and memory.
- Ensure that the economic potential of arts and culture is recognized and used to help boost tourism, economic growth and job creation.
- Ensure that the city's arts and creative assets are marketed locally, nationally and internationally for their excellence and/or local distinctiveness.
- Ensure mechanisms and partnerships to increase funding and support for arts and culture, acknowledging and rewarding projects of excellence.
- Optimize city-owned infrastructure of use of arts and culture bodies to support their development and that of the community they serve, and/or as spaces showcasing arts and culture at scale.
- Ensure mechanisms for coordination within the city and between the city and the arts and creative sector.
- Ensure empirical research that enables the city to improve its understanding of the sector and the delivery of services and support.

Creative economy fundamentals:
In July 2009, UNCTAD was invited by the South African National Arts Council, the ARTerial Network and the African Arts Institute to address a seminar by presenting the findings of the Creative Economy Report 2008. The aim was to consider expectations for the creative economy to improve development in South (United

Nations 2010). According to Wesgro (2016), the creative industries signify about 2% of the Western Cape province's GDP, with an estimated value added of R3 billion. The design sector employs approximately 18,000 people, which corresponds to 2% of Cape Town's employment. Also, the sector enjoys great support from different organizations, for instance, the Western Cape Clothing & Textile Service Centre (Clotex), Cape Craft and Design Institute, and the Creative Cape Town. As previously noted, national events like the advertising sector's Loerie Awards, the Design Indaba and the country's leading fashion week take place in Cape Town. The City of Johannesburg (2010) established that the Gauteng audio-visual industry contributes over R2.5 billion, with over 70% of the South African film and TV industry based in Gauteng. In the Cannes Lions Report 2009, Johannesburg was ranked the 10th most creative city in the world. Johannesburg has been home to the International Broadcast Centre and the sophisticated IP-based broadcasting platform. Also, Johannesburg, as the primary film shoot location, handles over 50% of all commercials shot in South Africa. Johannesburg is well positioned to become the centre of excellence in Africa in innovative and creative media and entertainment content generation, plus animation and computer-generated graphics (CGG).

Creative industries foreign direct investment (FDI) and contribution to urban development:
Recently, as Figure 2 shows, the inflow of FDI projects into South Africa has sufficiently contributed to the needed diversification of the urban economy. There is evidence indicating that FDI has the potential to contribute to growth through complementing domestic investment, as well as the transfer of skills for urban regeneration (United Nations 2013). The FDI in creative industries represents major potential for South African urban development. Thus, for the country to achieve its potential, there is a need to work hard on its foreign investment policy framework, infrastructure and the costs of doing business to

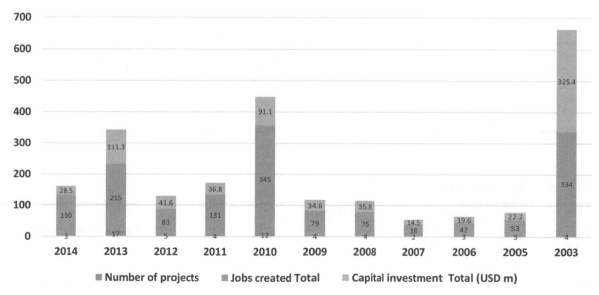

Figure 2: Number of FDI projects by year.
Source: fDi Intelligence from the *Financial Times* Ltd, 2014.

levels that will create opportunities for foreign investors. Furthermore, such actions will generate avenues for the country's creators and producers to compete globally and regionally (Joffe and Newton 2008).

The creative sectors have high economic growth potential, with the greatest prospects of contributing to exports, FDI and employment. Moreover, projects in the creative industries can create a positive influence on the youth because they provide them with many possibilities to be active in urban development schemes. Abisuga-Oyekunle and Fillis (2016) suggest that development strategies must identify the youth as a significant stakeholder in the development process. This will also contribute to social inclusion by attracting young people from the rural communities. The country has a vast and growing economy, and, if properly managed, could help attract considerable amounts of FDI (Pigato 2001).

Table 3 shows the Western Cape with both the highest average investment per project and the highest total investment of US$17.70 million and US$389.20 million, respectively. Gauteng has the highest number of jobs, while the Western Cape has the biggest project size with an average of 30 jobs per project.

Table 4 shows that out of a total of three destination cities, Johannesburg was the leading destination city, accounting for almost half of tracked projects. In this destination city, project volume increased in both 2010 and 2013. In each of these periods, eight projects were tracked.

Cape Town has the most jobs created (670), while Pretoria has the biggest project size on average, in terms of both job creation and investment.

Linking with foreign organizations can play an important role in learning and strengthening of local enterprises, assisting them to move into higher quality products and processes. Therefore, some strategy for FDI growth support must be provided around revitalizing local firms and putting together a different range of trade and projects policies designed to strengthen an active relationship between local and international enterprises. The strengthening of policy is of high significance to the creative

industries, especially to permit fair competition between highly focused foreign and national companies with prominent positions in creative services and value-added products.

Discussion

For the past 20 years, creative industries have been recognized as the main drivers of tourism, personal wealth, employment and foreign investment. These industries have likewise been employed for urban renewal projects. But, there is a widening gap between the developing and developed worlds, and a trend for the latter to consume creative products mainly from the former. That is why developing creative industries should be aligned with the objective of the SDGs to create new technologies and innovations that are available for urban development. The Gauteng province's Creative Industry Development Framework is one example that demonstrates the significance of the contribution to the success of the SDGs that development strategies can make. This framework makes clear the contribution of creative industries to social development goals. For example:

- community involvement in creative activities;
- regional integration and poverty alleviation across Africa;
- urban regeneration, especially in previously disadvantaged areas; and
- public-private/international partnerships in community-based creative programmes and projects such as indigenous music, fashion weeks, craft markets/festivals and FDI projects.

The exports of South Africa's creative industries increased by US$70 m between 2003 and 2012, attaining a total value of US$480 m. The export of creative goods grew from 2003 to 2012, with these average percentage increases: design 30%, audio-visuals 18%, jewellery 12%, visual arts 9%, recorded media 9% and new media 8%. In the new growth path for urban development, the South African government has made a commitment, which envisions the creation of five million jobs over

Table 3: FDI trends by destination state.

Destination state	No. of projects	No. of companies	Jobs created		Capital investment	
			Total	Average	Total (US$ million)	Average (US$ million)
Gauteng	31	31	755	24	318.20	10.30
Western Cape	22	22	670	30	389.20	17.70
Not specified	8	8	100	12	54.00	6.80
Total	61	58	1525	25	761.40	12.50

Source: fDi Intelligence from the *Financial Times* Ltd, 2014a.

Table 4: FDI trends by destination city.

Destination city	Projects		Companies		Jobs created	Capital investment (US$ million)
	No.	%	No.	%		
Johannesburg	28	45.90	28	48.28	621	211.70
Cape Town	22	36.07	22	37.93	670	389.20
Pretoria	3	4.92	3	5.17	134	106.50
Not specified	8	13.11	8	13.79	100	54.00
Total	61	100.00	58	100.00	1525	761.40

Source: fDi Intelligence from the *Financial Times* Ltd 2014a.

the next ten years (United Nations 2015). Furthermore, the creative industries have been identified as a significant contributor to the plan towards urban regeneration and the creation of jobs. The South African government has also recognized the TV and film industry as a sector that must be developed strategically in order to contribute more directly to economic development in the areas of employment, export and investment.

An analysis of FDI for creative industries suggests that there is a strong and positive connection between creativity, trade, FDI and economic development. However, South Africa's investment flows in the creative industries have not recorded a sharp increase in recent years. Figure 2 shows that the number of projects peaked in 2013 and then rapidly declined. Moreover, altogether, between January 2003 and September 2014, only 61 FDI projects were recorded nationwide. These projects represented a total capital investment of US$761.40 million, which is an average investment of US$12.50 million per project. During that same period, 1525 jobs were created. Average project size peaked in 2003 for both jobs created and capital investment.

Also, along with creative sector roles in urban regeneration strategies, there is a call for a debate on the role of the South African government in promoting creative activities within regional structural policy and in the same locations as urban regeneration initiatives. Particular attention should be given to addressing the strategies and policies carried out in major creative cities (such as Cape Town and Johannesburg). While the South African government is still committed to supporting small enterprises, as one of the drivers of economic growth required to make an impact on poverty and unemployment in the country, it is unable to successfully support such initiatives.

Furthermore, the particular effects of urban regeneration in South Africa are subject to the country's stage of development and income level. For example, as earlier noted, the Western Cape is well positioned to become a world-class destination for creative industries since Cape Town city was awarded the World Design Capital 2014 award (Wesgro 2013). South African creative imports experienced a significant increase in 2012, reaching a total of US$2,109 m, indicating the growing significance of cultural consumption. South African provincial and city authorities have combined efforts to encourage creative clustering in urban areas to showcase the role of design and the creative economy in urban regeneration, job creation and social cohesion. In connection with the Cape Town Partnership, the Creative Cape Town theme was established to encourage creative industries as a component of the economic strategy for urbanization. Likewise, the creative industries in Gauteng present important opportunities to contribute to the province's economic and social objectives. To realize this potential, a concise and practical strategic framework is needed to channel the implementation of creative industry projects across the province (South Africa 2006).

In summary, creative development in both Cape Town and Johannesburg show wealth creation, community development and job creation as major benefits that can develop from creative industry strategies for the urban economy. This study shows that there is a remarkable and productive interaction between creative economic activity, regeneration and inspiring quality of life in urban regions. Physical urban regeneration projects that adapt creative industries are profitable in many ways, not only for economic development but also for improvements to a city's image and reputation. Our analysis shows that the role of creative industries in economic development must be considered locally as well as nationwide, taking into account local contexts, characteristics, assets and resources. Regional or local regeneration strategies can make use of experiences from Johannesburg and Cape Town if problems, opportunities and situations are similar. However, any lessons from these cities must be 'interpreted' for local situations so that the particularities and local contexts are taken into account. Many instruments and policies used by the two leading cities mentioned in this study have proved to be valuable in other international cities. Examples include policies that relate to building networks, promoting creative institutions, encouraging collaboration, improving the exchange and supply of information, as well as education for creative entrepreneurship.

Concluding remarks

Although 'creative city' promotion has been successful and progressive in Cape Town and Johannesburg, it is inadequate for making sure local and regional economic development, wide-ranging job creation and creative diversity promotion occur. In other words, making Cape Town and Johannesburg more beautiful and marketable to visitors and foreign investors will not solve either city's challenges. When designing 'creative cities', social and economic development strategies should be considered. Thus, in order to develop the economy, an extensive policy framework for job creation and poverty alleviation should be mandatory. In addition, localized methodologies for regional economic development are required that understand the local creative industries' potential as a sector for urban development, job creation and local economic growth. This study observed a lack of such strategies that allow the creative industries to become an essential part of local sustainable urban development. To achieve this, it is necessary to deliberate on the multidimensional nature of the impacts that creative industries have on urban development, particularly, impacts on society, infrastructure, the economy and culture in each region.

Cape Town and Johannesburg are prospective creative mega-cities; therefore, they require specific policies and strategic measures to improve creative industries and make sure that the benefits extend broadly to other cities. Proper policy measures can help ameliorate the negative impacts of urban renewal and ensure social securities, such as job provision and livelihood for the people, social inclusion, basic infrastructure development, and the promotion of culture, traditions and heritage. Oyekunle (2014) argues that, in order to ensure social benefits, an urgent and radical policy action debate in South Africa is required. Oyekunle (2014) provides seven structured recommendations:

(1) Strengthening the industries' structure and policy framework.
(2) Capitalising on the creative sector's contribution to poverty reduction and economic development.
(3) Allowing capacity-building activities at different levels.
(4) Investing in sustainable creative industry development along the whole value chain.
(5) Acknowledging that despite its economic importance, the creative economy also produces non-monetary value that contributes to achieving people-centred, inclusive and sustainable development.
(6) Introducing a systematic communication system and awareness-building for the creative industries.
(7) Examining the relationships between the formal and informal sectors as central for modified creative industry policy development.

Urban committees should be established, and planners should be recruited who make sure that the policies and strategies of the government, policymakers and businesses can make cities more attractive to investors and meet local needs. The planners should also be responsible for ensuring that government provides basic services, improves infrastructure and develops a quality environment. Also, the government should intervene to allow the development of industrial areas to allow for creative production in such areas. In addition, policy responses and effective planning and regulation should be created to guarantee creative entrepreneurs and investors affordable spaces in the creative cities, so that they can contribute to local economic development.

Creative community-based programmes should also be established within cities. This is to enable communities' participation in decision-making for establishing long-term social sustainability and social inclusion. The programmes should provide opportunities for self-development through entrepreneurship skills. Local communities should be offered an opportunity to acquire skills and training in order to be absorbed into the national economy. To stimulate production, creative industries can be raised up to enhance their contribution to urban regeneration. This study also identified that higher-level skills development should be promoted, and training and education initiatives for the creative industries should be upgraded. Oyekunle (2015a) argues that to enhance the creative industries' competitiveness, the country needs industry-led skills training that facilitates growth in entrepreneurship and supports access to progress within the industries for individuals from all backgrounds.

To support creative cities, appropriate public policy is required. This consists of community development to allow for creative production and community engagement to promote diversity. Museums, libraries, theatres, schools, community centres and recreation facilities should be developed and upgraded. Also, events should be created to celebrate cultural diversity. Existing creative events in Cape Town tend to celebrate specific cultures and do not essentially promote creative diversity. Additionally, creative firms should be engaged in a programme to enhance diversity.

Finally, despite the fact that improvement has been made in relation to urban regeneration and promoting the creative industries in Cape Town and Johannesburg, some challenges still exist. To provide a dynamic and an enabling environment, action should be taken to guarantee adequate basic amenities and development of infrastructure. Creative diversity should be promoted, and inclusive development should be ensured by providing tangible opportunities for the whole community to both shape development goals and benefit from them. Also, debates about South African creative industries should be embedded in developmental goals, with the aim of involving as many inhabitants of cities as possible.

Disclosure statement
No potential conflict of interest was reported by the author.

ORCID
Oluwayemisi Adebola Oyekunle ⬥ http://orcid.org/0000-0001-7738-3132

References
Aber, J. 2009. *The Future of Shrinking Cities: Problems, Patterns and Strategies of Urban Transformation in a Global Context*. Pallagst Karina et al. eds. California: Institute of Urban and Regional Development.
Abisuga-Oyekunle, O. A., and I. Fillis. 2016. "The Role of Handicraft Micro-enterprises as a Catalyst for Youth Employment." *Creative Industries Journal*. Routledge: UK. doi:10.1080/17510694.2016.1247628.
Bontje, M., S. Musterd, K. Zoltan, and A. Murie. 2011. "Pathways Toward European Creative Knowledge City-Regions." *Urban Geography* 32 (1): 80–104.
Booyens, I. 2012. Creative Industries, Inequality and Social Development: Developments, Impacts and Challenges in Cape Town. [Online] Assessed October 2, 2014. http://download.springer.com/static/pdf/992/art%253A10.1007%252Fs12132-012-9140-6.pdf?auth66=1422016433_ae8ed5eb1234e2b27d2eb62aa4af95cc&ext=.pdf.
Canada, Policy Research Group. 2013. "The Creative Economy: Key Concepts and Literature Review Highlights." *Canadian Heritage*, May, 2013.
Duxbury, N. 2004. "Creative Cities: Principles and Practices." Background Paper F47 Family Network. Canada Policy Research Networks, Ontario.
Evans, G. 2005. "Measure for Measure: Evaluating the Evidence of Culture's Contribution to Regeneration." *Urban Studies* 42 (5/6): May: 959–983.
Evans, G. 2009. "Creative Cities, Creative Spaces and Urban Policy." *Urban Studies* 46: 1003–1040.
FDi Intelligence. 2014. "FDI into South Africa in Creative Industries." FDi Intelligence from the Financial Times Ltd. Trends Report, January 2003 to December 2014.
Florida, R. 2002. *The Rise of the Creative Class: And how It's Transforming Work, Leisure, Community and Everyday Life*. New York: Basic Books.
Florida, R. L. 2004. *Cities and the Creative Class*. London: Routledge.
Florida, R. L., and I. Tinagali. 2004. *Europe in the Creative Age*. London: Demos.
Hall, P. 2000. "Creative Cities and Economic Development." *Urban Studies* 37: 639–649.
Harris, N. 2007. "City Competitiveness." Originally drafted for a World Bank study of competitiveness in four Latin American cities, 2007 Unpublished. [Online]. Accessed May 1, 2016. http://www.dpu-associates.net/system/files/City+Competitiveness+09.pdf.
Hesmondhalgh, D. 2007. *The Cultural Industries*. London: Sage.

Hlobo, Z., and S. Sanan. 2014. City of Cape Town Draft Arts, Culture and Creative Industries Policy. Introduction to and Review of Policy Document. Cultural Policy Reading Group, 30th April.

Hutton, T. A. 2004. "The new Economy of the Inner City." *Cities* 21: 89–108.

Joffe, A., and M. Newton. 2008. *The Creative Industries in South Africa. Sector Studies Research Project.* Cape Town: HSRC Press.

Landry, C. 2000. *The Creative City. A Toolkit for Urban Innovators.* London: Sterling.

Lazzeretti, L., R. Boix, and F. Capone. 2008. ""Do Creative Industries Cluster? Mapping Creative Local Production Systems in Italy and Spain"." *Industry and Innovation* 15 (5): 549–567.

Lemanski, C. 2007. "Global Cities in the South: Deepening Social and Spatial Polarisation in Cape Town." *Cities* 24 (6): Jan. 448–461.

Lloyd, R. D. 2006. *Neo-Bohemia: Art and Commerce in the Postindustrial City.* London: Routledge.

Miszlivetz, F., and E. Markus. 2013. Creative cities, sustainable regions. ISES Working Paper Series, June, 2013.

Newton, M. 2003. *Joburg Creative Industries Sector Scoping Report.* Unpublished Report for the Economic Development Department, City of Johannesburg.

Oakley, K. 2009. "Getting Out of Place: The Mobile Creative Class Takes on the Local." In *Creative Economies, Creative Cities, Asian-European Perspectives*, edited by L. Kong and J. O'Connor. Dordrecht, Netherlands: Springer.

Oakley, K., and J. O'connor. 2015. *The Routledge Companion to the Cultural Industries.* New York: Routledge.

O'connor, J. 2004. "A Special Kind of City Knowledge: Innovative Clusters, Tacit Knowledge and the "Creative City'." *Media International Australia* 112: 131–149.

O'connor, J. 2007. Manchester: The Original Modern City. The Yorkshire and Humber Regional Review, special edition, 13–15.

O'connor, J. 2009. "Creative Industries: A New Direction?" *International Journal of Cultural Policy* 15 (4): 387–402.

O'connor, J., and L. Kong. 2009. *Creative Economy, Creative Cities: Asian European Perspectives.* Dordrecht: Springer.

Oyekunle, O. 2014. "Building the Creative Industries for Sustainable Economic Development in South Africa." *International Journal of Sustainable Development* 7(12): Dec. 47–72.

Oyekunle, O. A. 2015a. "Developing Creative Education in South Africa: A Case of Western Cape Province." In *Cultural Entrepreneurship in Theory, Pedagogy and Practice*, edited by O. Kuhlke, A. Schramme and R. Kooyman. Rotterdam: Eburon.

Oyekunle, O. A. 2015c. Management of Design and Innovation for Sustainable Development of South Africa Handicraft Sector. SARChi Working paper series, No. 2015-009.

Pigato, M. A. 2001. The Foreign Direct Investment Environment in Africa. Africa Region Working Paper Series No. 15, April.

Pratt, A. 2010. "Creative Cities: Tensions Within and Between Social, Cultural and Economic Development. A Critical Reading of the UK Experience." *City, Culture and Society* 1: 13–20.

Pratt, A. 2015. "Creative Industries and Development: Culture in Development, or the Cultures of Development?" In *The Oxford Handbook of Creative Industries*, edited by C. Jones, M. Lorenzen, and J. Sapsed. UK: Oxford University Press.

Rogerson, C. 2005. "Case Study Prepared for the World Bank-Netherlands Partnership Program Evaluating and Disseminating Experiences in Local Economic Development." Investigation of Pro-Poor (LED) in South Africa, 2005.

Rogerson, C. 2006. "Creative Industries and Urban Tourism: South African Perspectives." *Urban Forum* 17: 149–166.

Rutten, P. 2016. "Cultural Activities & Creative Industries, a Driving Force for Urban Regeneration." Urban Act [Online]. Accessed April 3, 2016. https://www.academia.edu/4276834/Culture_and_urban_regeneration._Findings_and_conclusions_on_the_economic_perspective.

Scott, A. J. 1997. "The Cultural Economy of Cities." *International Journal of Urban and Regional Research* 21 (2): 323–340.

Scott, A. J. 2000. *The Cultural Economy of Cities: Essays on the Geography of Image Producing Industries.* London: Sage.

Scott, A. J. 2007. "Capitalism and Urbanization in a new key? The Cognitive–Cultural Dimension." *Social Forces* 85: 1465–1482.

Shand, K. 2010. "Newtown: A Cultural Precinct – Real or Imagined?" *MA Thesis.*, University of the Witwatersrand, Johannesburg.

South Africa. 2013a. *Creative Industries Sector Fact Sheet.* The Western Cape Destination Marketing Investment and Trade Promotion Agency (WESGRO).

South Africa. 2013b. "Revised White Paper on Arts, Culture and Heritage." Version 2 (4 June 2013). [Online]. Accessed February 11, 2014. http://www.dac.gov.za/sites/default/files/REVISEDWHITEPAPER04062013.pdf.

South Africa, City of Cape Town. 2013. Draft Tourism Development Framework 2013 to 2017. March 2013.

South Africa, City of Cape Town. 2014. Arts, Culture and Creative Industries Policy – (policy number 29892). Approved by council on 3rd December 2014 C22/12/14.

South Africa, City of Johannesburg Metropolitan Municipality. 2004. Department of Finance and Economic Development. Final offering circular dated 13th, April.

South Africa, City of Johannesburg Metropolitan Municipality. 2011. *Joburg 2040: Growth and Development Strategy.* Braamfontein, Johannesburg: City of Johannesburg Metropolitan Municipality.

South Africa, Department of Sport, Recreation, Arts and Culture. 2006. Budget Statement 2, Vote 12.

South Africa, Western Cape Destination Marketing, Investment and Trade Promotion Agency. 2016. Sectors: Creative industries. Wesgro key facts.

United Kingdom, Department for Culture, Media and Sport. 2014. *Creative Industries Economic Estimates. Statistical Release.* UK: DCMS.

UNCTD. (United Nations Conference on Trade and Development). 2010. *Creative Economy Report 2010: a Feasible Development Option.* Geneva: United Nations.

UNCTD. (United Nations Conference on Trade and Development). 2013. "Strengthening Linkages between Domestic and Foreign Direct Investment in Africa." Fifty-seventh executive session, 26-28 June 2013.

UNCTD. (United Nations Conference on Trade and Development). 2015. "Creative Economy Outlook and Country Profiles: Trends in International Trade in Creative Industries." Unedited, United Nations.

UNCTD. (United Nations Conference on Trade and Development). 2016. "Creative Economy Outlook and Country Profiles: Trends in International Trade in Creative Industries." United Nations. [Online]. Accessed July 12, 2016. http://unctad.org/en/PublicationsLibrary/webditcted2016d5_en.pdf.

WESGRO. 2013. "Creative industries sector fact sheet. The Western Cape Destination Marketing Investment and Trade Promotion Agency 2013." [Online]. Available from: http://wesgro.co.za/publications/publications/2013-creative-industries-sector-fact-sheet. [Accesed: 13/7/2016].

WESGRO. 2016. "Western Cape Destination Marketing, Investment and Trade Promotion Agency. Sectors: Creative industries. Wesgro key facts." [Online]. Available from: http://wesgro.co.za/investor/sectors/creative-industries.

Wood, P., and C. Landry. 2007. *The Intercultural City: Planning for Diversity Advantage.* London, Sterlinga: Earthscan.

In-vitro diagnostics (IVDs) innovations for resource-poor settings: The Indian experience

Nidhi Singh and Dinesh Abrol

This article illustrates how the present institutional arrangements and the policy regime under perusal have not been able to support the development of an ecosystem for innovation-making for *in-vitro* diagnostics (IVDs) for resource-poor settings. Policies favouring trade liberalization and foreign direct investment, market deregulation, strong intellectual property rights, absence of stringent regulations of accreditation and quality control and limiting public R&D support to basic research and development of scientific and technical manpower have defined the dynamics of innovation-making for the IVDs in India since the year 2000. Investment in the IVDs for the management of priority diseases for resource-poor settings continues to be only a small fraction of the R&D investment, translational research and market formation in India. Seventy-five per cent of diagnostics needs are still met through imported and maladapted diagnostic innovations. Although with the help of public-funded policy initiatives undertaken by the government there are now some young start-ups that have emerged and are beginning to focus on some of the diagnostic needs and challenges facing resource-poor settings, but they have not been able to enter the market with fully developed products in any kind of significant way. Low levels of interest in innovation, making for resource-poor settings, is reflected in the system-building activities of public sector R&D institutions, industry and the healthcare system. Lack of collaboration between national R&D institutions and large domestic firms continues to be the defining feature of the national innovation system in the case of IVDs. Innovation-making for resource-poor settings is yet to become a priority for the challenge-based innovation system-building approach. Analysis confirms the impact of convergence of neo-liberal deregulation, trade liberalization and investment liberalization on the processes of fragmentation and underdevelopment of capabilities, preference for market calculations over social calculations and undervaluation of the needs of resource-poor settings. Since the relevant actors have been bidding a good-bye to the values of universality, equality and comprehensiveness, the neglect of non-market based social calculations is reflected in the innovation-making practice of R&D institutions and industry. The article suggests that the challenge of innovation-making for resource-poor settings cannot be tackled without shifting away from the path of deregulation, liberalization of trade and investment, and moving into the implementation of a challenge-based innovation system-building approach by the state, in the case of IVDs.

Introduction

In-vitro diagnostics (IVDs)[1] are currently the key to healthcare value-chain improvement as these directly influence the quality of patient care, health outcomes and downstream resource requirement.[2] In the last few decades, technological advancements and the evolution of genetic and molecular biology[3] have laid the foundations for the development of more advanced and sensitive IVDs. In addition, advances in information technology have led to the automation of many tests that were previously conducted manually. Because of the continuous process of innovation-making with the use of the above-mentioned technologies, IVDs have taken several forms, from large, bench-top instruments that can only be used in sophisticated laboratories located in urban areas to point-of-care tests (POC) that can be used in remote rural regions which lack basic healthcare facilities, i.e. resource-poor healthcare settings.[4]

In the field of IVDs, resource-poor settings demand that the actors give priority to the relevant context specific considerations in innovation-making. One, the requirements of treatment of infectious diseases is distinct. Two, innovations need to fit appropriately the resource-poor conditions of healthcare facilities. Innovations that can be easily performed and maintained by well-equipped labs do not work well in the resource-poor healthcare facilities of rural and small towns in India. Three, innovations imported from the industrialized world do not perform properly due to lack of basic infrastructure of power, cold-chain facilities and so on, and the absence of ambient temperature. Four, the system of capabilities for the use of innovations should have a close fit with the requirements of accuracy, accessibility and affordability and to be met at the point-of-care (POC) end.

IVDs at the POC end are certainly more reliable for use in resource-poor healthcare settings. These can be 'readily administered by a minimally trained healthcare worker or can be self-administered, do not require an ambient temperature and also, theoretically, offers the ultimate "test and treat' model which reduces the likelihood of losing a patient to follow up in resource-poor settings' (Drain et al. 2014). One of the most significant examples is the introduction of Rapid Diagnostic Test for HIV and Malaria in early 1990s which has had a major impact on the way diagnostic and treatment are performed in resource-poor settings. It is because these immune-chromatographic based assays provide information on disease status with more sensitivity and in lesser turnaround time as compared to laboratory based ELISAs.

Focus on the development of policy instruments and institution building practices capable of promoting innovations for resource-poor settings is a priority of Indian policymakers. It is necessary for the policy regime to make researchers, entrepreneurs and healthcare personnel prioritize appropriate social calculation based system-building activities to deal with the challenges of indigenization and the development of diagnostics for resource-poor settings. Studies undertaken have highlighted how the challenges facing the indigenous development of IVDs have not been tackled appropriately in India. Since 2000 these challenges have continued without being addressed sufficiently and, to date, still persist in an acute form, we need to examine the problems of the existing policy regime and focus on the failures experienced vis-à-vis the contribution of the public sector R&D organizations, private sector industry and healthcare to the system-building activities for the benefit of innovation-making for resource-poor settings to find out what kind of policy change is necessary.

The article is divided into seven distinct sections. The next section reviews the studies made since 2000 on the status of IVDs industry and the impact of policies under perusal on the development of indigenous innovations in the area of IVDs. The section following that describes the theoretical and analytical framework used to analyze the failures being experienced in respect of innovation-making for unmet diagnostic needs and diagnostic technologies in resource-poor settings of India. The section thereafter describes methodology, followed by the section that analyzes the performances of system-building activities within the science, translational and industrial bases for the promotion of IVDs development for resource-poor settings. The next section maps and provides an analysis of the extent of the performance of system-building activities for IVDs development relevant to resource-poor settings in India. The penultimate section identifies in brief the challenges facing the development of IVD technologies in India and analyzes the innovation system failures (directionality, demand articulation, policy co-ordination and reflexivity) that illustrate why the functions of innovation systems in the case of IVDs are underdeveloped and how the influence on the practice and policy of R&D, technology development and innovation-making can be ultimately traced to the level of system-building activities that are the result of a policy regime characterized by liberal and open policies of trade and investment and strong patents. The final section concludes the study.

Literature review

Past studies highlight how the challenges facing the development of IVDs innovations were not tackled on account of the distortions in import duty structures which persist even to the time of this study. These distortions have been an important problem for local manufacturers trying to contribute to the challenge of indigenization of imported IVDs in India since 2000. Visalakshi (2009) argued that the non-availability of ELISA plates and various raw materials from an indigenous source were a key challenge for the development and diffusion of indigenous immunodiagnostics during the early nineties. These raw materials had to be imported, but attracted high import duties. Thus, imported finished products were cheaper than and preferable to indigenous products. Furthermore, because of economic considerations, the

local manufacture business of diagnostics was characterized by low revenue and small returns, resulting in domestic firms not being interested in venturing into the field of diagnostics.

Similarly, Ghosh (1996) also pointed out that how the customs duty structures were favourable, even during the late nineties, for the bulk import of finished diagnostic kits rather than local production. Import duties on raw materials were higher than those on finished products in the case of medical diagnostics. Jarosławski and Sabarwal (2013) showed that 75–80% of medical diagnostic needs were being met through imports to India, but that for the vast majority of the Indian population these imported kits were unaffordable. They also noted that many of these imported diagnostic kits were maladapted. In many cases, resource constraints in healthcare settings were reported to be mainly responsible for maladaptation.

Ramani and Visalakshi (2001) attempted an analysis of the economic, scientific, technological and industry-wide issues in the case of biotechnology based innovations. Their case studies of biotechnology innovation also highlighted how the commercialization of indigenous innovations was being hindered by system-wide factors. They recognized that the problem was not limited to the challenge of trade policy. While they did identify that these system-wide factors lay in the lack of indigenous capabilities, lack of promotion of local manufacturers, lack of coordination between research institutions and industry, variations in their culture, existence of trade and investment barriers, lack of articulation of demand for indigenous products and lack of development of supporting regulatory institutions in India, they did not examine the sources of the persistent failures vis-à-vis the contribution of all the different actors to innovation system-building activities for the introduction and diffusion of IVDs suitable for resource-poor settings in India.

It is clear from this brief literature review that the policy regime and practice of institution building have not been favourable to the indigenous development of in-vitro diagnostics for quite some time. In this article, we propose to address the issue of the source of persistence of these problems vis-à-vis the policy regime under perusal for the last two decades. In this article, the main focus of our investigation is how the challenge of indigenization and innovation-making for resource-poor settings could not be appropriately tackled by policy-makers and health innovation system builders in India.

Theoretical and analytical framework

Scholars building on the innovation system approach suggest that the adoption of a challenge-based understanding of "innovation system" development is necessary for the strengthening of the performance of healthcare system in a country that is catching-up with the developed world. Cassiolato and Soares (2015) show how the development and diffusion of healthcare technological innovations require the contribution of a strong scientific and industrial base through the system-building activities of all the actors participating in the health innovation system. They also show that the successful making of health innovations significantly depends on the "building blocks" of health system governance, human resource development, development finance, informatics, technologies and service delivery.

It is our understanding that the health innovation system should be seen as a system where the performance of the system depends on the quality of the relationships among the components of the system, and where the quality of the relationships among the building blocks of the health innovation system is determined by the logic of the policy regime. For the realization of health innovation system-building, it matters whether the policy regime favours market or non-market calculations-based system-building activities. The challenge-based innovation system-building approach enables the pursuit of non-market social calculations when the actors participating in the building of the science, translation and industrial bases are trying to contribute to the structures and functions of the national innovation system for the benefit of innovation-making for resource-poor settings and indigenization, namely knowledge creation, knowledge transfer and diffusion, industrial enterprise formation, market creation, regulation, legitimization and so on.

The challenge-based, non-market social calculation-based steering and coordination, direction setting, demand articulation and creation and reflexivity are essential to promote the system-building contribution of the relevant actors to innovation-making for resource-poor settings. In theory, all the relevant actors, whether public sector R&D organizations, foreign firms, large domestic firms or young start-ups, can be made to respond favourably and contribute to the development of an innovation system for in-vitro diagnostics for resource-poor settings. A systematic assessment of the contribution of market calculations-based policy instruments and a challenge-based innovation approach favouring non-market social calculations to the building of science, translational and industrial bases is the focus of our theoretical contribution in this article.

System-building activities relating to interactive learning, competence-building for advanced technologies, etc., can be measured. It is possible to trace the impacts of policy instruments on the outcomes for the establishment of the relevant structures and functions of the national innovation system. Connections of innovation system-building with the policy reforms framework under perusal can be investigated for the impact on interactive learning, competence-building in relation to the development of advanced technologies and innovation-making for the benefit of local priorities. It is our understanding that even in the framework of market creation policy, instruments favouring market-based calculations can undermine the system-building contribution of innovation-making actors to the development of a national innovation system for the introduction and diffusion of indigenous innovations for resource-poor settings. Most of the problems with innovation system-building can be expected to persist under a policy regime that does not adequately incorporate non-market calculations, while incentivising and disciplining the relevant actors in the case of a health innovation system. These are theoretical postulates which need to be explored using a historical analysis based investigation into the contributions of all the relevant actors to innovation system-building. Figure 1 depicts the analytical framework of the study.

Methodology

Keeping the challenge of confirming the relevance of the proposed theoretical framework in view, the study proposes to measure the contribution of **system-building** activities of all the relevant active actors, with the aim of seeking an understanding of the failures underway vis-á-vis the building of structures and functions of a national innovation system. System-building is defined as 'the deliberate creation or modification of broader institutional or organizational structures in a technological innovation system carried out by innovative actors' (Musiolik, Markard, and Hekkert 2012, 1035, section 2.3, paragraph 1). It includes the creation or reconfiguration of value chains as well as the creation of a supportive environment for an emerging technology in a more general way.

System-wide failures also arise on account of the failures on the front of anticipatory governance, namely in relation to the challenges of envisioning, steering and coordination, directions setting, demand articulation and creation and reflexivity being addressed inappropriately by the policymakers. In the case of new and emerging technologies, these system-wide failures are known to adversely affect the building of the science, translational and industrial bases. Outcomes of the efforts undertaken by the government, industry and users of IVDs for system-building for the benefit of resource-poor settings are measurable. Analysis of the outcomes focuses in this study on the anticipated impact on the directions of R&D activities being undertaken in the public and private sectors, stage of indigenous technologies developed by young start-ups, contribution of foreign firms, industrial structure, articulation of demand for investment in relevant diagnostic products and the extent of the development of the industrial base available in respect of components and finished products introduced by foreign and domestic firms.

Outcomes from the implementation of **system-building** activities are assessed in this study in terms of how the pursuit of market and non-market calculations-based system-building activity has ultimately impacted on the development of the science base through research and development promotion, the translational base through knowledge transfer and exchange and the industrial base through enterprise formation and market creation for the benefit of innovation-making for resource-poor settings in the case of IVDs.

Innovation actors and the indicators of performance of their system-building activities that have been used in the study are listed in Tables 1 and 2. System-building activities of innovation actors for the development of the science base, the translational base the and industrial base under IVDs innovation system are analyzed in order to identify the effect of current innovation and the policy regime on:

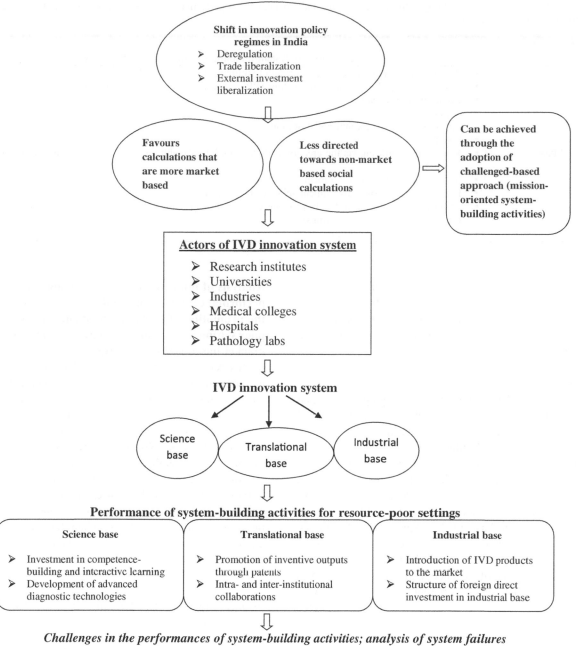

Figure 1: Analytical framework of the study.
Source: Authors own depiction

Table 1: Innovation actors analyzed in the study.

Innovation actors analysed in the study	
Categories	Name of actors
Research institutes	Institutes of CSIR, ICMR, DBT, DST active in IVD research
Universities	Jawaharlal Nehru University (JNU), Hyderabad Central University (HCU), Madras University, Banaras Hindu University (BHU), Indian Institute of Technology, Delhi (IIT, Delhi)
Medical colleges and hospitals	All India Institute of Medical Sciences (AIIMS), Post Graduate Institute of Medical Education and Research Chandigarh(PGIMER), Christian Medical College
Industries	
Large domestic diagnostic companies	Agappe diagnostics, Tulip diagnostics, Mediclone biotech, Bhat Biotech, J Mitra, Span Diagnostics Ltd.
Young start-ups	Xcyton Diagnostics, Achira labs, Reamatrix, Chromous biotech Bisen Biotech, Revelation Biotech, Bigtec Labs
Foreign diagnostic companies	Roche diagnostics, Qiagen India, Becton Dikinson India, Beckman Coulter India, Transasia Bio-Medicals, Siemens Diagnostics

Table 2: Indicators of system-building activities used in the study to analyze the performance of system-building activities in the science, translational and industrial bases.

Components of innovation	Indicators of system-building activities used in the study
Science base	• Promotion of financial resources through extramural research (EMR) • Promotion of public-private partnerships (PPPs) and product development partnerships (PDPs) • Collaborative research • G-finder analysis
Translational base	• Inter-institutional research collaborations • Patents
Industrial base	• Value-chain building via the creation of local suppliers capable of providing components of finished products • Indigenous product development • R&D activities using government schemes (PPPs & PDPs) • Value-chain development for the products embedded employing newer platforms and product development for resource constrained settings) • Collaborative R&D activities

1. *The innovation-making challenges specific to resource-poor settings.*
2. *The interventions that have been devised in the country and how adequate they are in dealing with those challenges.*

Analysis is done using the bibliometric method. Publications in IVD are mapped by using an indexed database *Web of Science*, which is a collection of online journals and academic citations provided by Thomson Reuters. G-Finder analysis was carried out on the publication activities in order to identify the type of research (basic, clinical and applied). The data was downloaded for the period 1991 to 2015. Information on extramural research projects (EMR) was collected from the database of the Department of National Science and Technology Management Information System (NSTMIS) for the period 2000 to 2015 available at Department of Science and Technology (DST) website. The datasets on patents in the IVD were obtained using the search engine http://www.ipindia.nic.in/, supported by Department of Industrial Policy and Promotion, Ministry of Commerce and Industry, Government of India, for the period 2000 to 2015. Further, a website search and study of annual reports of industries, research institutes, universities and government departments was done.

Analysis of innovation system functions and performance of system-building activities vis-á-vis the promotion of IVDs development for resource-poor settings using policy initiatives informed by non-market based social calculations

In this section, we highlight how far the past efforts of government policy initiatives for the promotion of innovation in the area of IVD for resource-poor settings have contributed to the innovation system-building challenge of innovation-making for resource-poor settings over the last two decades. Not many programmes and policy initiatives were embedded in non-market, social calculation-based system-building activities during this period in India. Below we analyze the extent of positive outcomes obtained vis-à-vis the formation of a science base, translational base and industrial base when the policy initiatives explicitly prioritized non-market social calculations in the building of innovation system functions in India.

Performance of system-building activities within the science base
Human resource development
Table 3 shows the contribution of the relevant system-building activities undertaken with the help of policy initiatives for human resource development in the field of diagnostics. The Department of Biotechnology

Table 3: System-building activities for Human Resource Development.

Departments	System-building activities for Human Resource Development
Department of Biotechnology	The Department of Biotechnology (DBT), Govt. of India has initiated a multi-institutional partnership programme on bio design called the National Bio design Alliance between Translational Health Science & Technology Institute (THSTI), Regional Centre for Biotechnology (RCB), International Centre for Genetic Engineering and Biotechnology (ICGEB), All India Institute of Medical Sciences (AIIMS), Indian Institute of Technology (IIT) Delhi, IIT Chennai and Christian Medical College (CMC) Vellore for the development of rapid point-of-care *In-vitro* diagnostic technologies. DBT in collaboration with Alagappa University and Bharthidasan University, Karaikudi, Tamil Nadu for post-M.Sc Advanced Diploma course in Molecular Diagnostics. DBT conducted an Indo-US workshop on 'Low-Cost and Therapeutic Medical Technologies' at the Centre for DNA Fingerprinting and Diagnostics (CDFD) on behalf of Department and National Institute of Biomedical Imaging and Bioengineering (NIBIB), USA for the development of low-cost biomedical imaging technologies and low-cost, point-of-care diagnostic technologies for disease areas of greatest need. Biodesign fellowship initiated in diagnostic and biomarker development through national biodesign alliance focusing on the development of point-of-care.

Table 4: Knowledge creation and development through publicly funded S&T institutions.

Departments	Institutes	Basic research	Experimental development, translational research and product development
Department of Biotechnology	Translational Health Science and Technology Institute (THSTI)	- Novel sample processing for the simple and rapid diagnosis of TB, MDR-TB and XDR-TB - Identification of novel protein biomarkers for early diagnosis of pregnant women at risk for preterm birth	Development of multiplexed, 'lab-on-chip' technologies to bring multiple biomarker tests on to a single, universal platform, instead of multiple diagnostic tests on multiple different instruments to be used in remote clinics and low-resource settings as well as secondary and primary hospitals.
Department of Health research (ICMR)	Centre for Research in medical entomology (CRME), Madurai	Detection of JE virus antigen in desiccated vector mosquitoes	–
	National Institute of Malaria Research (NIMR)		Indigenous production of monoclonal antibodies PfHRPII and pLDH achieved for improved diagnostics for malaria
	National Institute of Virology (NIV), Pune	A real-time RT-PCR useful for early diagnosis was developed for detection of dengue viral RNA	–
	Rajendra Memorial Research Institute of Medical Sciences (RMRIMS), Patna	–	Species-specific PCR developed for detecting Leishmania donovani, tested in the endemic area and widely utilized in referral labs as a confirmatory test
	National Tuberculosis Research Institute (TRC)	–	New rapid molecular methods for detection of rifampicin, isoniazid and ethambutol resistance in TB developed

Source: Compiled on the basis of information available on departmental websites as at July 2015

(DBT) pursued such activities in collaboration with universities like Alagappa University (Karaikudi) and Bharthidasan University (Tiruchirappalli) in Tamil Nadu initiated post M.Sc and the Advanced Diploma course in Molecular Diagnostics, and started a Biodesign fellowship for students through the National Biodesign Alliance, focusing on the development of point-of-care diagnostic technologies. A taskforce was developed in an Indo-US workshop conducted by DBT for the creation of human resources for the development of low-cost, biomedical imaging technologies and low-cost, point-of-care diagnostic technologies for disease areas of greatest need.

Knowledge creation and development through publicly funded S&T institutions
Table 4 shows the emerging contribution of public sector institutions to the building of the function of knowledge creation and development. In recent times, this function has been considerably strengthened by the activities of institutions such as the DBT and the Department of Health. Investment in basic and translational research projects is contributing to the emergence of young start-ups by focusing on the diagnostic needs of resource-poor setting and by helping the young start-ups to develop rapid, point-of-care

diagnostic tests and tests based on advanced diagnostic technologies. Further, thanks to this support, young start-ups are innovating in the area of reagents which can either persist in harsh environmental conditions or do not require ambient temperatures for storage.

Knowledge creation and development through public-private partnerships
Table 5 shows the strengthening of the function of knowledge creation and development through innovative funding mechanisms and public-private partnerships (PPPs). The DBT and the Council of Scientific and Industrial Research (CSIR) have been contributing through promotional funding schemes like the Small Business Innovation Research Initiative (SBIRI), the Biotechnology Industry Partnership Programme/Biotechnology Industry Research Assistant Council (BIPP/BIRAC) of the DBT and the New Millennium Indian Technology Leadership Initiative (NMITLI) of the CSIR. These schemes allocate funds to promote PPPs and young start-ups in the area of therapeutic, vaccine and diagnostic development for diseases of national importance. At present, in diagnostic development for resource-constrained settings, the SBIRI is supporting five projects, BIPP/BIRAP, three projects and NMITLI, two projects.

Table 5: Knowledge diffusion/transfer.

Departments	Knowledge transfer
Department of Biotechnology	CDFD has licensed its tuberculosis diagnostics technology to M/s ArkaNanomeds Pvt. Ltd. DBT supported AIIMS to develop rapid diagnostic test for leishmaniasis and a test for extra-pulmonary TB and this has been successfully licensed to Arbro Pharmaceuticals

Source: Compiled on the basis of information available on departmental websites as at July 2015

Table 6: Knowledge creation and development through public-private partnerships.

Departments	Funding schemes	Public-private partnerships
DBT	Biotechnology Industry Partnership Program (BIPP) in 2010 under the management of Biotechnology Industry Research. BIRAC has been set up with a vision to stimulate, foster and enhance the strategic research and innovation capabilities of the Indian biotech industry, particularly SMEs	- Supporting Big-Tech labs for the development of the point-of-care detection of infectious disease using handheld micro PCR - Supporting Revelations Biotech for the development of low-cost, rapid, quantitative PCR technology for molecular diagnosis - Supporting Chromous Biotech Multiplex Fast-PCR based diagnosis and prognosis of tuberculosis.
	Small Business Innovation Research Initiative (SBIRI) in 2005. SBIRI is managed by the Biotechnology Industry Research Assistance Council (BIRAC). SBIRI is set up to promote public-private partnerships	- Arbro Pharmaceuticals Ltd., New Delhi in collaboration with All India Institute of Medical Sciences and LRS Institute of TB and Respiratory diseases, developing diagnostic for TB - Bhat Bio-Tech India (P) Limited, Bangalore, in collaboration with the National Institute of Malaria Research, developing diagnostic for malaria - Bisen Biotech and Biopharma Pvt. Ltd., Gwalior, in collaboration with Jiwaji University, developing diagnostic for TB - Genomix Molecular Diagnostics (P) Limited, Hyderabad, in collaboration with BITS, Pilani, NIMR, New Delhi, NIMR, Jabalpur, Osmania University, developing diagnostic for malaria
CSIR	CSIR launched the programme of New Millennium Indian Technology Leadership Initiative (NMITLI) in 2001. NMITLI is set up to promote public-private partnerships	- Supporting Big-Tech labs for the development of novel molecular diagnostics for eye diseases and low vision enhancement devices - Supporting Big-Tech Labs for the development of point-of-care rapid diagnostic kit for TB

Source: Compiled on the basis of information available on departmental websites as at July 2015

Performance of system-building activities within the translational base

Knowledge diffusion/transfer

Table 6 describes the strengthening of the function of knowledge diffusion and transfer. This function is being strengthened with the contribution of the DBT through activities such as formation and growth of knowledge incubators like the Translational Health Science & Technology Institute (TSTHI), and initiated a multi-institutional partnership programme on Biodesign called the National Biodesign Alliance. This biodesign programme started with the alliance between THSTI, Regional Centre for Biotechnology (RCB), International Centre for Genetic Engineering and Biotechnology (ICGEB), All India Institute of Medical Sciences (AIIMS), Indian Institute of Technology (IIT) Delhi, IIT Chennai and Christian Medical College (CMC) Vellore. It focuses on the development of diagnostic technologies useful in resource-constrained settings. Centre for DNA finger printing and diagnostics (CDFD) and AIIMS have transferred their diagnostic technologies to ArkaNanomeds Pvt Ltd. and Arbro Pharmaceutical, respectively, contributing to knowledge diffusion.

Mobilizing users for the adoption of diagnostics services for resource-constrained settings and market formation

Table 7 shows the strengthening of the function of mobilizing users for the adoption of diagnostics services for resource-constrained settings. Market formation activities are being strengthened by the Ministry of Health and Family Welfare and the Department of Health Research. The challenge of expansion of the use of diagnostic services

in resource-poor settings can be met by increasing the number of mobile medical units in remote areas. Initiatives have been undertaken for the introduction of diagnostic technologies in several disease control programmes under which diagnostics for malaria, Japanese Encephalitis and the CD4 test for HIV have been successfully introduced in the National Disease Control Programmes (NDCPs). Initiatives, though small, have also been undertaken to enable workers to diagnose and effectively treat diseases in remote areas.

Performance of system-building activities within the industrial base

For the last few years, young start-ups have been playing a significant role in undertaking system-building activities for development of IVDs relevant for resource-poor settings. Analysis indicates that system-building activities undertaken by these young start-ups are mainly supported by government policy and initiatives.

Performance of system-building activities by young start-up firms: Role of government in supporting their entrepreneurial activities and resource mobilization

Most of these start-ups have been developed by science professionals. Converted into technocrats, they are playing a far more prominent role in the development of diagnostics for diseases related to resource-poor settings of India. They are making diagnostics cost-effective. They are developing rapid, point-of-care tests which do not require a sophisticated laboratory environment. Also, they are working on priority diseases. Table 8 shows the product profiles of young start-up firms and their

Table 7: Mobilizing users for the adoption of diagnostics services for resource constrained settings and market formation.

Departments	Action taken	Illustrative examples
Ministry of Health and Family Welfare	Ministry of Health and Family Welfare is undertaking initiatives to provide diagnostic services through various national level programmes to control communicable and non-communicable diseases	- Mobile medical units increased from 363 in 2010 to 442 in 2011 in order to provide diagnostic and outpatient care closer to hamlets and villages in remote areas under the national rural health mission programme - Initiatives have been taken for empowering grassroot workers in diagnosing and treating malaria cases in remote and inaccessible areas by scaling-up the availability of bivalent rapid diagnostic kits (RDK) and Artemisinin-based Combination Therapy (ACT) under NBVDCP
Department of Health Research	DHR has introduced diagnostics under several National Disease Control programmes	- A kit for Japanese encephalitis developed and supplied for the national programme. - Bivalent rapid diagnostic malaria kits tested, approved and successfully introduced into the national programme - CD4 count tests for HIV are introduced by NACO to provide this test free of charge in government hospitals

Source: Compiled on the basis of information available on departmental websites as at July 2015

focused disease areas. Table 9 shows the significant diagnostic technologies useful for resource-poor settings in which young start-ups have been granted patents.

Indigenous IVD product development by young, domestic start-up firms relevant for resource-poor settings

Coming to the indigenous product development by these young start-ups, 17 indigenous diagnostic products have been developed. Out of these 17 products, nine have been developed with the help of extramural research (EMR), government promotional funding schemes (NIMITLI, BIPP and SBIRI) and other international funding organizations. Further, out of the nine products, two products received EMR support while three benefited from government promotional schemes. Patents were granted only in the case of four products out of 17 indigenous products. TB and other viral and infectious diseases account for nine out of 15 indigenous products. Government support was absent in only the case of two indigenous products.

Big-Tech Labs, Bhat Biotech, Revelations, Bisen and Chromous used government schemes to develop indigenous products. Big-Tech Labs has been playing a major role. It is important to note that the promoter of Big-Tech Labs did his B. Tech and M. Tech at BITS, Pilani. It is a home-grown firm, which has significantly benefited from promotional schemes introduced by the government (Tables 10 and 11). As far as therapeutic area is concerned, TB, HIV and infectious diseases have been given greater priority (Table 12).

Table 8: Product/ services and diseases/ health conditions focused on by young start-ups involved in IVD development for resource-poor settings in India.

Young start-ups	Products / services	Diseases/ health condition focused
Mediclone Biotech Pvt. Ltd.	Monoclonal blood grouping antibodies, Immunoserological kits, Biochemistry kits, Rapid cards, Urinalysis strips	Infectious diseases and snake toxins
ABL Biotechnologies	Reagents and immunoassays	Anti-bacterials, anti-virals
Xcyton Diagnostics Ltd.	Molecular diagnostics and services (Syndrome Evaluation System (SES) for detection of infectious diseases	CNS infections, blood stream infections (sepsis, pneumonia, dengue, antibiotic resistance, pyrexia, chikungunya) Immunosuppressed infections, ophthalmic infections, HPV infections
Bhat Biotech	Rapid diagnostic kits, Elisa and contract research services	Pregnancy, HIV, hepatitis, malaria, Dengue, chikungunya, swine flu (H1N1), syphilis, TB, cardiac markers
ReaMetrix	Reagents, immunoassays	HIV/AIDS, autoimmune diseases, stem cell analysis, leukaemia/ lymphoma, immune phenotyping.
Achira labs	Reagents, instruments and rapid diagnostic kits	Thyroid disorders, fertility, diabetes and infectious disease
BigTech Labs	Microfluidic devices, Elisa, PCR	TB and infectious diseases
Bisen Biotech	Diagnostic reagents	Infectious diseases
Revelations Biotech Pvt. Ltd.	Molecular diagnostics	Infectious and non-infectious diseases, metabolic disorders and genetic horoscope
Chromous Biotech Pvt. Ltd.	Molecular diagnostics, Immunology reagents, kits and contract research services	TB and other infectious diseases

Source: Compiled by the authors on the basis of information available on companies' websites as at July 2015

Table 9: Important IVD technologies suitable for resource-poor settings developed by young start-ups in India.

Companies	Diagnostic technologies	Patenting Year
XcytonDiags	CheX: Elisa-based, in-vitro diagnostic (IVD) kit for HIV	1994
Bigtech	Micro-PCR device and reagents for low throughput, rapid, point-of-care in-vitro diagnosis of infectious diseases, suitable for harsh conditions	2006
ReaMetrix	Dry-tri: Cold-chain independent and easy to use CD4/CD8 assay reagent for HIV management; suitable for harsh conditions	–
Achira labs	Immunoassay-based microfluidic chips and point-of-care device for low throughput, rapid, in-vitro diagnosis (biochemistry); suitable for low-resource settings	2009
	Immunoassay-based fabric chips for device-free, point-of-care in-vitro diagnosis of infectious diseases	2010

Source: Compiled by the authors on the basis of information available on companies' websites as at July 2015

Table 10: Indigenous product development using extramural research funding by young start-ups.

Companies	EMR
Mediclone Biotech Pvt	Development and manufacture of immunodiagnostic kit and post-exposure prophylaxis of rabies using monoclonal antibody cocktail
ABL Biotechnologies	Development of point-of-care diagnostic for cancer

Source: Compiled by the authors on the basis of information available on companies' websites as at July 2015

Analysis shows how the performance of system-building activities within the science, translational and industrial bases has been improved inadequately through the pursuit of a challenge-based (mission-oriented) approach, based on policy initiatives of the government. Within the science base, government agencies (Council of Scientific

Table 11: Indigenous product development by young start-ups using patents, government promotional schemes for PPPs and funding from international donor organizations.

Companies	Patent	Govt. promotional schemes for PPPs	International donor organizations
Big-Tech Labs	Micro-PCR device and reagents for low throughput, rapid, point-of-care in-vitro diagnosis of infectious diseases; suitable for harsh conditions	- Novel molecular diagnostics for eye diseases and low vision enhancement devices - Point-of-care rapid diagnostic kit for TB - Point-of-care detection of infectious disease using handheld micro PCR	For the development of cost-effective Cepheid X-pert TB diagnostic test
ReaMetrix India	Dry-Tri: Cold-chain independent and easy to use CD4/CD8 assay reagent for HIV management; suitable for harsh conditions	–	–
Achira labs	- Immunoassay-based micro fluidic chips and point-of-care device for low throughput, rapid in-vitro diagnosis (biochemistry); suitable for low-resource settings - Immunoassay-based fabric chips for device-free, point-of-care in vitro diagnosis of infectious diseases	–	
Xycton	CheX: ELISA-based in vitro-diagnostic (IVD) kit for HIV	–	–
Bhat Biotech	–	HRP-II/ p-LDH based diagnostic kits for the differential detection of malarial parasites	–
Bisen Biotech	–	TB screen test for diagnosis of pulmonary and extra-pulmonary tuberculosis; evaluation of prototype kit at selected hospitals/ peripheral health centre/ research laboratories	–
Revelations Biotech Pvt. Ltd.	–	Development of low-cost, rapid, quantitative PCR technology for molecular diagnosis	–
Chromous Biotech Pvt. Ltd.	–	Multiplex Fast-PCR based diagnosis and prognosis of tuberculosis	–

Source: Compiled by the authors on the basis of information available on companies' websites as at July 2015

Table 12: Disease-wise indigenous diagnostic product development by young start-ups under different categories relevant for resource constrained settings.

Categories	Malaria	TB	HIV	Rabies	Cancer	Eye diseases	Other bacterial/viral infections	Total
Extramural Research				1	1			2
Patents			2				3	5
Govt. promotional Schemes for PPPs	1	3				1	2	7
Any other		1						1
Total	1	4	2	1	1	1	5	15

Source: Compiled by the authors on the basis of information available on companies' websites as at July 2015

and Industrial Research (CSIR), Indian Council of Medical Research (ICMR) and Department of Biotechnology (DBT)) have made an effort with the implementation of a few select niche products, aimed at bringing publicly funded research and innovation into public-private partnerships for commercial developments. Within the translational base, major efforts have been made by the DBT. For example, it set up THSTI in 2009 for the translation of scientific knowledge into healthcare. Development of diagnostics especially for resource-poor settings of the country is a mission under the programme of National Biodesign Alliance.

The DBT has conducted a limited number of mission-based training programmes over the last two decades for enhancing interactive learning and collaboration to foster system-building activities in respect of new knowledge diffusion to strengthen the science base and translational base for the development of IVDs innovation system. The industrial base for resource-poor settings in the area of IVDs is mostly built up by young-start-ups. Young start-up firms have developed by using mainly government initiatives in place for the pursuit of research and development. Today, young start-ups are the main actors in the industrial base insofar as taking interest in the innovation-making for resource-poor settings is concerned. System-building using the challenge-based innovation system approach favouring non-market social calculations to meet the diagnostic needs of resource-poor settings in India has not been responded to by large firms in a significant way. Demand articulation and reflexivity are yet to find their due place in the perusal of the challenge-based approach. Positive outcomes are limited to the area of tuberculosis.

Mapping the extent of system-building activities for resource-poor settings within the science base, the translational base and the industrial base in India

This section provides an analysis of the outcomes arising out of the system-building activities as a whole by all the actors active within Science base, Translational base and Industrial base. Below we investigate how much of the system-building activities in IVDs innovation system making have been devotedas such in practice to serve the diagnostic needs of resource-poor settings in India.

Performance of system-building activities within the science base

In this section, we analyze the contribution of EMR and publication activity (a system-building activity) under development for the benefit of resource-poor settings in

terms of the targeted disease pattern to find the gaps vis-à-vis local needs and applications.

Extramural research projects (EMR) for local priorities (2000–2015)

Analysis of EMR projects on IVDs collected from NSTMIS (National Science and Technology Management Information System) indicates that from the year 2000 to 2015, a total of 325 projects were undertaken by science base actors, of which only 76 focused on resource-poor settings (Figure 2). Figure 3 indicates an uneven pattern of attention to the development of innovations in respect of resource-constrained settings. Relatively speaking, cancers, viral/bacterial infections, TB, Malaria and HIV are better investigated in the case of resource-constrained settings. There is need to pay greater attention to the development of diagnostics for resource-constrained settings in the case of diseases like kala-azar, Japanese encephalitis, cholera, typhoid and diabetes.

Among funding organizations, only four government funding bodies in India are active in IVDs research funding, amongst which the DBT (Department of Biotechnology) and the ICMR (Indian Council of Medical Research) are leading funders, whereas the Department of Science and Technology (DST) and the Council of Scientific and Industrial Research (CSIR) have relatively less presence in the area of IVD research and development (Figure 4). Analysis of the composition of funding in terms of allocations made to different types of research performers suggests that research institutions (RIs), institutes of national importance (INI) and

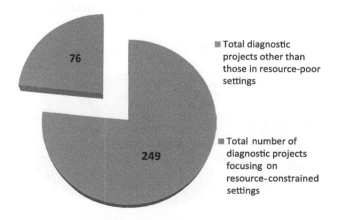

Figure 2: Total number of IVD projects vs. number of IVD projects focusing on resource-poor settings funded through EMR. *Source*: Data collected from NSTMIS database of DST as at July 2015

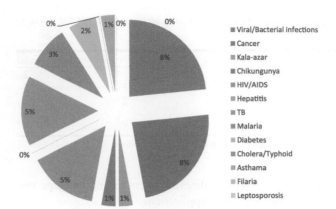

Figure 3: Ratio of disease-wise EMR funding for IVD research focusing resource-poor settings (2000–2015).
Source: Data collected from NSTMIS Database of DST as at July 2015

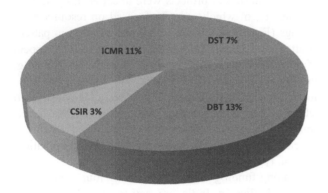

Figure 4: Funding organization involved in IVD research funding focusing on resource-poor settings (2000–2015).
Source: Data collected from NSTMIS database of DST as at July 2015

universities (UNIs) received a major part of the resources allocated for resource-poor settings under EMR funding (Figure 5).

Publication activities on IVD research for local priorities (1991–2015)

Research publication on IVDs published during the period 1991–2015 by science base actors indicates that

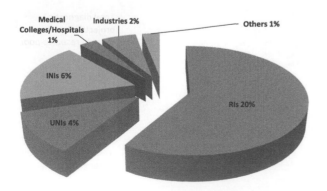

Figure 5: Institutional LOCI of IVD research funding focusing on resource-constrained settings (2000–2015).
Source: Data collected from NSTMIS Database of DST as at July 2015

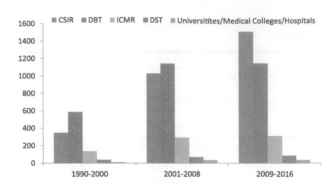

Figure 6: Publication activities on IVD research by different institutes.
Source: Publication data collected from web of science as at July 2015

significant work in the area of IVDs research was started after 2000 by different research and academic institutions (Figure 6). In comparison with other biomedical research areas like drugs and vaccines, IVDs research publications constitute only 4% of the total publications (Figure 7). This shows that the science base lacks capabilities in IVDs research relative to other biomedical areas. Therefore, the volume of knowledge produced in terms of publication activities is not very significant in the case of IVDs.

Classification of publications according to the focus on disease area (Figure 8) indicates that although in the case of the science base some of the public sector based R&D organisations have started working on IVDs of high-burden diseases, more emphasis needs to be given to R&D activities in these diseases, especially for communicable diseases (malaria, TB and HIV) which disproportionately affect the country.

Further, **G-finder**[5] analysis undertaken in case of the total number of IVDs publications published by the science base gives an assessment of R&D mismatches, that is, research is not embedded in the public health perspective because out of the total IVDs publications during 1991–2015, 75% are in basic research and only 16% and 9%, respectively, are focused on clinical research and applied research (Figure 9).

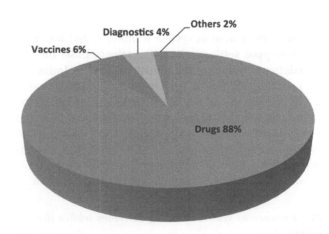

Figure 7: Ratio of publication activities on diagnostic research and other biomedical areas (drugs & vaccines).
Source: Publication data collected from web of science as at July 2015

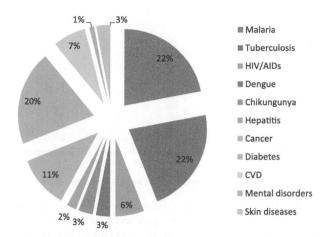

Figure 8: Ratio of IVD publications on focused disease areas. *Source*: Publication data collected from web of science as at July 2015

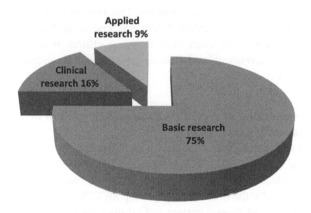

Figure 9: G-finder analysis on total IVD research publications. *Source*: Publication data collected from web of science as at July 2015

Performance of system-building activities within the translational base

In this section, we analyze the contribution of collaborative research (a system-building activity) under development for the benefit of resource-poor settings in terms of the targeted disease pattern to find the gaps vis-à-vis local needs and applications.

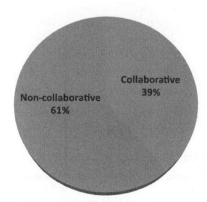

Figure 10: Ratio of collaborative and non-collaborative publications. *Source*: Publication data collected from web of science as at July 2015

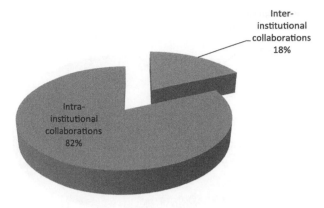

Figure 11: Ratio of intra- and inter-institutional collaborative publications. *Source*: Publication data collected from web of science as at July 2015

Patterns of research collaboration for the benefit of resource-poor settings

Analysis of the existing pattern of research collaboration through publication activities on IVDs from the period 1991 to 2015 illustrates that very few collaborative research programmes have been undertaken on IVDs (Figure 10), and most of the collaborations are intra-institutional rather than inter-intuitional (Figure 11). This means that there is an absence of knowledge sharing and coordination among the different actors, resulting in weak system-building for innovation and development. Collaborations are mostly concentrated in the area of research and development of simple basic IVD processes, or on the process of instrumentation rather than in research and development of advanced IVD technologies like PCR-based molecular diagnostics, point-of-care diagnostics, etc., which are the required solutions to overcome diagnostic challenges in the resource-poor settings of the country (Figure 12).

Patenting activities, resource-poor settings and new technology platforms

Analysis of patents in the area of IVD research (1991–2015) shows that Indian research institutions and firms lack in-house technological capabilities for the development of new patentable IVD technologies that are found

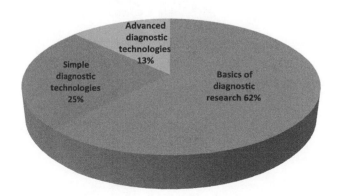

Figure 12: Technology-wise national collaborative publications pattern in diagnostics. *Source*: Publication data collected from web of science as at July 2015

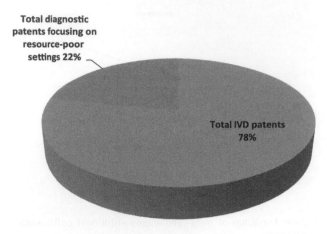

Figure 13: Ratio of total IVD patents vs. IVD patents focusing on needs of resource-poor settings.
Source: Data collected from Indian Patent Office Database as at July 2015

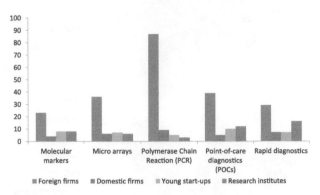

Figure 15: Technologies-wise patenting activities in the case of diagnostics relevant for resource-poor settings (1991–2013).
Source: Data collected from Indian Patent Office Database as at July 2015

to be useful for resource-poor settings. Analysis illustrates that there are very few IVD patents that focus on the needs of resource-poor settings (Figure 13). The number of patents owned by Indian research institutions and firms is insignificant, whereas foreign firms hold a huge number of IVDs patents (Figure 14), a factor that contributes to the high and unaffordable costs of IVDs for the vulnerable population residing in resource-poor settings. As a result, only some of these diagnostics have been introduced in National Disease Control Programmes.

The dismal patent scenario indicates a slowing down of the pace of indigenization in diagnostics for resource-poor settings. Further, it shows that Indian firms continue to lack in innovation capabilities in areas of advanced diagnostic technologies like molecular diagnostics and rapid diagnostic point-of-care tests, which hold the potential for the required solutions of diagnostics needs for resource-poor settings (Figure 15). Most of the patents in these advanced technologies are owned by foreign firms. Under the strong patent regime, getting into these research areas require high licensing fees or transaction costs, thus impeding or blocking Indian firms from undertaking further

innovations in these advanced IVD technologies. Therefore, Indian firms are primarily involved in developing IVDs reagents and instruments. The possibility of domestic firms and scientists being blocked from the realization of success from their ongoing efforts in research and development and innovation-making is already a reality.

Performance of system-building activities within the industrial base

In this section, we analyze the contribution of the industrial base to the development of FDI inflows, products introduced onto the market and indigenous product development for the benefit of resource-poor settings in terms of the targeted disease pattern to find the gaps vis-à-vis local needs and applications.

Market structure of IVDs for resource-poor settings in India

Analysis of the market structure of IVDs indicates that, at present, IVDs products relevant to resource-poor settings account for only 20%, which is very low to meet the demand, as out of the total 413 IVDs products in the market, only 71 are relevant to resource-poor settings (Figure 16). Further analysis indicates that the

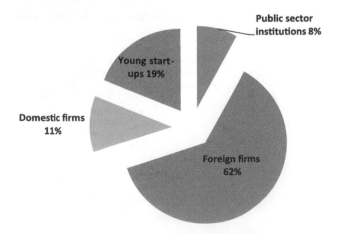

Figure 14: Ratio of IVD patents owned by different innovation actors focusing on needs of resource-poor settings.
Source: Data collected from Indian Patent Office Database as at July 2015

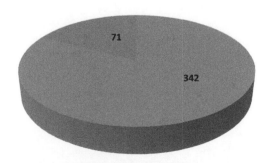

Figure 16: Total number of IVD products relevant for resource-poor settings vs. total number of IVD products on the Indian market.
Source: Compiled by the authors on the basis of information available on selected IVD industries websites and annual reports as at July 2015

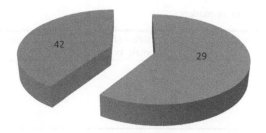

- No. of finished IVD products related to resource-poor settings
- No. of components of finished IVD products related to resource-poor settings

Figure 17: Number of finished products vs. components of finished products.
Source: Compiled by the authors on the basis of information available on selected IVD industries websites and annual reports as at July 2015

market is dominated by imported finished IVDs products, while the presence of components of IVDs products in the market is very low (Figure 17). This shows that demand for IVDs is mostly met through imported products, and firms lack interest in the development of indigenous IVDs products since they are mostly involved in business for profit. Analysis confirms that at present the IVDs market is dominated by foreign firms, constituting 68% of the market with 48 products, while large domestic firms have less presence in the market because of low revenue generated in this area (Figure 18). In the past few years, however, many young start-ups have emerged, which have been playing a significant role by focusing on the development of IVDs technologies relevant to resource-poor settings (as previously discussed).

Structure of FDI inflow in the IVDs sector
FDI inflow into the diagnostic sector in India indicates that a large amount of FDI goes to businesses providing diagnostic services, like Metropolis, SRL diagnostic, Thyrocare and Quest. The inflow is received through acquisition, non-acquisition or automatic routes. The main reason behind this is the large revenue generation from investment in this area; these labs have large business

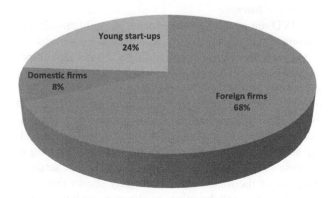

Figure 18: Ratio of type of firms' presence in Indian IVD product market.
Source: Compiled by the authors on the basis of information available on selected IVD industries websites and annual reports as at July 2015

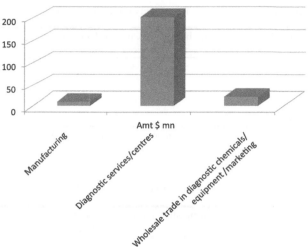

Figure 19: FDI inflow in the diagnostic sector from September 2004 to March 2013.
Source: Department for Industrial Policy and Promotion (DIPP), Government of India

chains at both the national and the international level. The manufacturing sector receives very little FDI. Some FDI goes to wholesale trade in diagnostic chemicals/equipment and marketing. This shows that only the diagnostic service sector business attracts FDI, while other diagnostic sectors lack FDI inflow. FDI inflow in the IVDs sector does not seem to be conducive to resource-poor settings, as it favours the expansion of diagnostic service businesses rather than the development of diagnostic products relevant to resource-poor settings (Figure 19).

Performances of IVDs firms in India
Analysis is made of the contribution of 19 companies (six domestic firms, seven young start-ups and six foreign firms) active in IVD development in India. Product profiles of the companies illustrate that domestic companies are behind in the development of finished diagnostic products for resource-constrained settings compared with foreign firms that have a larger share as far as the number of diagnostic products is concerned. Foreign firms dominate the market for finished diagnostic products relevant to resource-constrained settings. Many young start-ups have entered the finished product market in the last few years. Compared to domestic companies, young start-ups are more interested in the development of markets for diagnostics relevant to resource-poor settings.

Assessment shows that the growing presence of young start-ups is a distinguishing feature of the Indian innovation system. Compared with foreign firms, more young start-ups are showing interest in the segment of rapid point-of-care (POC) diagnostics relevant to resource-poor settings. The encouraging news from the market is about the growing involvement of young start-ups in finding cures for diseases such as TB, malaria and HIV – known to be neglected by the industry and professionals alike, these diseases are now on the radar of young start-ups for investment. Tables 13–15 analyze the current status of finished diagnostic products and components considered relevant to resource-poor settings that have been introduced. Table 16 analyzes, disease-wise, the

Table 13: Number of IVD products relevant to resource-poor settings in the Indian market.

Types of firms	Number of relevant products* for resource-poor settings	Total number of products available in market (finished products/ components of finished diagnostic products)
Domestic	6	51
Young start-ups	17	59
Foreign	48	303
Total	**71**	**413**

Notes: *Relevant products include finished products as well as components of finished diagnostic products available in the market.
Source: Compiled by the authors on the basis of information available on selected IVD industries websites and annual reports as at July 2015

Table 14: Number of finished IVD products relevant for resource-poor settings in the Indian market.

Types of firms	Total finished in-vitro diagnostic products available in the market	Total number of resource-poor settings related finished products
Domestic	23	1
Young start-ups	29	12
Foreign	134	29
Total	186	42

Source: Compiled by the authors on the basis of information available on selected IVD industries websites and annual reports as at July 2015

Table 15: Number of components of finished IVD products relevant for resource-poor settings in the Indian market.

Types of firms	Total components of finished in-vitro diagnostic products available in the market	Total number of resource-poor settings related components of finished products
Domestic	28	5
Young start-ups	30	5
Foreign	169	19
Total	227	29

Source: Compiled by the authors on the basis of information available on selected IVD industries websites and annual reports as at July 2015

number of products introduced to the market for resource-poor settings.

Analysis clearly indicates that the extent of performances of system-building activities with the Indian science, translational and industrial bases have not favoured much non-market based social calculations favouring innovation-making for resource-poor settings in the area of IVDs. Analysis of the system-building activities indicates the lack of a problem-solving mindset. Presently, the learning and competence-building activities are mainly driven by market calculations in the science and translational bases. Investment is also low in the area of advanced technologies like molecular diagnostics, rapid point-of-care tests, etc. It is obvious that the full potential of these technologies to solve diagnostic problems in resource-poor settings is not being realized. In the case of the industrial base, large foreign and domestic firms are almost absent from the performance of system-building activities favouring IVD development for resource-poor settings. Only young start-up firms are seen to be undertaking system-building activities relevant to resource-poor settings.

Emerging challenges of the IVDs innovation system for the development of system-building activities relevant for resource-poor settings
Tables 17–19 clearly show that performance of system-building activities for resource-poor settings in comparison to overall IVDs development is very low, which

confirms the persistence of system weakness arises due to lack in directionality, demand creation and policy coordination and reflexivity. In this section, we provide an analysis of the system-wide failures of innovation, denoted as system failures in this study, namely directionality failure, demand creation and policy coordination and reflexivity failures. Further, Table 20, based on the interviews with actors, confirms the persistence of system failures in IVDs innovation system of India.

Directionality failure
Insufficient investment in competence-building and interactive learning
The IVD innovation system lacks in problem-solving, goal-oriented research for resource-poor settings in India. Investment in system-building activities for competence building is lacking on the part of large firms as well as public sector R&D organizations. Although some efforts have been made by government departments and young-start-up firms, these are small, especially when considering the needs of competence-building. System-building activities for steering and co-ordination are lacking in the case of innovation-making for resource-poor settings. Analysis of the translation base clearly shows that there are very few collaborative publication activities. The number of actors involved in interactive learning is small (Figures 10–12). Another factor hindering competence-building and interactive learning is the interaction of the bench to bedside (i.e. from laboratory to clinical

Table 16: Disease-wise number of IVD products relevant for resource-poor settings in Indian market.

Types of Firms	Total diagnostic products available in market	Diseases-wise diagnostic products relevant for resource-poor settings									Total number of resource-poor settings related products
		TB	HIV	Malaria	Dengue/ chikungunya	Diabetes	Hepatitis	Cancer	Bacterial/ viral infections	Genetic diseases	
Domestic	51	–	1	–	–	1	–	2	2	–	6
Young start-ups	59	2	2	1	–	1	1	5	4	1	17
Foreign	303	2	4	2	1	5	-	15	14	5	48
Total	413	4	7	3	1	7	1	20	17	4	71

Notes: Relevant products include finished products as well as components of finished diagnostic products available in the Indian market.
Source: Compiled by the authors on the basis of information available on selected IVD industries' websites and annual reports as at July 2015

settings). Well-trained human resources as well as favourable institutional culture are presently lacking to link the bench to bedside. In India, there is shortage of adequately trained laboratory personnel. There is the problem of inadequate funding for training and education. Over 90% of India's medical colleges do not have training and education for laboratory diagnostic tests in their study courses (http://www.who.int/phi/en, 2011).

Demand articulation failure
Insufficient investment in indigenous innovation
Analysis shows that system-building activities for IVD innovation undertaken in the present innovation regime for resource-poor settings in the sphere of science, translational and industrial bases are highly inadequate on account of insufficient resources being devoted to demand creation. A lack of contribution from public procurement and advanced market commitments from the side of the government are evident from the analysis. Activities at the product development end are driven by market calculation. Products imported by original equipment manufacturers in finished form are still prevalent. Emphasis on indigenous research has been notably lacking amongst local players.[6] India must thus look to imports to cater for market demand. Diagnostic tests are costly and unaffordable. At all the levels, that is, the science, translational and industrial bases, investments in system-building activities are too weak to meet the diagnostics demands of resource-poor settings. Emphasis on the development of advanced IVDs like molecular diagnostics, rapid

point-of-care tests, etc. are too low to serve the emerging demands.

Policy co-ordination and reflexivity failure
Absence of the performance of system-building activities by large foreign firms and domestic firms
Liberalization of trade and investment has not stimulated the large domestic and foreign firms to undertake system-building activities. These firms are more oriented to profit-seeking market calculations. Further, the market environment and governmental support are not found conducive enough to provide young start-up firms with sufficient strength to make a mark. To some extent, the entrepreneurial activities of these young start-ups are being strengthened and supported by government support via the mobilization of financial resources through extramural funding and through the promotion of a limited number of public-private partnerships and product development partnerships. But market formation for these start-ups often weakens due to the prevalence of imported products. Indigenous products face a disadvantage in a market dominated largely by foreign firms.

Lack of regulation (issues of maladaptation of IVDs)
At present, almost 80% of imported diagnostics are maladapted to resource-poor settings of India because of deregulation. To be effective, IVDs must be analytically and clinically validated. These validations require accreditation from a clinical laboratory performing diagnostic tests in order to get certification from a regulatory body (NABL, CLIA and CAP) in India. Regulatory measures

Table 17: Ratio of performance of system-building activities within science base for resource-poor settings (comparison of system-building activities for the development of immunodiagnostics and molecular diagnostics within total IVDs system-building activities).

Performance of innovation actors	Indicators of innovation-making for resource-poor settings								
	1990–2000			2001–2010			2011–2016		
	Publications	Patents	EMR	Publications	Patents	EMR	Publications	Patents	EMR
Total in-vitro diagnostics	22.14	4.32	2.79	37.46	4.38	4.76	20.23	3.04	0.88
IVDs for resource-poor settings	0.17	0.04	0.04	0.40	0.15	0.23	0.23	0.13	0.29
Immunodiagnostics	24.01	5.17	3.29	38.69	5.19	4.81	14.99	3.27	0.58
Immunodiagnostics for resource-poor settings	0.20	0.03	0.05	0.23	0.13	0.08	0.08	0.08	0.20
Molecular diagnostics (MDs)	13.17	0.24	0.37	31.59	0.49	4.51	45.49	1.95	2.20
MDs for resource-poor settings	0.00	0.00	0.00	10.00	4.00	8.00	8.00	2.00	6.00
MDs for resource-poor settings	0.00	0.00	0.00	1.22	0.49	0.98	0.98	0.24	0.73

Source: Publication data are collected from web of science, EMR projects data are collected from NSTMIS database of DST and patent data are collected from Indian Patent office, as at July 2015

Table 18: Ratio of performance of system-building activities within Translational base for resource-poor settings (comparison of system-building activities for the development of immunodiagnostics and molecular diagnostics within total IVDs system-building activities).

Performance of innovation actors	Indicators of innovation-making for resource-poor settings								
	1990–2000			2001–2010			2011–2016		
	Total collaborations	Inter-institutional collaborations	Intra-institutional collaborations	Total collaborations	Inter-institutional collaborations	Intra-institutional collaborations	Total collaborations	Inter-institutional collaborations	Intra-institutional collaborations
Total IVDs	16.58	3.39	13.19	22.42	5.41	17.01	11.00	2.10	8.89
IVDs for Resource-poor Settings	0.34	0.04	0.30	0.60	0.17	0.43	0.47	0.13	0.34
Immunodiagnostics	19.01	1.89	17.11	22.74	3.48	19.25	8.25	1.47	6.78
Immunodiagnostics for Resource-poor settings	0.31	0.06	0.24	0.43	0.24	0.18	0.18	0.06	0.12
Molecular diagnostics (MDs)	10.84	6.94	3.90	21.68	9.97	11.71	17.49	3.61	13.87
MDs for resource-poor settings	0.43	0.00	0.43	1.01	0.00	1.01	1.16	0.29	0.87

Source: Collaborations are mapped through publications collected from web of science as at July 2015.

for diagnostics are poorly developed; at present, less than 10% of India's 20,000 clinical laboratories are accredited. The burden of cost for implementation of quality control programmes and a general lack of awareness are leading to a lack of pressure for appropriate innovation (http://www.nabl-india.org).

Of the laboratories participating in quality control programmes, 75% exist in only five states, accounting for 30% of the population. Since most of these laboratories are private laboratories, it has been noted that the population residing in resource-poor settings not only lacks access to private labs, but also is more likely to undergo maladapted and substandard testing which is subject to inadequate safety protection. The government has indeed been taking several steps to develop a clearly defined regulatory framework for the medical devices industry, but system-building activities for institutionalization and legitimation for regulating IVDs are lacking.

Although government bodies and associations such as the MoH&FW (Ministry of Health and Family Welfare) and the CDSCO (Central Drug Standard Control Organization) are working to promote indigenous production, the CDSCO works under the aegis of the MoHFW. It lays down rules, standards and approves the import and manufacturing of drugs, diagnostics, devices and cosmetics. All products must be registered with the CDSCO before they can be launched in the Indian market. Despite these efforts, a well-defined regulatory pathway for diagnostics products is awaited[7] (http://ehealth.eletsonline.com/2012/12/indian-diagnostics-a-leap-in-the-dark/).

Recently, the Government of India issued the revised draft of Medical Devices Rules 2016, introducing the "Quality Management System"[8] mandatory for notified medical devices and in-vitro diagnostics. But its effectiveness is still to be tested in the coming years.

Challenges of system-building activities for the development of IVDs for resource-poor settings: Experience of actors based on interviews (Table 20)
Conclusion

This article shows that the persistence of problems in the case of innovation-making for the IVDs lie in the pursuit of policies favouring and encouraging actors that limit themselves to the pursuit of the logic of economic profit and scientific reputation. This encourages actors participating in the building of the science, translational and industrial bases to practise market calculations-based system-building activities. The element of the contribution of the logic of market-based calculations is the main contribution of this article. In the analytical framework of the national system of innovation, this element allowed the authors to trace the connections of system-building failures to the policies of deregulation, trade and investment liberalization, and stronger intellectual property, all favouring market calculation-based decision-making in India over the period of last two decades.

The article shows that the system failures are taking place because non-market, social calculation-based policy correctives are not being pursued adequately vis-

Table 19: Ratio of performance of system-building activities within Industrial base for resource-poor settings (comparison of system-building activities for the development of immunodiagnostics and molecular diagnostics within total IVDs system-building activities).

| Performance of innovation actors | Indicators of innovation-making for resource-poor settings | | | | | |
| | 1990–2000 | | 2001–2010 | | 2011–2016 | |
	Total products introduced in market	Total indigenous products	Total products introduced in market	Total indigenous products	Total products introduced in market	Total indigenous products
Total in-vitro diagnostics	21.06	5.73	44.17	6.75	19.22	3.07
IVDs for resource-poor settings	1.84	0.82	7.98	1.64	4.70	0.61
Immunodiagnostics	27.59	8.05	46.55	6.03	10.92	0.86
Immunodiagnostics for resource-poor settings	2.30	1.15	9.20	1.72	5.17	0.29
Molecular diagnostics (MDs)	4.96	0.00	38.30	8.51	39.72	8.51
MDs for resource-poor settings	0.71	0.00	4.96	1.42	3.55	1.42

Source: Websites and annual reports of the companies assessed as at July 2015

à-vis the activities of envisioning, steering and coordination, direction setting, demand articulation and reflexivity. Since almost all the actors participating in the building of the science, translational and industrial bases are responding to the policies informed primarily by the logic of either economic profit or scientific reputation, the national system of innovation persists with its rigidities and structural limitations. The article argues that immaturity will continue to be shown vis-à-vis the challenges of indigenization and innovation-making for resource-poor settings. These shortcomings are the consequence of an insufficient contribution by non-market, social calculation-oriented policies.

In addition, the article confirms that the contribution of public-sector research organizations, large firms of foreign and domestic origin and young start-ups is not guided by challenge-based, non-market calculations. Their contribution to the strengthening of the science, translational and industrial bases is guided by market, calculation-based decision-making. The share of their respective contributions to interactive learning and competence-building for advanced technologies for the

Table 20: Challenges of system-building activities experienced by different innovation actors.

Actors	Challenges experience in the performance of system-building activities	Type of failures
Biomedical Technology Unit, Sree Chitra Tirunal Institute for Medical Sciences and Technology, Trivandrum	Interviews with the scientists working on IVD diagnostic technologies have revealed that funding is the major challenge for their research. Extramural funding is very low in the area of diagnostics as compared with other biomedical research.	Directionality failure
Translational Health Sciences and technology Institute (THSTI), Faridabad	Scientists working in the area of advanced IVD diagnostic platforms, like point-of-care diagnostics and molecular diagnostics, have shared their experiences that the introduction of advanced diagnostics in resource-poor settings is challenging because of a lack of infrastructure and social barriers, including the mental makeup of those doctors and patients who continue to follow traditional diagnostic tests.	Policy co-ordination and reflexivity failure
Sir Gangaram Hospital, New Delhi	Interviews with the doctors handling genetic diseases have shown that IVDs for the detection of genetic disorders are too expensive. In the case of genetic disorders at present, there is a shortage of doctors and a lack of awareness of diseases and tests.	Demand articulation failure
Dr P K Ghosh (Former Advisor of Department of Biotechnology)	Dr Ghosh shared his insights on the management of IVD research in India. He said that this research area lacks in incentives, competence, interactive learning and competence-building; therefore, there is prevalence of imported diagnostics in India. Large firms do not want to invest in diagnostic research because of low revenue returns and high uncertainties prevailing.	All the systemic failures persist (directionality, demand articulation, policy co-ordination and reflexivity)

Source: Based on the interviews conducted in October 2016

development of innovations suitable for resource-poor settings is therefore quite small. Initiatives taken by the government to encourage them to practise non-market or social calculation-based institution building needs to be accelerated for the pursuit of innovation-making for resource-poor settings.

Further, this article confirms that the underdevelopment of capabilities of local manufacture for resource-poor settings is the result of the myopic conduct of large foreign and domestic firms and due to the lack of sufficient public investment in the process of learning, competence-building and innovation-making. Claims of the advocates of external liberalization, stronger IPRs and more freedom for larger-size firms are not confirmed, even in short measure. Emerging capability gaps are connected to the absence of systematic technological indigenization by large foreign and domestic firms populating the market.

Finally, the article shows that the problem of how to tackle the gaps and needs of resource-poor settings in the field of IVD technologies is not a challenge which can be taken care of by policies on trade and investment alone. It is ultimately a challenge of the formulation of an appropriate science, technology, innovation and industrial policy in India. As markets, governments and professions are actively contributing to these persistent failures. Innovation system scholars thus need to focus on the emerging institutional voids. Interventions of the state, industry and users must eliminate the rigidities persisting within the innovation system ecosystem. Inappropriate incentives will have to be done away with to ensure the formation of socially responsible systems of innovation.

Acknowledgement

We acknowledge that this work was undertaken in 2014 as a collaborative research programme between the Institute for Studies in Industrial Development (ISID) and the Public Health Foundation of India (PHFI). This paper is a revised and updated version of Working Paper No-166, titled 'Challenges of In-Vitro Diagnostics for Resource-poor Settings: An Assessment, ISID-PHFI Collaborative Research Programme'.

Disclosure statement

No potential conflict of interest was reported by the authors.

Notes

1. In-vitro diagnostics are the technologies which diagnose patients using their body fluids or tissue samples to confirm the state of disease or other healthcare conditions (FDA).
2. Diagnostics has penetrated deeply into the healthcare system. Almost 60–70% of medical treatments are based on laboratory diagnostic tests. Diagnostics affect 60% or more of downstream decision-making in disease-management, resulting in improved health outcomes and net cost saving for the healthcare industry. Accurate diagnostics influence patient care in many ways: early risk assessment for disease; targeting disease long before the symptoms occur, targeting disease more specifically, often with less invasive treatment; estimating prognosis more accurately; and, managing chronic disease more effectively (The Lewin Group report 2005).
3. Several breakthrough technologies which significantly contributed to the development and transformation of diagnostic technologies are: 1) Invention of monoclonal antibodies (MAb) by Koehler and Millstein in 1975, which made it possible to detect minute amount of disease-related protein biomarkers when combined with fluorescent labelling and chemistry in blood, cells, or tissue samples of a patient's body. 2) Invention of the Polymerase Chain Reaction (PCR) in the mid-1980s by Kary Mullis, that allowed the amplification of a few molecules of DNA or RNA by many orders of magnitude into diagnostically measurable quantities. 3) The success of the Human Genome Project in 2001, which paved the way for the determination of the complete genomes of many globally important disease-causing micro-organisms, that provide a vast array of potential target sequences that could be used to measure the onset and progression of diseases.
4. "Resource-poor settings" are the conditions lacking in infrastructure for basic healthcare facilities, cold chains, power, reagents, expert staff, etc.
5. The G-FINDER survey tracks global public, private and philanthropic investment into product research and development for neglected diseases. These are diseases that predominantly affect developing countries and for which products are needed but there is insufficient commercial pull to stimulate R&D.
6. Owing to the deficiency of required skills, technical expertise and the infrastructure to support the elevated costs for conducting research and development for products, the share of domestic players has been considerably lowered. (KEN research report, 2013).
7. Due to lack of regulatory legislation, there is no clarity on the classification of and requirements for approval of diagnostics products and novel medical devices in India. The time of approval, diagnostics products and medical devices are categorised into critical and non-critical diagnostics. Any diagnostics tool developed by an indigenous company or academic institution has to go through a validation process conducted by an independent DCGI-approved testing lab such as National Institute of Biologicals (NIB), apart from in-house validation procedures conducted by the company itself. For other products, it needs to be proven that there are more if the same products approved in India. Despite huge technological differences between the two, diagnostics are still treated as drugs by the DCGI. Currently, the DCGI recognises primarily HIV, Hepatitis B and C blood grouping, and Syphilis kits, as these need serious licensing and regulations. Regarding the rest of the kits, there are neither guidelines nor any institutes, with known positives and negatives to be tested. [AQ10] This results in Indian companies that innovate new diagnostics facing great difficulties in catering for overseas market opportunities, as the country of origin has no properly laid-out procedures.
8. As per the new Rule, the licensee of medical devices and in-vitro diagnostics shall comply with the requirements of the "Quality Management System" as laid down in Schedule M-III. The provisions of this Schedule M-III shall be applicable to manufacturers of finished devices, in-vitro diagnostics, which specifies requirements for a quality management system that shall be used by the manufacturer for the design and development, manufacture, packaging, labelling, testing, installation and servicing of medical devices and in-vitro diagnostics. If the manufacturer does not carry out design and development activity, the same shall be recorded in the quality management system. The manufacturer shall maintain conformity with this Schedule to reflect the exclusions. It is emphasized that the quality management system requirements specified in this Schedule are in addition to complementary technical requirements for products. Manufacturers of components or parts of finished devices and in-vitro diagnostics are encouraged to use appropriate provisions of this regulation as guidance (Phramabiz, June 2016).

References

Cassiolato, E. Jose, and C. Maria Clara Soares. 2015. "Innovation Systems, Development and Health: An Introduction." In: *Health innovation Systems, Equity and Development*, edited by Cassiolato, José E and Soares, Maria Clara C, 145–172. E-papersServiços Editoriais, Rio de Janeiro, Brazil.

Drain, P.K.; Hyle, E.P.; Noubary, F.; Freedberg, K.A.; Wilson, D.; Bishai, W.R.; Rodriguez, W.; Bassett, I.V. 2014, "Diagnostic point-of-care tests in resource-limited settings." *Lancet. Infectious Diseases*, 14 (3): 239–249.

Ghosh, P. K. 1996. "Indian Experience in Commercializing Institutionally Developed Biotechnologies." *J. of Scientific and Industrial Research*, 55: 860–872.

Jarosławski and Saberwal. 2013. "Case Studies of Innovative Medical Device Companies From India: Barriers and Enablers to Development." *BMC Health Services Research* 13 (199): 1–8.

Musiolik, J., J. Markard, and M. P. Hekkert. 2012. "Networks and Network Resources in Technological Innovation Systems: Towards a Conceptual Framework for System-building." *Technological Forecasting and Social Change* 79 (6): 1032–1048.

Ramani, Shyama V., and S. Visalakshi. 2001. "The Chicken or Egg Problem Revisited: The Role of Resources and Incentives in the Integration of Biotechnology Techniques." *International Journal of Biotechnology* 2 (4): 297–312.

The Lewin Group. 2005, July. "The Value of Diagnostics Innovation, Adoption and Diffusion Into Health Care." Prepared for *AdvaMed*. Available at https://static.aminer.org/pdf/PDF/000/247/915/is_implementation_adoption_and_diffusion_in_healthcare.pdf

Visalakshi, S. 2009. "Role of Critical Infrastructure and Incentives in the Commercialization of Biotechnology in India." *Asian Biotechnology and Development Review* 11 (3): 63–78.

Possibilities of Russian hi-tech rare earth products to meet industrial needs of BRICS countries

N. Yu. Samsonov, A. V. Tolstov, N. P. Pokhilenko, V. A. Krykov and S. R. Khalimova

The prospects of the creation in Russia of a new scientific and technological sector of production from rare earth materials from the unique niobium rare earth Tomtor deposit (Russia, Yakutia) are discussed in this article. The authors show that an effective innovative technological chain 'ore processing – getting highly liquid REM-products' could be created. This could make it possible in the near future to begin the integration of Russian products into the global market of highly liquid rare and rare earth metals where today two BRICS countries – China and Brazil – are in the leading positions.

Introduction

In modern economies, innovations are the vital source of competitiveness and economic development. One of the ways to promote innovative development is the stimulation of hi-tech industries, which is not possible without development of hi-tech materials and a country's own domestic resource base. In this article, we discuss perspectives of Russian innovative materials to meet the needs of hi-tech industries in the BRICS countries.

Statement of the problem

The global rare elements market is controlled by just two BRICS countries (Brazil and China). About 90% of the world's niobium is produced in Brazil based on Araxá carbonatite deposits and about 90% of the rare earth elements (REE) are produced in China from the giant field Bayan-Obo deposit.

The Tomtor niobium rare earth deposit, which is located in the Arctic zone in the north-west of Yakutia (Russia), is a virtually inexhaustible source of highly liquid minerals that will be used in the hi-tech sectors of industrial production, military industrial complexes and the nuclear industry for the foreseeable decades. For example, at least 40% of the critical technologies which are required for innovative dominance in developed and developing countries (such as the BRICS countries) – from the development of advanced weapons and nuclear energy to the melting of special steels and alloys, and the creation of major structural nanomaterials – is not realizable without the rare earth metals (REM) and related hi-tech materials and products based on them (Tolstov et al. 2014).

There are many applications of REM and the corresponding final products based on them. Rare metals and rare earths are included in the technological chains of weapons and military equipment production for the Russian Military forces and for export, as well as for a wide range of other products produced by the State Corporation 'Rosatom' enterprises, the State Corporation 'Rostech' and other companies and corporations strategic to the Russian economy (Pokhilenko et al. 2014).

Research questions and research objectives

The research question (hypothesis) of this article is the following: Is it possible to provide Russian industry with a wide range of rare earth products in the form of the entire lanthanides line of different purities for a long period of time with the development of the Tomtor deposit and consequent processing of its ore and obtaining collective REM carbonates (i.e. containing unseparated metals) from the ore? Moreover, is production of niobium, yttrium and scandium of high processing degrees (oxides of the individual metals, pure and high-pure metals and their compounds) ensured? Furthermore, is it possible to export the rare earth metals and hi-tech pig products for the needs of industries in the BRICS countries, especially China and India?

Literature review

The research subject is the economic and technological balance of rare earth products production, which allows for the optimization of obtaining of rare earth metals in quantity and value. This is a research paper, and its methodology is based on the analysis of the competence of possible manufacturer of final REM products, of the unique geological and industrial and technological features of rare earth raw materials and of the special economic and geographic characteristics of the deposit location, as well as of the prospective routes of the ore transportation to the place of processing. Methods of analysis are based on estimated economic models of evaluation of investment projects of solid minerals exploitation. Economic, product and technological data used for calculations rely on projected production plans and the price situation on the REM market. We also use system and industrial economic analysis.

In the first part of the paper we demonstrate the possible development of a wide range of competences in rare earth products production in Russia. The second part deals with the characteristics of Russian REE deposits as well as the problems of technological cycle of the Tomtor deposit development and peculiarities of processing of its ore. In the third part, calculations showing real perspectives of Russian hi-tech rare earth products to meet industrial needs of the BRICS countries are given.

In the literature on the economy of REM, the question of the optimization of REM distribution between the consumers and the problem of the Chinese and Brazilian monopoly in the market (REM and niobium correspondingly) are discussed (Dadwal 2011; Manceheri, Sundaresan, and Chandrashekar 2013; He 2014). Another important issue is the search for and implementation of new technologies for the reproduction of REM or technologies that could make it possible to use 'excess' REM (e.g. cerium, samarium, gadolinium, etc.) instead of scarce metals on the market (neodymium, dysprosium, terbium), (Gscheider 2014; Nicoletopoulos 2014; Rare Earths ... 2014). At the centre of these problems and solutions is the analysis of the role of the government in the market and its influence on the demand dynamics as the main mechanism for the development of the REM industry (European Commission ... 2011; Binnemans 2014; Cassard et al. 2014; Endl and Berger 2014).

Object of the research

The State Corporation 'Rosatom' (SC 'Rosatom') can play a significant role in the formation and revival of the Russian rare earth industry, and help take it to the world market. In this regard, it is worth mentioning what the State Corporation is and what its current spheres of competence are, as well as its technological and economic activities. The State Nuclear Energy Corporation 'Rosatom' ensures the implementation of governmental policy and the effective coordination of the use of nuclear energy, the stable operation of nuclear power and nuclear weapons sectors, nuclear and radiation safety.

'Rosatom' unites about 400 enterprises and scientific organizations, including all civil companies in the Russian nuclear industry, enterprises of the nuclear weapons complex, research organizations and the only nuclear icebreaker fleet in the world. The Corporation holds a leading position in the world nuclear technologies market, ranked first in the world in the number of simultaneously constructed nuclear power plants abroad, second in uranium reserves and the third in its production. It is also ranked the second in the world in terms of nuclear electricity generation, providing 36% of the world market for uranium enrichment services and 17% of the nuclear fuel market.

The goal of this paper is to show the possibilities and mechanisms of including SC 'Rosatom' enterprises into the production chain of REE of different purities and the final product based on them, as well as the resulting economic effect.

The data for the analysis (such as ore reserves, the content of valuable components, degrees of extraction, production and processing volumes) are given by the geological reports on the Tomtor deposit (the Blizzard site) and the parameters of the preliminary feasibility study of its development.

Combining the competencies of several Siberian divisions and enterprises of the SC 'Rosatom' referred to below, allows for the creation of an effective technological chain: REM processing – obtaining REM carbonate –

obtaining pure REM – obtaining innovative products of deep processing – producing products based on REM and, thus, to provide:

1. Formation of the segment, which is new for the SC 'Rosatom', of highly liquid innovative REM products (oxides and high-purity metals, and, in the future, the final rare earth products based on them, for example, for nuclear power) to fully meet the needs of the Russian hi-tech industry and for export, including to the other BRICS countries.

2. Development, concentration and accumulation of scientific and technological competencies in the sphere of innovative processing technologies of the unique Tomtor ores and subsequently obtaining individual rare earth metals from them.

3. Increase of scientific and technological competencies in decontamination of radioactive ores and products of chemical and metallurgical processing (thorium, uranium and their decay products).

4. Effective utilization of production and technological capabilities in the disposal of radioactive waste produced during processing, including the modernization of existing facilities and the creation of new REM complexes.

5. Creation and application of the production and economic model for the processing cycle regulation, storage and selective production of high-purity rare earth products subject to changes of the conditions in metal markets (optional balancing model of ore processing and producing REM-components) in the REM business (Binnemans 2014).

6. Training of highly qualified specialists in the REM ore processing and obtaining rare earth products.

Results and discussion

Currently, an effective research and industrial technology has been developed. It is adapted for one of the SC 'Rosatom' enterprises, Zheleznogorsk Mining and Chemical Combine located in the Krasnoyrsk region. The technology allows for the transformation of more than 75% of Tomtor ores into market products and the obtaining of products from both the first process stage (carbonates of rare earth elements) and the second process stage (the individual oxides), as well as high-purity products (pure REM, including heavy and the most expensive lanthanides, and their compounds). This technology was developed by the Institute of Chemistry and Chemical Technology SB RAS in Krasnoyarsk (by V. I. Kuzmin) and was used as the basic technology in the entering of the Tomtor deposit Blizzard site reserves on the State balance in 2000 (Tolstov, Konoplev, and Kuzmin 2011; Pokhilenko et al. 2014).

The proposed technology provides enormous economic effectiveness in terms of product processing degrees. Each subsequent stage of the technological chain significantly increases the value added – from 1.5 to 10 times. Scandium, europium, terbium, dysprosium, praseodymium and neodymium constantly hold the first places in the value and liquidity among the REM in the world market. Unique natural concentrations of the

Tomtor ores allow obtaining up to 1.0 kg of scandium, 0.8 kg of europium, 0.2 kg of terbium, 1.5 kg of dysprosium, 6 kg of praseodymium and more than 20 kg of neodymium from one ton of the ore. These are currently the scarcest, most expensive and highly liquid metals on the world's rare earth metals markets and will be so in the future as well (Kryukov et al. 2016).

Further, new alternative pyrometallurgical technology – liquation melting (Joint Institute for High Temperatures, Moscow, CJSC 'Lanthanum', Novosibirsk region) – has a serious potential. This technology allows for the obtaining of easily separable phosphate saline alloy with rare earths (slag) and, separately, alloy with niobium, significantly cheaper compared with the accepted alkaline hydrometallurgical method (up to 25–30%) (Delitsyn et al. 2015). With the use of the liquation technology, a set of laboratory tests has already been carried out and the corresponding products of the electroslag casting have been obtained from the Tomtor ores transferred by the Institute of Geology and Mineralogy SB RAS (Novosibirsk) to the organizations mentioned here.

Extraction and processing from 10 thousand tons of the Tomtor ore per year, with a gradual increase to 100 thousand tons, will be required to meet fully the REM needs of Russian industry.

The remoteness of the deposit and the complete lack of infrastructure, as well as the extraction volumes of the Tomtor ore arising from the needs of the economy (10 to 100 thousand tons), suggest the creation of a compact mining enterprise with the transportation of raw materials (transportation costs will amount to 3–5% of operating costs) into two possible centres of its primary processing. Options for the ore transportation are presented in Figure 1.

1. Zheleznogorsk MCC (SC 'Rosatom', Krasnoyarsk region). During winter, prepared ore in big-bags is delivered by trucks by winter snow road to the Anabar River mouth, stored on the ore site and then,

during navigation season, transported by water on the Yenisei to the MCC pier.

2. Krasnokamensk Hydrometallurgical Combine (private enterprise), uses territory and radioactivity removal technology of Priargunsk Industrial Mining and Chemical Combine (SC 'Rosatom', Zabaykalsky region). Similarly, the ore is delivered to the river mouth, then with trans-shipment on the Lena to Yakutsk (see Figure 2) and then by the railroad to the combine.

Table 1 presents the initial ore processing obtaining volumes of collective carbonate, dependent on increasing processing capabilities with access to full capacity (11.4 thousand tons) within five years, which is possible on the Zheleznogorsk MCC industrial site or on the Krasnokamensk MCC-Priargunsk MCC.

Further, it is advisable to organize separation and extraction of individual rare earth oxides, material purification, transformation of oxides into the high-purity metal alloys on the SC 'Rosatom' regional Siberian enterprises and (or) partly on the capacities of independent companies which have competencies in the production of rare metal and rare earth products:

- Novosibirsk Chemical Concentrates Plant (Novosibirsk);
- Siberian Integrated Chemical Plant (Seversk, Tomsk region);
- Manufacturing group 'Electrochemical plant' (Zelenogorsk, Krasnoyarsk region);
- Electrolytic chemical combine (Angarsk, Irkutsk region);
- CJSC 'Lanthanum' (Berdsk, Novosibirsk region);
- CJSC 'Rare metals plant' (Baryshevo, Novosibirsk region).

Table 1 represents possible volumes of primary REM products and ferroniobium and secondary raw materials (aluminum, phosphates, etc.) production when processing

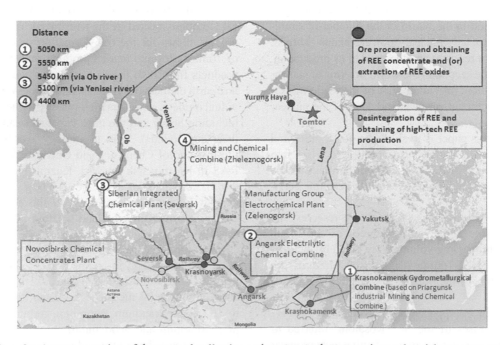

Figure 1: Options for the transportation of the ore and collective carbonates to the processing and enrichment enterprises. *Source*: Kryukov et al. 2016

Figure 2: Loading of the raw material in the big-bags onto the bulk carrier.
Source: Kryukov et al. 2016

capacities are between 10 to 100 thousand tons of the Tomtor ore per year.

The intermediate industrial products supplied to the SC 'Rosatom' enterprises for separation, extraction of individual rare earth oxides and transformation of the oxides into metal alloys are:

1. REM carbonate, including yttrium oxide (from 1.14 to 11.4 thousand tons per year);
2. Scandium oxide concentrate (0.02–0.2 thousand tons per year).

By-products (titanium powder, alumina, trisodium phosphate, ferroniobium) are secondary for the rare earth industry and are either sold to consumers by the owner of the initial raw material after the primary stage of ore processing or stored.

The possibilities for the SC 'Rosatom' to produce the corresponding nomenclature and volumes of REM oxides (including yttrium and scandium) on the basis of carbonates are presented in Table 2. At the first stage, processing of 1.14 thousand tons of carbonates allows for production of up to 582 tons of REM oxides (including yttrium and scandium) per year. Thus, a tenfold capacity increase

(full capacity) will allow for production of up to 5.82 thousand tons of oxides.

At the first stage, US\$21.4 million[1] per year total revenue from oxides sales is guaranteed and, at full capacity, US\$214.4 million (see Table 3). It is not difficult to calculate that if SC 'Rosatom's' total revenue is US\$9 billion (about 600 billion rubles in 2015), the share of the new prospective REM production segment could be between 0.3 and 2.3%.

This approach should be called a gradual rare earth products market launch, a launch without high expectations and illusions. Moreover, the cost of enrichment of some (up to 30%) of produced oxides to create innovative products – pure and high-purity metals – will be significantly higher.

The unique composition of the ores, the tremendous deposit resources and the flexible deposit ore processing technological scheme allow for the obtaining of a wider range of highly liquid and innovative products, including high-purity and the scarcest heavy REM, and consumer products on their basis, after full transformation of the ores. It should be mentioned here that the market

Table 1: Volumes of primary REM products and secondary components obtained in Tomtor ore processing.

Components, thousand tons	Period of development, year							Total
	Year 1*	Year 2*	Year 3	Year 4	Year 5	Year 6	Year 7–15	
Ore processing	0	0	10.0	30.0	52.0	77.0	100.0	**1069.0**
REM carbonates, including yttrium oxide (composite product of the 1st stage of processing)	0	0	1.14	3.42	5.92	8.78	11.4	**121.86**
Ferroniobium	0	0	0.51	1.52	2.64	3.91	5.08	**54.3**
Trisodium phosphate	0	0	7.7	23.1	40.04	59.29	77	**823.13**
Alumina	0	0	1.5	4.5	7.8	11.55	15	**160.35**
Scandium oxide concentrate	0	0	0.02	0.06	0.1	0.15	0.2	**2.13**
Titanium powder	0	0	0.28	0.84	1.45	2.15	2.8	**29.92**

*The investment period for the enterprise construction.
Source: Authors' calculations; Kryukov et al. 2016; Pokhilenko et al. 2016

Table 2: REM oxides produced on the basis of REM carbonates (sales plan).

| Components | Period of development, year | | | | | | | Total |
	Year 1	Year 2	Year 3	Year 4	Year 5	Year 6	Year 7–15	
Ore, ths. tons	0	0	10.0	30.0	52.0	77.0	100.0	**1069.0**
REM oxides, tons								
Lanthanum	0	0	120.0	360.0	624.0	924.0	1 200.0	**12 828.0**
Cerium	0	0	246.0	738.0	1 279.2	1 894.2	2 460.0	**26 297.4**
Praseodymium	0	0	25.0	75.0	130.0	192.5	250.0	**2 672.5**
Neodymium	0	0	98.0	294.0	509.6	754.6	980.0	**10 476.2**
Samarium	0	0	12.3	36.9	64.0	94.7	123.0	**1 314.9**
Europium	0	0	3.9	11.7	20.3	30.0	39.0	**416.9**
Gadolinium	0	0	12.0	36.0	62.4	92.4	120.0	**1 282.8**
Terbium	0	0	0.9	2.8	4.9	7.2	9.4	**100.5**
Dysprosium	0	0	7.3	21.9	38.0	56.2	73.0	**780.4**
Holmium	0	0	0.9	2.8	4.9	7.2	9.4	**100.5**
Erbium	0	0	1.9	5.7	9.9	14.6	19.0	**203.1**
Thullium	0	0	0.5	1.5	2.5	3.8	4.9	**52.4**
Ytterbium	0	0	1.9	5.7	9.9	14.6	19.0	**203.1**
Lutecium	0	0	0.5	1.4	2.4	3.6	4.7	**50.2**
Yttrium	0	0	30.7	92.1	159.6	236.4	307.0	**3 281.8**
Scandium oxide concentrate	0	0	20.0	60.0	104.0	154.0	200.0	**2 138.0**

Source: Authors' calculations; Kryukov et al. 2016; Pokhilenko et al. 2016.

situation for rare earth products – oxides and high-purity metals – with China controlling the prices, makes the question of developing and using the Russian REM business production and economic model for the processing cycle regulation, storage and selective production of high-purity rare earth products urgent. Moreover, Brazil has a significant impact on the global REM market, operating the Araxá high-quality pyrochlore deposit which contains niobium and rare earths as the secondary components.

The Tomtor ore field is located in Russia, in the northwestern part of the Republic of Sakha (Yakutia) in the territory of Olenek ulus (locality) 400 km from the coast of the Laptev Sea (see Figure 3) and is associated with the unique size carbonatite massif of alkaline ultrabasic rocks and carbonatites of the same name (Entin et al. 1990; Tolstov et al. 2014).

On an area of 30 square kilometres, there are three isolated sites (North, Blizzard and South), within which there is the ore bed of redeposited and epigenetically altered weathering crust (see Figure 4).

The Blizzard site has been explored; its resources of the first stage are entered on the State balance. Geological reserves and average components content of the Blizzard site are shown in Table 4. The content of harmful additions and radioactivity causing low environmental risks are the following: uranium – 0.0092%, thorium – 0.11%, radium – 0.157%.

The North and the South sites have been explored to a much lesser extent; exploration works are complete and prospecting works are being performed. Projected measured reserves were tested and evaluated as 1.1 million ton of ore with 1% niobium pentoxide, 5% total rare earth oxides and 0.04% scandium oxide concentrations (Tolstov et al. 2014).

High concentrations of rare and, in most importantly, the scarcest and the most expensive yttrium-earth metals in the Tomtor ores together with real possibilities of

Table 3: Sales of REM oxides, including yttrium and scandium (excluding related products), in US$ millions.

| Indicator | Period of development, year | | | | | | | Total |
	Year 1	Year 2	Year 3	Year 4	Year 5	Year 6	Year 7–15	
REM oxides sales, US$ millions								
Lanthanum	0	0	0.6	1.7	3.0	4.4	5.8	61.9
Cerium	0	0	1.1	3.2	5.6	8.3	10.8	115.4
Praseodymium	0	0	2.6	7.9	13.7	20.2	26.3	281.1
Neodymium	0	0	5.8	17.3	30.1	44.5	57.8	617.9
Samarium	0	0	0.1	0.4	0.6	0.9	1.2	12.8
Europium	0	0	2.7	8	13.8	20.4	26.5	283.4
Gadolinium	0	0	0.5	1.4	2.4	3.6	4.7	50.2
Terbium	0	0	0.6	1.7	2.9	4.3	5.6	59.9
Dysprosium	0	0	2.5	7.4	12.9	19.1	24.8	265.1
Holmium	0	0	0.5	1.4	2.4	3.6	4.7	50.2
Erbium	0	0	0.1	0.4	0.8	1.1	1.5	15.9
Thullium	0	0	0.7	2.2	3.8	5.7	7.4	79.0
Ytterbium	0	0	0.6	1.7	3.0	4.4	5.7	61.0
Lutecium	0	0	0.7	2.1	3.7	5.4	7.1	75.8
Yttrium	0	0	0.5	1.4	2.4	3.5	4.6	49.2
Scandium oxide concentrate	0	0	2	6	10.4	15.4	20.0	213.8
Total	**0**	**0**	**21.4**	**64.3**	**111.5**	**165.1**	**214.4**	**2291.9**

Source: Authors' calculations; Kryukov et al. 2016; Pokhilenko et al. 2016

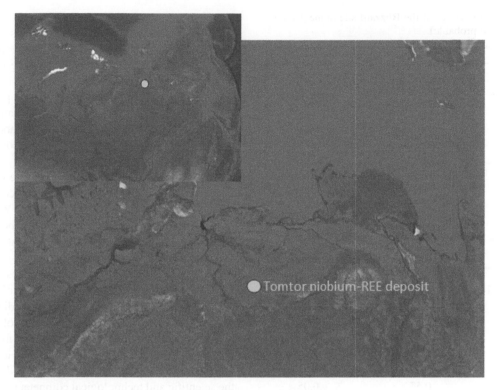

Figure 3: The location of the Tomtor ore field. Coordinates: NL 71°03′; HP 116°30′.
Source: Authors' calculations

infrastructural support (resource, energy, transport and logistics), despite the location in a remote area of north-west Yakutia, make it possible to recommend the deposit as a new and priority source of raw material for full range production of REM, ferroniobium, yttrium, scandium, alumino phosphates and other liquid products (Delitsyn et al. 2015). The geological reserves of the Blizzard site of the Tomtor deposit are given in Table 4 and the structure and the absolute content of the REM oxides in the Blizzard site ores are shown in Table 5.

The project of the deposit development and production of REM products from a high degree of processing of the Tomtor raw materials is a part of the State Programme 'Development of Russian industry and increase its competitiveness 2020', into which it has been included as a number one priority.

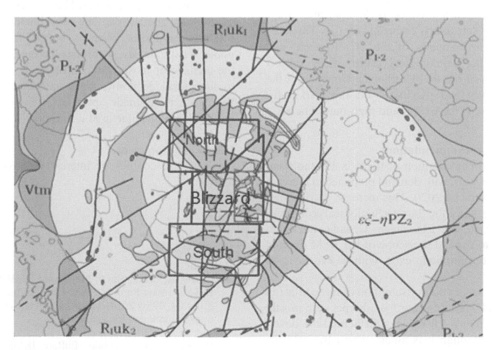

Figure 4: A fragment of the geological map of the Tomtor ore field.
Source: Entin et al. 1990; Kravchenko et al. 1990; Tolstov 2005; Lapin and Tolstov 1993; Tolstov and Samsonov 2014

Table 4. Geological reserves of the Blizzard site of the Tomtor deposit (proved and probable).

Ore (dry)		897.63 ths. tons
	Average oxide content, %	Oxides reserves, ths. tons
Niobium	6.06	54.36
Yttrium	0.50	4.27
Scandium	0.05	0.32
Rare earth elements	8.19	73.47

Source: Entin et al. 1990; Kravchenko et al. 1990; Tolstov 2005; Lapin and Tolstov 1993; Tolstov and Samsonov 2014

Table 5. The structure and the absolute content of the REM oxides in the Blizzard site ores, %.

Oxides	Relative content in the 100% amount of lanthanide oxides	Absolute content in the ore
Lanthanum	23.52	1.96
Cerium	47.7	3.97
Praseodymium	4.51	0.38
Neodymium	14.3	1.19
Samarium	1.63	0.14
Europium	0.57	0.05
Gadolinium	1.42	0.12
Terbium	0.22	0.02
Dysprosium	1.01	0.08
Holmium	0.18	0.01
Erbium	0.38	0.03
Thullium	0.07	0.01
Ytterbium	0.23	0.02
Lutecium	0.06	0.005
Total	100	7.98
Yttrium	–	0.50
Scandium	–	0.05

Source: Entin et al. 1990; Kravchenko et al. 1990; Tolstov 2005; Lapin and Tolstov 1993; Tolstov and Samsonov 2014

The Blizzard site of the Tomtor ore field was licensed in 2014. The owner of the license is 'Vostokengineering' LLC which is a subsidiary company of the 'ThreeArc Mining' (the joint company of the SC 'Rostech' and 'ICT group'). The technological chain of the project implies obtaining of the REM carbonate from one of the Russian enterprises and placing it on the market for further production of the oxides and high-purity metals.

Conclusion and recommendations

At present, global consumption of REM is 130 thousand tons of oxides per year, with demand predicted to increase to 180 thousand tons by 2020. This increase in consumption is reasonable, since the 'big' technological economies of the USA, China, the EU, Japan, Canada and South Korea continue to expand the scope of the REM use and increase their consumption, even with lower rates of GDP growth.

At the same time, in the consumption of rare earths (up to 3 to 5 thousand tons of rare earth oxides per year, mostly imported from China), Russia lags far behind the developed countries, although it is assumed that by 2025 the needs of the Russian economy will reach 15 thousand tons of rare earth oxides.

Obtaining the REM products from the Russian enterprises of the SC 'Rosatom' on the basis of domestic sources of raw material, unique in the mineralization scale (reserves and resources) and in the content of the rare earth components, allows Russia to avoid dependence on the current structure of global monopoly-producing countries and suppliers (China, Brazil) and the risk of unfair competition in the global REM market, as well as being subjected to global price fluctuations of the rare elements.

The research hypothesis, whether rare earth products including niobium, yttrium, scandium could meet the needs of Russian industry, is confirmed. With consumption volumes of Russian industry being about 2 thousand tons of rare earth oxides, the development of the Tomtor deposit guarantees REM deliveries to the domestic market. The surplus capacity makes it possible to export Russian REM to world market, including the other BRICS countries, primarily China and India.

Our estimations show that the value of highly liquid REM products made using concentrates from the ore of the Tomtor deposit reaches US$215 million per year. It is found out that this value enables the development of the scientific and technological competences in innovative ore processing technologies of the unique ores like Tomtor and the subsequent obtaining of individual rare earth metals from them. It is shown that combining competences within SC 'Rosatom' enable the creation of an effective technological chain to obtain highly liquid REM-products that could meet the industrial needs of the other BRICS countries as well.

As a result, BRICS could strengthen its role as a tool of competitive confrontation between countries that are focused on the development of basic and classic industries, mining and processing, and countries that are more or less focused on the collection of rent from the processes of globalization.

In conclusion, we would like to mention that Russia may hop on the last carriage of technologies and hi-tech products producers, even if in the form of primary goods. That is to say, it still has the opportunity to board the carriage of those who contribute to the development to the global economy, rather than receiving it in the form of ready-made products, especially products invented and manufactured elsewhere.

Disclosure statement

No potential conflict of interest was reported by the authors.

Note

1. 2015 prices for metal oxides were used in calculations.

References

Binnemans, K. Economics of Rare Earth: The Balance Problem// ERES2014:1st European Rare Earth Resources Conference/ Milos/04–09/09/2014. http://www.eurare.eu/docs/eres2014/ firstSession/koenBinnemans.pdf.

Cassard, D., F. Tertre, G. Bertrand, F. Schjoth, J. Tulstrup, T. Heijboer, and J. Vuollo. EuRare IKMS: An Integrated Knowledge Management System for Rare Earth Element Resources in Europe // ERES2014:1st European Rare Earth

Resources Conference/Milos/04–09/09/2014. http://www.eurare.eu/docs/eres2014/fifthSession/GuillaumeBertrand.pdf.

Dadwal, S. R. 2011. "The Sino-Japanese Rare Earths Row: Will China's Loss be India's Gain?" *Strategic Analysis* 2 (35): 181–185.

Delitsyn, L. M., G. B. Melent'ev, A. V. Tolstov, L. A. Magazina, A. E. Samonov, and S. V. Sudareva. 2015. "Tomtor's Technological Problems and Their Solution." *The Rare Earth Magazine* 2 (5): 164–179. (In Russian: Delitsyn, L.M., Melent'ev, G.B., Tolstov, A.V., Magazina, L.A., Samonov, A.E., Sudareva, S.V. (2015). Tehnologicheskie problemy Tomtora i ih reshenie. Redkie zemli, 2 (5) 164–179.)

Endl, A., and G. Berger. A Comparative Analysis of National Policy Approaches – With Focus on Rare Earth Elements in Europe// ERES2014:1st European Rare Earth Resources Conference/Milos/04–09/09/2014. http://www.eurare.eu/docs/eres2014/fourthSession/AndreasEndl.pdf.

Entin, A. R., A. I. Zajtsev, N. I. Nenashev, V. B. Vasilenko, A. I. Orlov, O. A. Tjan, Ju. A. Ol'hovik, S. I. Ol'shtynskij, and A. V. Tolstov. 1990. "On the Geological Sequence of Events Related to the Intrusion of Tomtor Alkaline Ultrabasic Rocks and Carbonatites (N-W Yakutia)." *Geology and Geophysics* 12: 42–45. (In Russian: Entin, A.R., Zajtsev, A.I., Nenashev, N.I., Vasilenko, V.B., Orlov, A.I., Tjan, O.A., Ol'hovik, Ju.A., Ol'shtynskij, S.I., Tolstov, A.V. (1990). O posledovatel'nosti geologicheskih sobytij, svjazannyh s vnedreniem Tomtorskogo massiva ul'traosnovnyh shhelochnyh porod i karbonatitov (S-Z Jakutiya). Geologiya i geofizika, 12: 42–45.)

European Commission. 2011. *Tackling the Challenges in Commodity Markets and on Raw Materials*. Brussels: European Commission, COM. (2011) 25 final.

Gscheider, K. The Rare Earth Crisis and the Critical Materials Institute's (CMI'S) Answer // ERES2014:1st European Rare Earth Resources Conference/Milos/04–09/09/2014. http://www.eurare.eu/docs/eres2014/firstSession/karlGschneidner.pdf.

He, Y. 2014. "Reregulation of China's Rare Earth Production and Export." *International Journal of Emerging Markets* 9 (2): 236–256.

Kravchenko, S. M., A. Yu. Belyakov, A. I. Kubyshev, and A. V. Tolstov. 1990. "Scandium-yttrium-rare Earth-Niobium Ores – A New Type of Rare Metal Raw Materials." *Geology of Ore Deposits* 1 (32): 105–109. (In Russian: Kravchenko, S.M., Belyakov, A.Yu., Kubyshev, A.I., Tolstov, A.V. (1990). Skandievo-redkozemel'no-ittrievo-niobievye rudy – novyj tip redkometall'nogo syr'ya. Geologija rudnyh mestorozhdenij, 1 (32): 105–109.).

Kryukov, V. A., A.V. Tolstov, V. P. Afanasiev, N. Yu. Samsonov, and Ya. V. Kryukov. 2016. "Providing the Russian Industry with High-Tech Products Based on the Raw Material From the Giant Deposits in the Arctic – Tomtor Niobium-Rare Earth and Popigai Superhard Abrasive." In *North and the Arctic in the New Paradigm of Global Development. Luzin Readings – 2016*, edited by E. P. Bashmakova and E. E. Toropushina, 204–206. Apatity: IEP KSC RAS. (In Russian: Kryukov, V.A., Tolstov, A.V., Afanasiev, V.P. Samsonov, N.Yu., Kryukov, Ya.V. (2016) Obespechenie rossijskoj promyshlennosti vysokotehnologichnoj syr'evoj produktsiej na osnove gigantskih mestorozhdenij Arktiki – Tomtorskogo niobij–redkozemel'nogo i Popigajskogo

sverhtverdogo abrazivnogo materiala in Bashmakova E.P. and Toropushina E.E. (eds.) " Sever i Arktika v novoj paradigme mirovogo razvitija. Luzinskie chtenija – 2016″ (pp. 204-206). Apatity: IEP KSC RAS.)

Lapin, A. V., and A. V. Tolstov. 1993. "New Unique Deposits of Rare Metals in Weathering Carbonatite Crusts." *Prospect and Protection of Mineral Resources* 3: 7–11. (In Russian: Lapin, A.V., Tolstov, A.V. (1993). Novye unikal'nye mestorozhdenija redkih metallov v korah vyvetrivaniya karbonatitov. Razvedka i ohrana nedr, 3, 7–11).

Manceheri, N., L. Sundaresan, and S. Chandrashekar. Dominating the World. China and the Rare Earth Industry// International Strategy & Security Programme (ISSSP), National Institute of Advanced Studies. – Bangalore, April 2013. – 61 p.

Nicoletopoulos, V. European Policies on Critical Materials, Including REE// ERES2014:1st European Rare Earth Resources Conference/Milos/04–09/09/2014. http://www.eurare.eu/docs/eres2014/fourthSession/VasiliNicoletopoulos.pdf.

Pokhilenko, N. P., V. A. Kryukov, A. V. Tolstov, and N. Y. Samsonov. 2014. "Tomtor as a Priority Investment Project of Ensuring Russia's Own Source of Rare Earth Elements." *ECO* 2: 22–35. (In Russian: Pokhilenko, N.P., Kryukov, V.A., Tolstov, A.V.,Samsonov, N.Y. (2014). Tomtor kak prioritetnyj investitsionnyj proekt obespecheniya Rossii sobstvennym istochnikom redkozemel'nyh elementov. ECO, 2: 22–35.).

Pokhilenko, N. P., V. A. Kryukov, A. V. Tolstov, and N. Yu Samsonov. 2016. "Creating a Strong Rare Earth Industry in Russia: State Corporations Without not Overpower." *ECO* 8: 25–36. (In Russian: Pokhilenko N.P., Kryukov V.A., Tolstov A.V., Samsonov N. Yu. (2016). Sozdanie sil'noj redkozemel'noj promyshlennosti Rossii: bes goskorporatsij ne osilit'. ECO, 8: 25–36.).

Rare Earths Market Prices. 2014. News and Analysis. Argus Rare Earths (1).

Tolstov, A. V. 2005. *The Main Ore Formations of the North of the Siberian Platform*. 2005. Moscow: IMGRE, 200. (In Russian: Tolstov, A.V. Glavnye rudnye formatsii Severa Sibirskoj platformy. / IMGRE. – M., 2005. 200)

Tolstov, A. V., A. D. Konoplev, and V. I. Kuzmin. 2011. "The Peculiarities of Forming the Unique Rare Metal Deposit Tomtor and Estimation of Perspectives of its Industrial." *Prospect and Protection of Mineral Resources* 6: 20–25. (In Russian: Tolstov, A.V., Konoplev, A.D., Kuzmin, V.I. (2011). Osobennosti formirovaniya unikal'nogo redkometall'nogo mestorozhdeniya Tomtor i otsenka perspektiv ego osvoeniya. Razvedka i ohrana nedr, 6: 20–25.)

Tolstov, A. V., N. P. Pokhilenko, A. V. Lapin, V. A. Kryukov, and N.Yu. Samsonov. 2014. "Investment Appeal of Tomtor Deposit and Prospect of its Increase." *Prospect and Protection of Mineral Resources* 9: 25–30. (In Russian: Tolstov, A.V., Pokhilenko, N.P., Lapin A.V., Kryukov, V.A., Samsonov N.Yu. (2014). Investitsionnaya privlekatel'nost' tomtorskogo mestorozhdeniya i perspektivy ee povysheniya. Razvedka i ohrana nedr, 9: 25–30.)

Tolstov, A. V., and N. Y. Samsonov. 2014. "Tomtor: Geology, Technology, Economy." *ECO* 2: 36–44. (In Russian: Tolstov, A.V., Samsonov, N.Y. (2014). Tomtor: geologiya, tekhnologiya, ekonomika. ECO, 2: 36–44.)

Developing an entrepreneurial mindset within the social sector: A review of the South African context

MME 'Tshidi' Mohapeloa

Aim of paper: This paper explores how an entrepreneurial mindset can be developed using a South African case due to its complexities of high social challenges in a middle-income country. These complexities have restricted opportunities at which the social sector operates.

Methodology: Empirical research was used to collect research papers on social entrepreneurs, with reference to social entrepreneur (ship) in South Africa. A search using Scopus with key words 'social entrepreneur (ship)' and 'South Africa' yielded 63 documents. The core question was to determine how the development of the entrepreneurial mindset has been documented in South Africa's social sector with specific reference to the non-profit sector to create social value.

Key findings: Developing an entrepreneurial mindset means influencing not only ways of thinking, skills and knowledge, but also a reflection through attitudes and an observable set of behavioural patterns.

Practical implications: The development of a conscious awareness that incorporates social and environmental issues and a contribution of 'Ubuntu' principles with a combination of resources for a mix of innovation to target social needs.

Conclusion: A targeted educational curriculum (both at school and institutions of higher learning) is required to integrate entrepreneurial development with a social focus

Introduction

A prominent feature that has characterized the social sector is diminishing donor funds, especially in the non-profit sector. This has forced leaders of non-profit organizations to develop creative and innovative ways to take their organizations beyond survival towards sustainability. Failure to do this could devalue the important contribution made by this sector in curbing social issue. Linked to this is the demand for improved effectiveness and sustainability as well as increased competition for scarce resources (Johnson 2000). On the other hand, international non-profit organizations are criticized for 'patronage, dependency, pathological institutional behaviour and financial malpractice' instead of effectiveness or improved sustainability (Johnson 2000; Fowler 2000, 638). A reduction in grant dependency has affected the survival of non-profit organizations thus putting them at risk (Urban 2010), whilst social issues are still on the increase. Hence, there is a need for financial independence in a limited funding space that requires an application of unequivocal methods, especially when traditional government initiatives are unable to satisfy the entire social deficit (Ferri 2010). The reduction of donor funding has made leaders in the public, private and non-profit sectors feel uncomfortable in the face of social challenges that have still not been dealt with.

South Africa has triple complexities (poverty, inequalities and unemployment) that affect the economy. These are still far from being resolved, yet the country is still viewed as a middle-income country. According to South Africa's National Development Plan (NDP) the country has high inequalities, a high unemployment rate for young graduates, redundancy and closure of major employment sectors due to economic recession. A growing number of communities live from hand to mouth, the ever-increasing disparities between rich and poor are some of the social challenges faced by the country (SA 2012/13). However, entrepreneurs could contribute to increasing the country's gross domestic product (GDP) and economic wealth.

If the non-profit social sector is to add value through its contribution of social wealth value creation, then a change in mindset is required, to make social impact. The guiding research question thus is to determine how an entrepreneurial mindset can be applied and utilized to develop strategies that could enhance the achievement of social goals. This deals effectively with the extent to which an entrepreneurial mindset is being developed within the social sector, with specific reference to the non-profit sector when creating social value. This means an exploration of how an entrepreneurial mindset can be developed using a social focus.

South Africa does not have a legal definition of entrepreneurs with a social focus. Thus, a wide range of clusters is used to define those with double or triple bottom-lines, such as non-profit organizations (NPOs), companies or co-operatives, in terms of market-related income generation and social benefits (Stead and Claasen 2009). Entrepreneurship has been used to mitigate challenges and curb social ills, such as the high unemployment rate, especially amongst the economically viable youth. However, a shift towards small- and medium-enterprise (SME) development and entrepreneurial opportunities with a social purpose being explored through the informal sector is further enhance by family owned or survivalist entrepreneurs. There seems to be a link between immigrant entrepreneurs, as they contribute to mitigating unemployment and poverty whilst participating in the economic growth of their host countries (Fakoti 2014).

Review of literature

The term 'social entrepreneurs' encapsulates and generates an idea whilst attempting to develop that idea into an opportunity which responds to social issues (Guclu, Dees, and Anderson 2002). However, the concept defines an individual as the founder or initiator of the initiative (Urban 2010). But it fails to acknowledge the contribution of and the role played by communities, organizations and teams within the organizations to develop innovative and creative methods to solve social problems (Mair and Marti 2006; Yunus and Weber 2010). Definitions look at social and entrepreneurial elements (Mair and Marti 2006; Yunus and Weber 2010). The core theme of this concept seems to highlight the social aspects which relate to non- economic outcomes and the application of entrepreneurial activities, such as trading, the pursuit of profits, income generation and employment opportunities (SEL 2005). With no universal definition adopted (OECD 1999; SEJ 2005), an acknowledgement that social goals are the main outcomes to be achieved is emphasized. Peredo and McLead (2005) agree with Dees (2008) when they acknowledge that for the social entrepreneur, wealth is a means to an end. Pomerantz (2003, 26) further elaborates that NPOs, with their goal to maximize revenue, should apply principles from for-profit businesses without neglecting their core social mission.

Markets and the social economy

It is essential when dealing with the social sector to also look at the social economy. According to Deraedt (2009), the social economy involves organizations where economic justice and democratic participation are typically defining features. Further to this, as observed by Guclu, Dees, and Anderson (2002), is the emphasis on the operating environment where markets, industry structure, the political environment and culture have a direct impact on the social economy. Thus, for social entrepreneurs to achieve social value, they need to address basic human needs by targeting untapped or market failures, filling gaps through partnerships and deploying new business models (Seelos and Mair 2005; Seelos and Mair 2008; Kerlin 2010; Hackett 2010; Yunus and Weber 2010).

An acknowledgement of diverse factors motivating social value creation is needed. These factors include the drive for increased stakeholder benefit, market forces that promote social improvements, serving the poor and engaging the poor in market activities (Dees and Anderson 2003; Mair and Marti 2006; Seelos and Mair 2008; Yunus, Moingeon, and Lehmann-Ortega 2010). However, globally, donor-directed philanthropic programmes can divert these needs. Thus, locally focused markets within this social economy should be the focus, reflected through interventions. For instance, such interventions need to grow the total market size for the industry to be structurally attractive, whilst meeting recognized social needs and demand. This implies that interventions should not only address market failures that guarantee a more-than-sufficient market size, but should also have social impact (Austin, Stevenson, and Wei-Skillern 2006).

Sustainable outcomes

Marshall, Coleman, and Reason (2011, 3) define sustainability as 'finding ways to live together within the carrying capacity of the planet, with equality and justice for all human communities' while allowing vibrant space for other life forms. Core sustainability drivers focus on ensuring that there is a sense of agency, availability of resources, a consciously created awareness, a sustainable approach and the crafts of practice to take action in the service of a more environmentally sustainable and socially just world. Urban's (2010) construction of the term 'social entrepreneur' combines simpler concepts for complex measures (Cooper and Emory 1995); however, Pomerantz (2003, 25) looks at activities and processes done within the social space.

This allows opportunities for the development of innovative, mission-supporting activities or non-profit activities, undertaken by individual social entrepreneurs and nonprofit organizations in association with for-profit companies focusing on financial, sustainable and social outcome. This ensures that strategic decisions strike a balance where social and environmental outcomes are never reduced at the expense of financial sustainability. Ensuring sustainability outcomes means acknowledging that development is not done at the expense of future generations, but that sustainability and development are met as determined in the context of the sustainable development goals (SDGs). Indicators that that help measure sustainability focus on empowerment (targeted communities/groups), using financial returns as a means towards social and environmental value creation, and look at the space where empowerment occurs, to determine whether economic, social and environmental factors are being considered, making this an approach that creates sustainable wealth.

Measures for sustainable wealth require a combination of the elements indicated in Table 1 which gives new paths and solutions, based on local needs. These identify priorities for wealth creation that need to be achieved effectively. These sustainability components could include social and economic dimensions (e.g. promoting health and combating poverty); decision-making based upon environmental development and/or conservation and management of resources (e.g. combating pollution and protecting forests and other fragile environments); strengthening the role of major groups (e.g. children, women and workers); and means of implementation such as education and technology (UN resolution *Agenda 21* on Sustainable Development, UN Department of Economic & Social Affairs, 1992).

The entrepreneur: An economic vs. a social focus

Four key determinants that separate social entrepreneurs from corporate entrepreneurs are their responses to: (a) market failures − opportunities for social entrepreneurs; (b) mission − social value creation vs. profit creation; (c) resource mobilization − how resources are effectively tapped into and; (d) performance measurement − use of financial indicators vs. non-quantifiable indicators. People, context, deal and opportunity (PCDO) are the domains that Austin, Stevenson, and Wei-Skillern (2006) took from commercial entrepreneurs but applied to

Table 1: Suitability measures.

Economic elements	Social elements	Environmental elements
Relies on effective application of economic outcomes.	Uses non-economic outcomes to achieve goals, such as education and health to alleviate poverty.	It incorporates all other relevant features of the environment that are present but could be lost if not taken care of.
Uses trading, accessing markets, sales of goods and services and exchange of monetary values.	Accessibility and the quality and availability of social services become the outcome.	The operating environment through markets, industry structure, political environment and culture (Guclu, Dees, and Anderson 2002).
Relies on contracts and service level agreements as the essence for any transaction.	Usually not-for-profit organizations, cooperatives or associations that strives to occupy the social economy space, as it is neither privately or publicly controlled (Social Enterprise Journal 2005).	Seeks the right environment, created by governments together with the private sector, to support social entrepreneurs and help them flourish (Johnson 2000).
Has the ability to access funds through micro finance so as to deal with social issues such as health or education.	Social value creation is the primary objective, done through the process of economic value creation as the by-product (Seelos and Mair 2005).	Incentives, such as tax laws for an investor to invest in innovative community initiatives.

social entrepreneurs. These highlight that the 'opportunity dimension of the framework is perhaps the most distinct owing to fundamental differences in missions and responses to market failure'. For corporate entrepreneurs, an opportunity must have a large or growing total market size and the industry must be structurally attractive. For a social entrepreneur, however, a recognized social need, demand or market failure usually guarantees a more-than-sufficient market size (Austin, Stevenson, and Wei-Skillern 2006). This means that a social entrepreneur must have a special breed of leadership qualities with specific character traits and behavior, similar to those of an economic entrepreneur, to enhance proactive, risk-taking and innovative measures. However, social entrepreneurial behaviour is deeply influenced by the concurrent requirements of the environment, the need to build a sustainable organization and the need to achieve the social mission (Weerawardena and Mort 2006).

Social entrepreneurs must manage and balance the social and financial aspects of their enterprise for the benefit of customers, especially when the enterprise is neither purely philanthropic nor purely commercial but has a social mission (Clark and Ucak 2006). It is important to note that not all enterprises with a social mission are non-profit organizations or all businesses, but can be mainly in the non-profit sector or as business doing good (Peredo and McLead 2005). At a global level, donor-directed programmes are locally focused and needs based. This is also reflected in social entrepreneurship interventions (Peredo and McLead 2005).

Key elements of an entrepreneurial mindset
From the literature review, it could be concluded that there are five core issues common to social entrepreneurs: (1) the effective use of a business model for sustainability; (2) tapping into economic, social and environmental approaches to create social wealth; (3) innovative use and combination of resources; (4) exploiting or pursuing opportunities (5) for social change or to address social needs. These can be grouped into the approach, the intersection, contributing factors and the social impact as the outcome (Figure 1).

As indicated in Figure 1, a socially focused entrepreneur looks at the social and/or the environmental aspect of the business objective. The bottom line is not only linked to profit, but responds effectively to social problems through the use of business principles and business models to address sustainability issues. However, social wealth tends to be measured through community empowerment where financial returns are seen as a means towards social and environmental value creation (Schwab Foundation for Social Entrepreneurs). These cannot be achieved by looking only at financial sustainability to create wealth, but also require consideration of the social and environmental factors. Thus, a combination of elements gives new paths and solutions, based on local needs that identify priorities for wealth creation to be achieved effectively. The economic element relies on effective application of economic outcomes such as trading, accessing markets, sales of goods and services and exchange of monetary values. Other economic elements such as contracts and service level agreements are the essence of any transaction, with the ability to access funds through micro finance. These help deal with social issues such as health or education.

The social element uses non-economic outcomes to achieve goals, such as education and health to alleviate poverty. Thus, accessibility and the quality and availability of these services become the outcome. The non-profit sector, cooperatives and associations strive to occupy the social economy space, as they are neither privately or publicly controlled (Social Enterprise Journal 2005). They make social value creation the primary objective, to be achieved through the use of economic value creation process as the by-product (Seelos and Mair 2005). The environmental element incorporates all other relevant features of the environment that are present, but could be lost if not taken care of. The right environment, created by governments together with the private sector, could support entrepreneurs with a social focus, and enable them to flourish (Johnson 2000). Incentives through tax could enhance investors to invest in innovative community initiatives.

Businesses utilize resources and technologies to tap into opportunities. However, if the needs met are social, then a

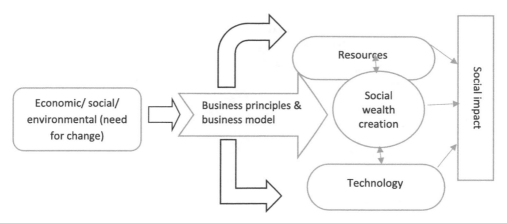

Figure 1: A model for developing a socially focused entrepreneurial approach.

social wealth outcome occurs based on how innovatively resources and technologies have been adapted. The new creation could be new, improved ideas where an innovative outcome brings change that responds to social needs, transferring wealth to expand resources and create opportunities for social issues. According to the Schwab Foundation profile for social entrepreneurs, it acknowledges innovation, sustainability and direct social impact as outcomes that transform traditional ideas for social change using pattern-changing ideas. This means that searching for opportunities to exploit or pursue requires time, the ability to identify social needs, use societal challenges and/or unmet gaps to pursue change, and apply new technologies as venture for social value outcomes. Urban (2010) sees this innovativeness as a proactive and risk-taking ability, where path breakers integrate business ethics with moral judgement. A view point that strengthens existing resources or taps into new avenues tends to enhance new ideas. Thus, creative measures such as ventures that are hybrid can leverage the effective use of minimal resources for a business with a social focus. Through private-public partnerships such hybrid social business ventures could be achieved. This includes ventures with an investor where minimal returns can be leveraged, or an investment with no interest where all profits are injected into the business. Social impact as the final product can be measured when the poor and marginalized are beneficiaries and stakeholders of entrepreneurial interventions with no negative externalities (Schwab Foundation for Social Entrepreneurs). This benefit extends to those in need, as they are usually most severely impacted by the social needs/ills. This does not mean that if a person uses this approach, it will lead to social wealth creation; instead, this is a rather radical approach that makes markets work for people (Johnson 2000). It is the ability to see opportunities in every (problem) situation, instead of a problem in every situation.

Methodology

The methodology used was similar to that of Gras, Mosakowski, and Lumpkin (2005) in their study. It focused on a content analysis for future research topics in social entrepreneurship, which included, the following steps: (1) collecting research papers on social entrepreneurs; (2) collecting papers referring to social entrepreneur (ship) in South Africa; (3) applying exclusion criteria; (4)

cleaning the data; and (5) analyzing the data. A search of the database, using Scopus and the key words 'social entrepreneur (ship)' and 'South Africa' was done. The search yielded 63 documents dating as far back as 1985, peaking between 2012 and 2015, and with major publications coming from the humanities, economics, business and social sciences. The latter did not exclude natural sciences such as environmental science and medicine engineering, etc.

Exclusion criteria looked at data that focused on economic entrepreneurs without any reference to social impact or social entrepreneur(ship) and the non-reference to South Africa. References to the southern Africa region or the continent, Africa, were also excluded, as they failed to capture a South African context. The cleaning process meant selecting content that spoke directly to South African issues, contexts and case studies. The final analysis step was done and revealed that the *Mediterranean Journal of Social Sciences* had more publications than Management or South African journals. However, there has been a marked increase in academic journals on social entrepreneurship over the last three years. Despite this, a consistently low number of entrepreneurship journals focusing on Southern Africa has been published in comparison with the number of management journals.

Figure 2 indicates publications by faculty. It reveals that the social sciences, economics, and the arts and humanities are the main sources of publications. The rest is shared with management and natural sciences publications. The latter seems to be an emerging trend.

Although the main publications are from South Africa as seen in Figure 3 below, not all were published from this country, as the US and the UK also had major publications that looked at the South Africa social sector, focusing on the non-profit sector.

South African higher education institutions have documented these issues, as indicated in Figure 4. However, business schools and centres of innovation that respond to training and development of entrepreneurs with a social focus in their entrepreneur development programmes seem to be low on publications.

Institutions such as the University of Johannesburg's Centre for Social Entrepreneurship, the University of Pretoria's Business School (GIBS) with its Social

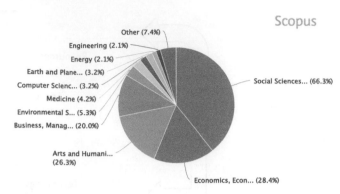

Figure 2: Analysis by publication faculty.

Entrepreneurship Certificate Programme, and the University of Cape Town's Bertha Centre for Innovation & Entrepreneurship are all special units dealing with entrepreneurship with a social focus (Steinman 2009).

Discussions

The development of an entrepreneurial mindset is a process that requires diverse activities and interventions that will stimulate and influence thinking processes whilst imparting skills and knowledge at the same time. However, social impact gets reflected through attitudes and an observable set of behavioural patterns that reflect the core mindset elements as discussed earlier: (1) the use of a business model for sustainability; (2) tapping into economic, social and environmental aspects; (3) innovative use and combination of resources, (4) exploiting or pursuing opportunities; and (5) addressing social needs.

Enabling and financial realities

Sustainability using business principles does not only look at a financial model as a core aspect for any business model, but incorporates the value proposition and the value constellation (Yunus, Moingeon, and Lehmann-Ortega 2010; Ostenwalder, Pigneur, and Tucci 2005). Thus, a look at aspects such as revenue stream and costs for entrepreneurs with a social focus acknowledges the financial limitations they face, especially when profits are not measured in terms of social value or social impact. This also adds a further difficulty linked with accessing funding from the banking sector due to an absence of fixed income or surety and limited alternative micro financing credits, a similar situation found in the Grameen model for the poor (Yunus and Weber 2010). Thus, alternative financing models available in South Africa, such as the 'stokvel[1]'and the loans from micro lenders have been useful, although they have own limitations (Yunus, Moingeon, and Lehmann-Ortega 2010).

Globalization has impacted on South Africa, yet its enabling opportunities in South Africa are still a far cry from those in developed countries such as the UK that have: (a) fostered a culture of social enterprise; (b) ensured that the right information and advice are available to those running social enterprises; (c) enabled social enterprises to access appropriate finance; and (d) enabled

social enterprises to work with government (Steinman 2009).

As a developmental country, South Africa uses a market-based approach to focus on market forces as a distribution tool for (and redistribution of) scarce resources. However, this approach lacks scaling-up opportunities that enable social enterprises to thrive. A need to foster conducive environments through investment opportunities and resources to address social needs could easily be achieved. The country has more than half of Africa's dollar millionaires, with a combined fortune of about US $390 billion, living in South Africa, but contributions to philanthropic causes that focus on social justice to support a transformation agenda on human rights and social justice for all cannot access government and/or hybrid funding (Ritchie 2011). However, those responding to human rights causes through advocacy activities have limitations in funding resources, as CSI tends to funds 'soft issues' (feeding-schemes and primary education, etc.) rather than 'hard issues' (advocacy and lobbying, and human rights development), with little focus on rights-based and social justice initiatives.

Perception of entrepreneurship with a social focus

When entrepreneurs tap into economic, social and environmental aspects they tend to respond to social ills such as health and education without harming the environment. This gets achieved through access to markets, trading and adapting with the changing environments. Traditional government initiatives have been unable to satisfy the entire social deficit. This is seen through ongoing service delivery protests, dependency on social welfare/grants (communities and non-profit sector), massive inequalities, poor education, inadequate housing, the HIV/AIDS pandemic, high unemployment and poverty rates (Rwigema and Venter 2004; Urban 2010). Although the focus has been mainly on education and human health, there have been some pockets of development in entrepreneurial approaches in dealing with environmental issues such as ecotourism and climate change (Walker 2014; Chirozva 2015).

Benefits for communities

Addressing social needs benefits communities. The significance of an entrepreneurial role with a social focus at community level can be indicated by the meaningful

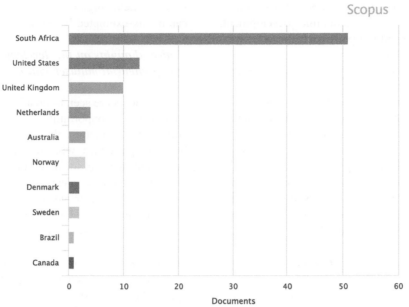

Figure 3: Documentation from country.

contribution of community development. However, cultural dimensions such as power, distance and collectivism do limit entrepreneurial activities (Takyi-Asiedu 1993; Hofstede 2001; Urban 2010). In South Africa, however, the contributions of Ubuntu principles have led to the achievement of high community involvement that is especially enhanced when benefits for the group are more important than for the individual (Ncube 2010). This gets supported by collective, socially-orientated, strength-based individuals embracing socio-structural remedies for human problems, whilst creating an environment that removes impediments and expands opportunities' (Bandura 1997; Urban 2010).

Innovation mix where entrepreneur(ship) focus on social issues

The mix of innovation when combined with resources targeting social needs creates opportunities to invest in scarce resources for future returns. Diverse social responses to opportunities require resources, which enhance serving basic, long-standing needs more effectively through innovative approaches (Austin, Stevenson, and Wei-Skillern 2006). South Africa's National Development Plan (NDP) that looks at different sectors for innovative approaches tends to be encouraging. The Minister of Economic Affairs, Ebrahim Patel, elevated interventions that support business with a social mission when he stated

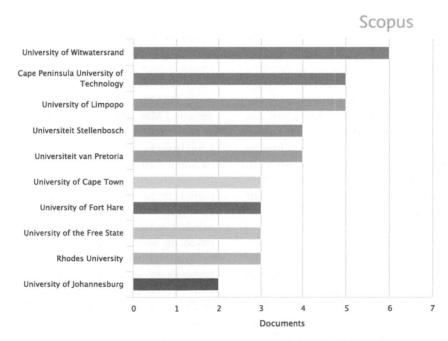

Figure 4: Higher institution affiliation.

that a 'percentage of pension fund investments would be earmarked for investment in economic development, proving a potential R70 billion pot for businesses with a social mission' (Fury 2010).

Innovation mix means exploitation of opportunities and resources in a strategic manner to maximize social impact. Corporate entrepreneurs as decision-makers who have been able to use the innovative mix should not be limited to CSI or enterprise development, but should consider alternatives for creating social change. Thus, an entrepreneurial mindset development process with a social focus needs to include change-making, tolerance for uncertainty and a willingness to pursue social goals, even where there are uncertainties full of obstacles or a lack of resources environments (Fury 2010).

Policies encouraging a new generation of entrepreneurs with a social focus

South Africa has 'existing policies, regulations and initiatives with direct bearing on enterprise development with a social focus. These include such as registration and obligation of companies and Non-profit organizations; tax law; B-BBEE, CSI, Enterprise development, and provision of business development services' (Fury 2010, 3). Contributory legislation includes:

(a) The NPO Act accommodates social enterprises that devote their income entirely to a public social purpose, thus creating an enabling environment in which NPOs can flourish and, by extension, to establish an administrative and regulatory framework within which organizations can conduct their affairs (Steinman 2009).

(b) The New Companies Act (Act No. 71 of 2008) requires that a non-profit company register without any members, auditing requirement are replaced by another form of independent verification, (except in the case of non-profit companies that exceed thresholds) determined by the minister.

(c) The Co-operatives Act (Act No. 14 of 2005) supports co-operatives work together as a business whose purpose is to reach a common goal, based on the value of self-help, self-responsibility, democracy, equality, equity and the ethical values of honesty, openness, social responsibility and caring for others.

(d) Close Corporations under the Close Corporations Act (Act No. 69 of 1984) (amended by the Companies Act, No. 71 of 2008) allows Close Corporations to be formed for social purposes and need not even pursue gain and can thus restrict dividends or payments to their members, so devoting most of their income and assets to achieving a social purpose.

The broad-based black economic empowerment (BBBEE) codes of ethics through their status, verification, ownership and codes is amongst the most favourable enabling factors, especially for procurement purposes where 15 scorecard points are allocated to companies which invest 3% of post-tax-profit for this purpose, three times the number of points allocated to CSI spend under the socio-economic development category (Fury 2010), 15 points to companies investing in 100% black-owned SMMEs and an extra 5 points for CSR contribution.

Public-benefit organization (PBO) status organization can be tax-exempted based on specific criteria.

Critical thoughts on the development of an entrepreneurial mindset with a social focus in South Africa

Entrepreneurs are seen as creative and innovative individuals who see opportunities in a need or a challenge. However, stimulating this process requires integrating business principles with social issues to create social value and achieve social wealth. This means the social mission should be clear and social enterprises should strive for ongoing triple (economic, social and environmental) outcomes. The ability to respond to challenges such as limited resources, no specific legal frameworks for social entrepreneurs and unfavourable opportunities requires an integration of diverse but relevant approaches such as partnerships with the profit sector, using hybrid approaches and service level agreements with the public sector. Entrepreneurial mindset development, as a conscious awareness, should incorporate social and environmental issues that form part of the educational curriculum (both at school and institutions of higher learning). The integration and responding to gaps should be part of the new generation of entrepreneurs with a social focus.

Initiatives could enhance access to social services and advocacy for issues such as environment, gender, poverty, hunger and partnership for development. Determining and measuring social impact can be a gauge that can be used to measure long-term social returns on investment. A triple bottom line for sustainability looks at financial, social and environmental goals, meaning measures should incorporate finance through a mixture of income streams and opportunities utilized. Linked to this is the ability to enter into contracts (economic element) and deliver a professional service (social element) in a competitive market (environmental element) for self-sufficiency. Thus performance is not measured in terms of financial performance (such as profitability – return on assets, return on equity and sales growth) but in less standardized and more idiosyncratic mechanisms that help to alleviate social issues, which legitimize the entrepreneur's role and the social mission as an outcome (Sullivan Mort, Weerawardena, and Carnegie 2003; Mair and Marti 2006; Weerawardena and Mort 2006; Certo and Muller 2008; Dees 2008).

Conclusion

In South Africa, entrepreneurs with a social focus search for highly effective and diverse funding models even though there is no incentive in focusing on hard cause. However, the development of the entrepreneurial mindset within the social sector is gaining momentum based on identified gaps being seen as opportunities to be explored using diverse and alternative resources as a response to social needs/mission. It can be concluded that the non-profit sector should overcome its not-for-profit mentality through embracing an entrepreneurial strategy while maintaining a social focus.

Disclosure statement

No potential conflict of interest was reported by the author.

Note

1. In South Africa, an informal savings society to which members contribute regularly and receive payouts in rotation.

References

Austin, J., H. Stevenson, and J. Wei-Skillern. 2006. "Social and Commercial Entrepreneurship: Same, Different, or Both?" *Entrepreneurship Theory and Practice* 30 (1): 1–22.

Bandura, A. 1997. *Self-efficacy: The exercise of control.* New York: W.H. Freeman.

Certo, T., and T. Muller. 2008. Social entrepreneurship: Key issues and concepts, Mays Business School, Texas A&M University, College Station, U.S.A.

Chirozva, C. 2015. "Community Agency and Entrepreneurship in Ecotourism Planning and Development in the Great Limpopo Transfrontier Conservation Area." *Journal of Ecotourism* 14 (2-3): 185–203.

Clark, C. H., and S. Ucak. 2006. RISE For-Profit Social Entrepreneur Report: Balancing Markets and Values, *Social Enterprise Program* Eugene M. Lang Centre for Entrepreneurship Columbia Business School.

Cooper, D. R., and C. W. Emory. 1995. *Business Research Methods.* Chicago: Richard D. Irwin.

Dees, J. 2008. "Philanthropy and Enterprise: Harnessing the Power of Business and Social Entrepreneurship for Development." *Journal of Innovations Technology Governance Globalization* 3 (3): 119–132.

Dees, J. G., and B. B. Anderson. 2003. "2. For-Profit Social Ventures." *International Journal of Entreprenuership Education* 2 (1): 1–26.

Deraedt, E. 2009. Social Enterprise: A Conceptual Framework, Conceptual Discussion Paper for the ILO Social Enterprise Development Targeting Unemployed Youth in South Africa (SETYSA) project, Katholieke Universiteit Leuven, Compiled for ILO.

Fakoti, O. 2014. "The Entrepreneurial Alertness of Immigrant Entrepreneurs in South Africa." *Mediterranean Journal of Social Sciences* 5 (23): 722–726.

Ferri, E. 2010. Social Entrepreneurship and Institutional Approach A Literature Review.

Fowler, A. 2000. "NGDOS as a Moment in History: Beyond aid to Social Entrepreneurship or Civic Innovation?" *Third World Quarterly* 21 (4): 637–654.

Fury, B. 2010. Social Enterprise Development in South Africa - creating a virtuous circle, Tshikululu Social Investment.

Gras, D., E. Mosakowski, and G. T. Lumpkin. 2005. Future Research Topics in Social Entrepreneurship, A Content-Analytic Approach. *Independent Research.*

Guclu, A. J., G. Dees, and B. Anderson. 2002. The Process of Social Entrepreneurship: Creating Opportunities Worthy of Serious Pursuit, *Centre for the Advancement of Social Entrepreneurs*, Duke The Fuqua School of Business.

Hackett, M. T. 2010. "Challenging Social Enterprise Debates in Bangladesh." *Social Enterprise Journal* 6 (3): 210–224.

Hofstede, G. 2001. *Culture's Consequences: Comparing Values, Behaviors, Institutions, and Organizations across Nations.* Thousand Oaks, CA: Sage Publications.

Johnson, S. 2000. Literature Review on Social Entrepreneurship Research Associate Canadian Centre for Social Entrepreneurship.

Kerlin, J. 2010. "A Comparative Analysis of the Global Emergence of Social Enterprise." *VOLUNTAS: International Journal of Voluntary and Nonprofit Organizations* 21: 162–179.

Mair, J., and I. Marti. 2006. "Social Entrepreneurship Research: A Source of Explanation, Prediction and Delight." *Journal of World Business* 41 (1): 6–44.

Marshall, J., G. Coleman, and P. Reason. 2011. *Leadership for Sustainability: An Action Research Approach.* Oxford: Greenleaf Publishing.

Ncube, L. B. 2010. "Ubuntu: A Transformative Leadership Philosophy." *Journal of Leadership Studies* 4 (3): 77–82.

Ad Hoc Task Force on Corporate Governance. 1999. *OECD Principles of Corporate Governance.* Paris: OECD.

Ostenwalder, A., Y. Pigneur, and C. L. Tucci. 2005. "Clarifying Business Models: Origins, Present, and Future of the Concept." *Communications of Association for Information System* 15: 1–43.

Peredo, A. M., and M. McLead. 2005. "Social Entrepreneurship: A Critical Review of the Concept." *The Journal of World Business* 41 (1): 56–65.

Pomerantz, M. 2003. "The Business of Social Entrepreneurship in a "Down Economy"." *Business* 25 (3): 25–30.

Ritchie, G. 2011. Social Justice Philanthropy Needs Firm Local Footing: *Http//*www.ngo *pulse.co.za* Wednesday, October 26, 2011 - 10:12.

Rwigema, H., and R. Venter. 2004. *Advanced Entrepreneurship.* Cape Town: Oxford University Press, South Africa.

Schwab Foundation for Social Entrepreneurs Profile: *Http://* www.schwabsfoundation. *com/.*

Seelos, C., and J. Mair. 2005. "Social Entrepreneurship: Creating New Business Models to Serve the Poor; Indiana University Kelley Business School." *Business Horizons* 48: 241–246. www.elsevier.com/locate/bushor.

Seelos, C., and J. Mair. 2008. "Corporate Strategy and Market Creation in the Context of Deep Poverty." IESE Business School: University of Navarra, Spain.

SEJ 2005 Social Enterprise Journal

SEL. 2005.

South Africa. National Planning Commission 2012/2013. *Our Future – Make It Work, National Development Plan 2030.* Pretoria Government Printers.

Stead, R., and R. Claasen. 2009. Business Development Services for Social Entrepreneurs, *Research paper on BDS for Social Enterprise*, South African Institute for Entrepreneurship.

Steinman, S. 2009. An Exploratory Study Into Factors Influencing An Enabling Environment For Social Enterprises In South Africa, University of Johannesburg Centre for Social Entrepreneurship Faculty of Management, Commissioned by ILO.

Sullivan Mort, G., J. Weerawardena, and K. Carnegie. 2003. "Social Entrepreneurship: Towards Conceptualisation." *International Journal of Nonprofit and Voluntary Sector Marketing* 8 (1): 76–88.

Takyi-Asiedu, S. 1993. "Some Socio-Cultural Factors Retarding Entrepreneurial Activity in sub-Saharan Africa." *Journal of Business Venturing* 8 (2): 91–98.

United Nations. 1992. UN Department of Economic & Social Affair. UN resolution Agenda 21 on Sustainable Development.

Urban, B. 2010. "Social Entrepreneurship Activity and Different Skills Associated with Successful Social Entrepreneurship in South Africa." Faculty of Management/Department of Entrepreneurship, University of Johannesburg: Johannesburg.

Walker, G. 2014. Communicating Science to the Public: *Opportunities and Challenges for the Asia-Pacific Region.*

Weerawardena, J., and G. Sullivan Mort. 2006. "Investigating Social Entrepreneurship: A Multidimensional Model." *Journal of World Business* 41 (2006): 21–35.

Yunus, M., and K. Weber. 2010. *Building Social Business: A new Kind of Capitalism That Serves Humanity's Most Pressing Needs.* New York: Publication Affairs.

Yunus, M., B. Moingeon, and L. Lehmann-Ortega. 2010. "Building Social Business Models: Lessons From the Grameen Experience." *Long Range Planning,* 43: 308–325.

Index

Printed and bound by CPI Group (UK) Ltd, Croydon, CR0 4YY

18/10/2024

01776250-0020